AF780569

Being Algae

Critical Plant Studies

PHILOSOPHY, LITERATURE, CULTURE

Series Editor

Michael Marder
(*IKERBASQUE / The University of the Basque Country, Vitoria*)

VOLUME 8

The titles published in this series are listed at *brill.com/cpst*

Being Algae

Transformations in Water, Plants

Edited by

Yogi Hale Hendlin
Johanna Weggelaar
Natalia Derossi
Sergio Mugnai

BRILL

LEIDEN | BOSTON

This is an open access title distributed under the terms of the CC BY-NC 4.0 license, which permits any non-commercial use, distribution, and reproduction in any medium, provided the original author(s) and source are credited. Further information and the complete license text can be found at https://creativecommons.org/licenses/by-nc/4.0/

The terms of the CC license apply only to the original material. The use of material from other sources (indicated by a reference) such as diagrams, illustrations, photos and text samples may require further permission from the respective copyright holder.

Cover image by Claes Bech-Poulsen.

Library of Congress Cataloging-in-Publication Data

Names: Hendlin, Yogi Hale, editor. | Weggelaar, Johanna, editor. | Derossi, Natalia, editor. | Mugnai, Sergio (Botanist) editor.
Title: Being algae : transformations in water, plants / edited by Yogi Hale Hendlin, Johanna Weggelaar, Natalia Derossi, Sergio Mugnai.
Description: Leiden ; Boston : Brill, 2023. | Series: Critical plant studies, 2213-0659 ; volume 8 | Includes bibliographical references and index.
Identifiers: LCCN 2023034542 (print) | LCCN 2023034543 (ebook) | ISBN 9789004683303 (hardback) | ISBN 9789004683310 (ebook)
Subjects: LCSH: Algae—Social aspects. | Human-plant relationships. | Human ecology. | Ecocriticism.
Classification: LCC QK566 .B45 2023 (print) | LCC QK566 (ebook) | DDC 579.8—dc23/eng/20230906
LC record available at https://lccn.loc.gov/2023034542
LC ebook record available at https://lccn.loc.gov/2023034543

Typeface for the Latin, Greek, and Cyrillic scripts: "Brill". See and download: brill.com/brill-typeface.

ISSN 2213-0659
ISBN 978-90-04-68330-3 (hardback)
ISBN 978-90-04-68331-0 (e-book)
DOI 10.1163/9789004683310

Copyright 2024 by Yogi Hale Hendlin, Johanna Weggelaar, Natalia Derossi and Sergio Mugnai. Published by Koninklijke Brill NV, Leiden, The Netherlands.
Koninklijke Brill NV incorporates the imprints Brill, Brill Nijhoff, Brill Schöningh, Brill Fink, Brill mentis, Brill Wageningen Academic, Vandenhoeck & Ruprecht, Böhlau and V&R unipress.
Koninklijke Brill NV reserves the right to protect this publication against unauthorized use.

This book is printed on acid-free paper and produced in a sustainable manner.

Contents

Figures

Notes on Contributors

Yogi Hale Hendlin
Ph.D., is an environmental philosopher and public health scientist. Hendlin is Editor-in-Chief of the journal *Biosemiotics*, writes on interspecies communication, critical plant studies, and the role of industrialism affecting our environment which in turn determines our health. As a professor at Erasmus University Rotterdam, the Netherlands, as well as researcher at the University of California, San Francisco, Hendlin's interdisciplinary work regularly appears in the academic and popular press. Also co-edited by Hendlin, *Food and Medicine: A Biosemiotic Perspective* appeared in 2021.

Johanna Weggelaar
is an independent project leader and researcher working at the crossroads of science and the creative sector. With a background in general engineering from the École Centrale de Lyon in France and in Cultural History from the University of Utrecht in the Netherlands, she conducts multidisciplinary projects where expertise from various backgrounds are necessary to tackle complex problems. She has been working for the last 5 years on the role algae have in the environment and in our societies, researching how human-algae relations may be interpreted.

Natalia Derossi
graduated Summa Cum Laude from Erasmus University College, Rotterdam, the Netherlands. She has tutored Ecology at Erasmus University College. She then pursued a Research Master Degree from the École des Hautes Études en Sciences Sociales, Paris, France, with a thesis dealing with the promises of infinite growth unleashed by recent biotechnology advancements, written starting from an ethnographic fieldwork conducted at the largest European biotechnology facility in Wageningen, the Netherlands. Currently, she lives in Paris working as an interpreter and conducting field research for a Franco-Swiss Environmental Justice coordination framework fighting against the devastating effects of fossil finance.

Sergio Mugnai
Ph.D., is a plant biologist and currently Senior Lecturer at the Erasmus University College (Rotterdam, The Netherlands), with a major interest in environmental sustainability. He was previously a researcher at the University of Florence (Italy) where he conducted research on plant response to abiotic

stressors as well as plant internal communication. He also worked at the European Space Agency (ESA) as Biology Project Scientist for almost four years. In this position he followed several biology experiments on board the International Space Station (ISS).

Andrew Lopez
is a Ph.D. candidate in Philosophy at Queen's University. Andrew specializes in animal philosophy, feminist philosophy, and political philosophy, with research interests in philosophy of biology and Latin American philosophy. He brings these fields together to understand the ethical, political, and philosophical relevance of the rich cognitive, social, and cultural lives of nonhuman animals. His dissertation focuses on the cognitive and epistemic lives of nonhuman animals and how these should inform political philosophy concerned with interspecies relations. His work has been published in *Synthese* and the *Journal of Ethics and Social Philosophy*.

M. Polo Camacho
is an Applied Philosopher at the Center for Practical Bioethics, where his work emphasizes the need for community-based approaches to end-of-life ethics and the ethics of emerging technologies. In addition to bioethics, he has interests in philosophy of science, philosophy of biology, and environmental ethics. His academic work has been published in *PTPBio, History and Philosophy of the Life Sciences, Biosemiotics*, and *Studies in History of Philosophy of Science Part A*.

Jesse D. Peterson
(he/him) is an American from the traditional homelands of the Shoshone, Paiute, Goshute, and Ute tribes near Salt Lake City, UT, and now works as a Lecturer with the Radical Humanities Laboratory and School of the Human Environment at University College Cork. He researches societal relationships to ecological challenges using transdisciplinary methods, having focused on issues related to oceans and biodiversity. His research addresses topics such as the production of ocean health and pollution, more-than-human relationships, socio-ecological death, citizen science, biodiversity data, and innovation in research methods. His publications on these topics are scheduled with or have been published as: "Sea Farming and Feminist Blue Humanities" (special issue for *Journal of Agricultural and Environmental Ethics*); "Can the Baltic Sea Die? An Environmental Imaginary of a Dying Sea" (in *Death's Social and Material Meaning Beyond the Human*); "Storying Toxic Time-Scape Trajectories: Intersections among Algal Toxins and More-than-Human Bodies" (in *Toxic Time-Scapes: Examining Toxicity across Time and Space*); "The Metaphor

of 'Ocean' Health is Problematic" (*Frontiers in Marine Science*); "Doing Environmental Humanities: Inter/Transdisciplinary Research through an Underwater 360° Video Poem" (*Green Letters*); *Excessive Seas: Waste Ecologies of Eutrophication* (PhD dissertation), and elsewhere.

Nina Lykke

Professor Emerita, Dr. Phil., Gender Studies, Linköping University, Sweden, and Adjunct Professor, Aarhus University, Denmark. She has participated in the building of Feminist Studies in Scandinavia and Europe since the 1970s. She is also a poet and writer, and co-founder of the international network for Queer Death Studies. Current research interests: human-diatom-relations; queering of death and mourning in posthuman, queerfeminist, materialist, decolonial and eco-critical perspectives; intersectionality; autophenomenographic and poetic writing. Author of over 200 publications among others monographs such as *Cosmodolphins. Feminist Cultural Studies of Technology, Animals and the Sacred* (with M. Bryld, 2000), *Feminist Studies* (2010), and *Vibrant Death. A Posthuman Phenomenology of Mourning* (2022), edited volumes such *as Writing Academic Texts Differently* (2014), *Assisted Reproduction Across Borders* (2016, w M. Lie) and *Pluriversal Conversations on Transnational Feminisms* (2023, w. R. Koobak, P. Bakos, S. Arora and K. Mohamed, and journal articles such as "Queer Death Studies: Death, Dying and Mourning from a Queerfeminist Perspective". *Australian Feminist Studies* (2020) (with M. Radomska and T. Mehrabi); "Decolonising Mourning. World-Making With the Selk'nam People of Karokynka/Tierra del Fuego" *Australian Feminist Studies* (2020) (with H.M. Vargas and C. Marambio).

Ole G. Mouritsen

Ph.D. DSc, is a physicist and professor emeritus of gastrophysics and culinary food innovation at the University of Copenhagen. He is an elected fellow of the Royal Danish Academy of Sciences and Letters, the Danish Gastronomical Academy, Sigma Xi, and Gastronomische Academie Deutchlands. He is president of the Danish Gastronomical Academy. He is the author of several scientific books and about 400 scientific papers and reports, in addition to being recipient of a number of prestigious science and science communication prizes. He is passionate about Japanese food culture and is appointed Japanese Cuisine Goodwill Ambassador. In his spare time, he cooks, collaborates with chefs, furthers his knowledge of all aspects of food, and writes articles and books about the science of food and taste, including English editions *Sushi: Food for The Eye, the Body, and the Soul* (2009), *Seaweeds: Edible, Available & Sustainable* (2014), *Umami: Unlocking the Secrets of the Fifth Taste* (2014), *Life-As a Matter of Fat* (2015), *Mouthfeel: How Texture Makes Taste* (2017), *Octopuses, Squid and Cuttlefish: Seafood for Today and Tomorrow* (2021), and *The Science and Art of Pickled Vegetables: Tsukemono* (2021).

J. Lucas Pérez Lloréns
Ph.D., is a biologist and Professor of Marine Ecology at the Faculty of Marine and Environmental Sciences (University of Cádiz, Spain). He is president of the Spanish Phycological Society. He is author of four scientific books as well as nearly 100 scientific papers. His field of research encompasses from seagrass and seaweed ecology and ecophysiology to aquaculture. He is very much passionate of culinary ethnology of marine produce (mostly seaweeds), including its use in avant-garde cuisine. He loves cooking and he has collaborated with several chefs and cooking schools. He was awarded in several culinary contests. Two of his books: *¿Las algas se comen?* (2016) and *Those Curious and Delicious Seaweeds* (2018) were awarded as "Best of the World" in the categories of Nutrition and Health Institutions and translation, respectively at the "Gourmand World Cookbook Awards" (2016 and 2018). Together with colleagues from the University of Cadiz, I have advised the restaurant Aponiente (three Michelin stars) in the experimental cultivation of "marine rice", which has allowed the chef Ángel León to win the National Award for Gastronomic Innovation 2021 granted by the Royal Spanish Academy of Gastronomy.

Kathryn Larsen
is an architect in Denmark, and graduate of TU Delft's architecture program. Through her creative practice Studio Kathryn Larsen, she works with biodesign and natural materials in construction, including seagrass and algae. Her work focuses on historical and futuristic applications.

Mustafa Yavuz
has BSc degree on Biology Education and has served as a Teacher of Biology between 2002 and 2015, at government schools. He has MSc. and Ph.D. degrees in Biology, in his both degree studies, he has focused on the taxonomy, ecology and biomonitoring of the lichens. His second Ph.D. degree study is entitled "De Plantis Tradition from Aristotle to Avicenna" which he defended in December 2022. Mustafa Yavuz worked at Istanbul Medeniyet University, History of Science Department as Assistant Professor between 2015 and 2020. In April 2020, He has obtained the Assoc. Prof. Dr. degree in History of Science (Philosophy) whence he continues to serve in the same department. His studies are related to the history and philosophy of biology as well as taxonomy and ecology of lichens.

Gustaaf Hallegraeff

(Ph.D. University of Amsterdam, D.Sc, University of Tasmania) is recognised internationally for his marine biosecurity work on harmful algal blooms impacting on human health, the fish farm and shellfish industries, their stimulation by coastal eutrophication, relationships with climate change and global spreading via ship's ballast water. He has published over 250 research papers and 12 books, including a stage setting UNESCO Monograph on Oceanographic Methodology on Harmful Marine Microalgae and the first Australian Marine Phytoplankton Flora. He supervised 36 Ph.D. graduands. He was honoured with a Fellowship of the Australian Academy of Technological Sciences and Engineering, Eureka Prize for Environmental Research, President's Prize of the International Institute for Marine Engineers, University of Tasmania Research Medal, and Lifetime Achievement Award from the International Society for the Study of Harmful Algae.

Jon L. Pitt

is Assistant Professor of Japanese Environmental Humanities at the University of California, Irvine. His current book project, *Becoming Botanical: Rethinking the Human through Plant Life in Modern Japan*, looks to bring insights from Critical Plant Studies into the study of modern Japanese literature and cinema. He is also the translator of poet Hiromi Ito's *Tree Spirits Grass Spirits* (*Kodama kusadama*, 2023 via Nightboat Books), and the host of the environmental humanities podcast Nature : Mono. He has contributed chapters to the *Handbook of Modern and Contemporary Women Writers* (MHM Limited) and *The Coronavirus Pandemic in Japanese Literature and Popular Culture* (Routledge).

Melody Jue

is Associate Professor of English at the University of California, Santa Barbara. She is the author of *Wild Blue Media: Thinking Through Seawater* (Duke University Press, 2020), which won the 2020 Speculative Fictions and Cultures of Science book award, and is co-editor with Rafico Ruiz of *Saturation: An Elemental Politics* (Duke Press, 2021). Professor Jue has published articles in journals including *Grey Room, Configurations, Women's Studies Quarterly, Resilience*, and *Media+Environment*.

Brenda Parker

is an Associate Professor in Sustainable Bioprocess Design at the UCL Department of Biochemical Engineering.

Marcos Cruz
is an Architect and Professor of Innovative Environments at the Bartlett UCL. Both are co-founders and directors of Bio-Integrated Design (Bio-ID), a unique interdisciplinary research and teaching platform that brings together the fields of biotechnology, computation and design for applications within the built environment. In order to address our ongoing climate emergency, Bio-ID merges a range of expertise by employing laboratory research in applied science of biochemical engineering; novel design as a research tool to encompass new products, multifunctional and bioreceptive materials; ecology and space. Project outcomes vary from being grown objects, bioreactors, components or landscapes, all of which emerge from the complex relationship between the environment, specific socio-cultural contexts, and the interfacial properties of materials and organisms. Research from the Bio-ID programme has led to publications, a patent filing and numerous international awards. Design outputs have been exhibited and presented at the Centre Pompidou, London Design Festival, ACADIA, the London Design Museum, Ars Electronica, and the Venice Biennale.

Julia Lohmann
is Associate Professor in Contemporary Design at Aalto University and the founder of the Department of Seaweed, a transdisciplinary community of practice investigating the potential of macro algae as a design material with a regenerative eco-systemic impact. She uses her artistic practice as research through design, to explore the ethical and material value systems underlying our relationship with flora and fauna. Lohmann is developing empathic, collaborative and co-speculative approaches to design. She promotes more-than-human-centric, regenerative practices benefitting socio-ecological systems. Lohmann studied at the Royal College of Art, where she has also taught and completed an AHRC-funded collaborative PhD scholarship between the RCA and the Victoria & Albert Museum. At Aalto University, she teaches MA courses on critical design practices, materials and living systems. Julia Lohmann researches biomaterials and the role of design in enabling eco literacy, promoting marine conservation and attaining UN Sustainable Development Goals. Julia Lohmann's work is part of major public and private collections worldwide, including the Museum of Modern Art in New York, and has received awards, bursaries and support from the Esmée Fairbairn Foundation, the British Council, Jerwood Contemporary Makers, D&AD, Stanley Picker Gallery, Arts Foundation, Wellcome Trust and Cooper Hewitt Smithsonian Design Museum.

Soo Jung Ryu

is a Ph.D. fellow at the Aarhus School of Architecture in Denmark. Her research is called "Urban Seascaping – How to live not just by the sea but *with* the sea". It is design-based research that investigates the potential role of integrating the sea and seaweed as a catalyst for alternative waterfront development in coastal cities in light of rising sea levels. Her current work focuses on the need for more transdisciplinary collaboration between creative disciplines and natural sciences in designing nature-based solutions as part of coastal adaptation. Soo is a Korean-New Zealander who graduated with a Master of Architecture from the University of Auckland and Victoria University of Wellington in Aotearoa, New Zealand (NZ) and a Master of Science from Aalborg University, Denmark (specialising in Urban, Energy and Environmental Planning). Before pursuing her Ph.D., she also worked with various architecture firms across Aotearoa, New Zealand and Great Britain. Notably, she was involved in the rebuilding process of Christchurch, NZ, after the devastation of the February 2011 earthquake.

Cintia Organo Quintana

has obtained her Ph.D. in Biological Oceanography in 2009 at University of Sao Paulo, Brazil. She became Associate Professor at the Department of Biology, University of Southern Denmark (SDU) in 2022. Her research interests focus on marine ecology, and on the links between marine biodiversity and ecosystem functioning, investigating the impacts of eutrophication, pollution, habitat loss and climate change. She works actively on protection, conservation and restoration of marine ecosystems and their potential to provide mutual benefits to biodiversity and society.

Introduction

Algal mor(t)ality

> What happens to a diatom in the upper, sunlit strata of the sea may well determine what happens to a cod lying on a ledge of some rocky canyon a hundred fathoms below, or to a bed of multicolored, gorgeously plumed seaworms carpeting an underlying shoal, or to a prawn creeping over the soft oozes of the sea floor in the blackness of mile-deep water.
>
> RACHEL CARSON, *The Sea Around Us* (1951)

Aquatic environments are as networked together by algae as the terrestrial world is by plants. Microalgae (unicellular organisms, both prokaryotic and eukaryotic) and macroalgae (eukaryotic multicellular organisms) regulate oxygen in water, provide the alimentary fundament for the rest of sealife, and metabolize nutrients. Algae breathe life into our seas. And despite – or perhaps, because of – algae being the ancestors of plants, phylogenetically, some algae groups are not plants at all. In fact, algae is an informal term for a vast polyphyletic group of photosynthetic organisms that includes species from multiple distinct clades. Some algae groups don't even belong to the Viridiplantae clade (to which land plants belong, Ruhfel et al., 2014). Microplankton are the original photosynthesizers; but because they include cyanobacteria, diatoms and dinoflagellates among others – the mix of life making up the plankton biome – it makes more sense to classify them according to their ecological niches rather than their phylogeny. The queering of categories 'algae' proliferate, both through their variety and effects, produce an oblique lens through which we can help explain the mediums in which we live and think vegetal life.

Rachel Carson's marine-centric perspective conveyed its wonder at a time when the ocean still posed a world of unknown mysteries. In her trilogy *The Sea Around Us*, the ocean is the central character, reminding us of the interdependence and vibrancy of life unbeknownst to the casual observer of brine. Her capacity to relate a narrative of the sea as our primordial origins, delicate yet vast, inspired humility and reverence in the face of increasing isolation. Like many of the great ecologists, the urgency Carson felt is even more tangible now that the effects of the climate crisis have become codified in the modern

© Y. H. HENDLIN, J. WEGGELAAR, N. DEROSSI & S. MUGNAI, 2024 | DOI:10.1163/9789004683310_002
This is an open access chapter distributed under the terms of the CC BY-NC 4.0 license.

liturgy of science and local experience. A litany of decay is codified in an infrastructure of resignation both built and biopsychosocial, signals incapacitating collective action, a failure to follow the flow of algae and their symbionts. Like so many other aspects of our reality – too big to fail, beyond perturbation from the affairs of mere mortals – the ocean suffers at our iron hand unprecedentedly hasty acidification and biodiversity collapse; oceanic transformations remaining mostly hidden to us terrestrial beings. Predictably, as human populations despair at the downstream consequences of our ecological callousness, we nonetheless continue treating most complex dynamics as solvable through the coercion of techno-scientific, militaristic grit. Suddenly, the ocean and its composition are regarded as a new tool of control, a new battleground for the quixotic war on the chaotic consequences of the Anthropocene, itself only further driving its cruelest consequences.

Some manifestations of planetary dysbiosis are rather tangible: the depletion of fish populations, sea level rise, sudden blooms of – sometimes toxic – algae. Due to weather extremes and polluted waters, algae blooms are multiplying, leading to the sorts of environmental and economic nuisances that finally get social systems designed to ignore such ecological repositionings to notice: precipitating eutrophication events, hindering boat navigation, creating inconveniences for tourism, causing severe health risks. Due to our social systems reacting most when algae respond to our disrespect of their ecological preferences rather than attending to the myriad 'ecosystem services' they provide, wild algae receive bad press, popularly labeled as a kind of new plague. Episodic blooms of cyanobacteria, of *Ulva* on the French Brittany coast, or of Sargassum in the Caribes, now punctuate the summer season. It's tempting for many to approach these occurrences as a kind of intruding phenomena to be endured, despite their cascading effects on the lives and comfort of innumerable maritime and terrestrial organisms. But such a route bypasses the ecological interconnectedness Carson and other lovers of the sea so poetically describe.

At once toxic life-killing response to our elemental pollution and simultaneously savior substance to be engineered to buy oxygen, food, cow methane alleviation, climate control, and coveted energy to sustain our unsustainable industrial culture, the current demonization and deification of algae prevents seeing algae for what they are. Detoxing from reducing algae to their megamachine use-value and projects (inflation and deflation) of our hopes and fears is no easy feat.

Perhaps because they occupy a medium so foreign to our sensibilities – the oceanic – algae have received scant attention in Critical Plant Studies (for an exception, see for instance Lawrence, 2022). The ancestors of all plants on

Earth seem forgotten from the recent eruption of academic research attempting to question the place of plants in our human frameworks. This volume aims at giving algae the importance they deserve in the field of Critical Plant Studies and nuancing the perspectives on algae we nurture.

Another reason why water plants in general, and algae in particular, have been overlooked by Critical Plant Studies, is due to their foreignness. Humans, even seafaring ones, ultimately are a terrestrial species (although our ability to consciously breathe may have resulted from our propensity to dive in water and hold our breath to forage for seafood). In western biology's convenient if lazy shorthand of importance-by-proximity, we tend to paint importance in concentric circles from our own standards, based on models of ourselves and our needs. Oil, for example, has become so important because it is the input for the hierarchies of control we've grown used to. While seaweed plays a much more central role in some cultures, such as Japan, where not only food, but also materials traditionally are dependent on various seaweeds, algae for the most part have proven recalcitrant to industrialization. Harvesting and processing algae requires an artisan's pace, rather than the rapacious speed of factories (though efforts are afoot, also, to enfold algae into the enclosed linearity of industrial process). The academic illegibility of algae has been symmetrical to its previous economic and political marginalization.

This more or less conscious exclusion of algae, particularly microalgae, from academic research is not new. Red tides due to algae blooms have been recorded since Antiquity, yet algae for centuries occupied a neglected field of botany, ignored from most plant classification until the 16th century (Ragan, 1998), even though both Theophrastus and Aristotle earlier mentioned and discussed algae in their work. Based on Aristotelian philosophical deductions and the conclusion that only the fructification organs were suitable for the construction of the most natural system in the field of botany, Italian botanist Andrea Cesalpino (1524–1603) was the first to establish a classification in which algae were considered together with fungi, moss and ferns, as part of a fifth group gathering "seedless plants," characterized by their imperfection:

> Some plants have no seed; these are the most imperfect, and spring from decaying substances; they have only therefore to feed themselves and grow, and are unable to produce their like; they are a sort of intermediate existences between plants and inanimate nature (cited in von Sachs, 1890 chapter 2)

Considered as the lowest level of vegetal life in a taxonomy based on the flowering capacity of plants, imperfect and flowerless algae for centuries were therefore relegated to the rank of primitive and underdeveloped plants. Unicellular algae were not yet even part of the picture, as they were observed for the first time at the end of the 17th century. Anton van Leeuwenhoek (1632–1723), a Dutch textile merchant who invented the modern microscope in the 17th century to observe the quality of the microfibers of textiles, was the first to observe bacteria and protozoa in a volume of water, likely observing microalgae which he mistook for tiny animals, which he termed *animalcules*. We need to wait until the 19th century to see algae being recognised as a major plant category. Here, Stephen Endlicher (1805–1849) developed an upper taxon classification which became accepted by many botanists at that time, and placed algae in an independent division called Thallophyta, together with lichens and fungi (Kaufmann, 1965).

The slipperiness of algae's form and function and their classification based on a terrestrial model of perfection explain how for centuries they were partly omitted from most classification attempts. Only with the development of DNA-based research in the 20th century did the whole classification system require rethinking based on the main dichotomy between prokaryotes and eukaryotes, revealing at the same time a third group, the Archaebacteria. With different forms of algae interspersing all three domains of life, it is no wonder they escaped demand and control classification schemes. Their diversity blurs frontiers, the original queer plant that is not one. Even now with phycology – the branch of botany covering seaweeds and algae – as a recognized discipline, debates about how to classify algae remain part of academic discussions. As Mark Ragan (1998: 11) puts it: "textbooks of algae describe organisms that are living and fossil, aquatic and non-aquatic, autotrophic, heterotrophic and mixotrophic, flagellate and non-flagellate, diverse in form, size, color, toxicity, habitat and geographic distribution," concluding that our understanding of algae might be better described as sociological application.

Also the obvious aptitude of algae for conquering specific environments and developing sometimes extremely sophisticated symbiosis strategies is questioning our conventional perceptions of a species habitat. Algae are colonial organisms which serve as food and habitat for countless other species. In fact, much like how endophytic fungi inhabit most terrestrial plants, algae colonize most aquatic environments, such as corals. *Zooxanthellae* algae provide corals with their brilliant colors, synthesize their waste, and provide the synergy which enables endless forms of other livings based in coral reefs. Algae and coral symbioses also occur, however, only under larger ecologically conducive circumstances. Ocean acidification and warming cause corals to expel *Zooxanthellae*, which turns them white, brittle, without sufficient food, and susceptible to pathogens. Algae are the guardians of corals but are only accepted as symbionts under the right temperature, pH, and other beneficial environmental conditions. The quality of the water determines the cooperative capacities for the coral-algal symbiont, and the profusions of life around algal gathering helps determine the composition of the water. Transformations in water create and affect transformations in water plants, and vice-versa.

The movement of plants from sea to land some 500–700 million years ago presented additional evolutionary challenges, which allowed plants to diversify in new ways. At the origin of this terrestrialization of life, we find a few algae species who started to colonize land environments by acquiring specific genes from soil bacteria making them resistant to drought and other extra-aquatic stress (Cheng, 2019). Subaerial algae are now widespread (Kharkongor and Ramanujam, 2014) although remaining one of the most overlooked groups of all algal groups in academic research. Here again, algae have gotten involved in striking symbiotic phenomena, such as lichens, and colonized a huge variety of habitats. Despite their ecological and evolutionary importance, this book will not include these species and will predominantly focus on marine/freshwater algal species.

This ecological approach to understanding critical plant studies can be helpful for honing synecological treatments of terrestrial plants as well. Understanding plants as meeting places for interspecies confluences, based on their thoroughgoing interspeciality themselves, creates productive ambiguities as to what the relevant unit of plant analysis is.

1 Rehabilitating Algae

The sudden interest in algae and seaweed for instrumental extractive purposes deserves an equal examination of why water plants have so long gone suppressed as core constituents of our biosphere, responsible for making up much of the composition of our current atmosphere, as well as playing an integral part in seawater's concoction.

In the Western tradition, algae have tended to be seen, with few exceptions, as slimy weeds from the sea, phlegmy masses to be avoided. The persisting name "seaweed" suggest we still consider them as an unwelcome and inedible species, a dispensable member of our forest gardens. Middle English variations included *sechaf* (sea-chaff) while Old English pronounced seaweed to be *fleotwyrt* (float-wort). Conjuring up almost Conradian notions of the aquatic realm as the deep/dark/excluded, in Europe seaweed held an ambiguous place in the cultural imaginary as a substance more to be feared than nourished by. Seafarers' legitimate unease at the sometimes real prospect of being pulled to their watery deaths by seaweed catching their boats with volitional tentacles can be contrasted with other European sailors and communities historically supplementing their diet with various seaweeds (O'Connor, 2017).

The ambivalent relation local communities held with seaweed is also illustrated by its indigenous uses. In Ireland, Brittany (France) or Denmark, seaweed traditionally has been used for food, as a natural fertilizer, to produce soap (through burning and using its ash), or as a construction material. The very fecundity of seaweed, however, seems to have been its downfall as industrialization held sway. Unable to reliably source and harvest seaweed in the same spot, but instead relying on seasons, tides, and complex ocean dynamics, seaweed harvesting – in the west, at least – proved disobedient to the exigencies of commodification and scale. Thus, with industrial culture's disregard for that which cannot be subsumed easily into mechanization, indigenous European seaweed knowledge and practices have largely disappeared.

Seaweed suddenly gained in popularity in the 19th century in Europe when natural history developed as a discipline. Darwin already described the importance of kelp in his "Voyage of the Beagle":

> The number of living creatures of all orders, whose existence intimately depends on the kelp, is wonderful. A great volume might be written, describing the inhabitants of one of these beds of sea-weed. […] Often as I recurred to a branch of the kelp, I never failed to discover animals of new and curious structures. […] I can only compare these great aquatic forests of the southern hemisphere with the terrestrial ones in the intertropical regions. Yet if the latter should be destroyed in any country, I do not believe nearly so many species of animals would perish, as, under similar circumstances, would happen with the kelp. (Charles Darwin, *The Voyage of the Beagle*, Volume III, Journal and Remarks 1832–1836, p. 305).

During the century, natural history exceeded the academic circles to infiltrate the amateur spheres. Seaweed raised a real enthusiasm among natural history lovers, and especially among women during the Victorian era in England. Hunting animals was perilous and therefore reserved for men only. Furthermore, the study of botany codified by the Linnaeus classification of plants was based on their sexual organs, which made it inappropriate for women. Luckily, studying seaweed was a safe and accessible hobby, leading to a real craze among Victorian women going "seaweeding" on the shoreline. What originated as a hobby eventually generated an impressive amount of knowledge gathered in detailed scrapbooks permitting the reappraisal of seaweed, then called "flowers of the sea". During the same period, Anna Atkins (1799–1871) experimented with the photography cyanotype process to record local seaweed species. She produced the very first book illustrated with photography ever printed. The volume entitled *Photographs of British Algae: Cyanotype Impressions* (ca. 1853) was considered as a true innovation in botanical photography and gained the recognition of the scientific community.

Outside of Europe, records of the relations of communities to seaweed go much further back in time. Seaweeds were first utilized for human consumption in Japan at least 1500 years ago, according to early written records. During the Asuka Era (600–700 AD) and Nara Era (700–800 AD) seaweeds were only consumed by the Japanese aristocracy, since it was seen as a special dish served in sacred ceremonies or bestowed upon nobility as an offering. Until the Middle Ages only wild seaweeds were used, which limited them as a food source. During the Tokugawa Era (1600–1800 AD), Ieyasu Tokugawa, the Shogun at that time, ordered fishermen in the small town of Shinagawa (now part of the Tokyo metropolitan area) to bring him fresh fish every day. The fishermen built an offshore fence and started a fish farm. They also found that seaweed preferred growing on this fence. Thus seaweed cultivation was created (The Seaweed Site, 2022).

The lack of attention for algae in western cultures is striking in comparison with other cultures but also with human history. A recent hypothesis of migration archeology throws into question the popular evolutionary theory according to which the first humans reaching the American continent crossed the then frozen Bering Strait between Siberia and Alaska. The first inhabitants in the Americas (at least from Asia) might have instead followed via boat a "Kelp Highway" along the shores from Asia to North America, well before the Bering Strait land bridge became accessible (~17,000 years ago) (Erlandson et al. 2007; Braje et al. 2017). This Kelp Highway offered travelers a nutrient-rich diet, from the kelp itself, but also via the variety of fish life it attracted. Kelp may have both nourished and guided, allowing human expansion into the Americas, underlying how entangled our histories are with the life water plants provide.

Paddling further down into Earth's history, 3.8 billion years ago the first unicellular cyanobacteria started photosynthesizing and producing oxygen from water as a by-product of this process. Known as the Great Oxidation Event, this shift signals the move from anaerobic to aerobic life with the oxygen-byproducts of cyanobacteria metabolism and respiration outpacing the absorption of oxygen into the water. And around 2.1–2.4 billion years ago, oxygen from these cyanobacteria eventually displaced the methane-heavy atmosphere, creating the oxygen-rich atmosphere we currently thrive on, which allowed the prospects for terrestrial life. Cyanobacteria-created oxygen molecules (O_2) in the air eventually were broken apart by UV radiation into free oxygen which recombined to form ozone, the gaseous sunscreen which prevents mutation-inducing UV rays from reaching Earth's surface. From an evolutionary perspective, we owe our lives to algae.

Indeed, all terrestrial life and the great complexity of organisms was empowered by the originators of photosynthesis on Earth: cyanobacteria, also called blue-green algae (West, 2022). From an ecosystemic perspective, these algae provided a crucial role in the nutrient cycle plus the basis of the food chain. The catalyst of photosynthesizing microscopic prokaryotes enabling the conditions for eukaryotic life is testament to the power of life processes to produce emergent events. This generosity characteristic of algal life produces the conditions for other forms of life, progenitors of diversity, the opposite of colonization. For instance, the smallest known photosynthetic organism is a microalgae – a marine cyanobacterium called *Prochlorococcus* – which is believed to be the most abundant producer of oxygen in the oceans (Tetu et al., 2019; West, 2022). But this oxygen only became available once all the rest of the oxygen sinks were filled – reactions, mineralization, terrestrial rusting. This slow, multi-billion year process, which enabled oxygen to slowly fill up every possible reservoir before accumulating enough to oxygenate the ocean and then fill the atmosphere, presents a lesson of patience and fortitude. It also paints a picture of an organism without agenda, dedicated to its unique ability to alchemize sunlight and water into energy and oxygen. In order to live, algae provide life.

2 Taxonomic Ambiguities

While resembling plants, algae overflow taxonomic kingdoms. From the 60 meter long Pacific Ocean giant kelp (*Macrocystis pyrifera*) to the micro ur-plant green algae (treated sometimes as plants (division *Chlorophyta*) and sometimes as protozoans (phylum *Chlorophyta*, kingdom Protista)), photosynthetic aquatic organisms of all sizes have escaped the sights of critical plant studies. In Brill's series, for example, algae receive but an excerpt on speculative *algaerobot* colonies in *The Plant Contract, Art's Return to Vegetal Life* (2018). Despite being the origins of all terrestrial plants, the invisible promiscuities of algae may at first glance appear even more foreign a subject for philosophical investigation than (terrestrial) plants. The reason for our scarceness of knowledge in respect of algae is almost certainly due to the struggle of finding a unique and univocal definition of what algae are and finding suitable methods for their classification. Since the term 'algae' includes a gigantic number of species, with ubiquity on the earth, and often a miniscule size that requires sophisticated techniques to even visualize let alone isolate species, any attempt at an ultimate taxonomic classification will be limited. Furthermore, algae are not a specific taxonomic group, but a category of living organisms defined by a term used similarly to the way in which people refer to "trees," "grass," or "herbs." From a botanical perspective algae are classified as thallophytes, meaning that they are considered as plants lacking roots, stems and leaves. They use chlorophyll-*a* as their primary photosynthetic pigment, or other photosynthetic pigments, such as phycobilins (red algae). They also lack the sterile cover of cells around reproductive cells, although they can have defined tissues containing specialized cells. Compared to terrestrial vascular plants, however, the degree of specialization or differentiation of cell types is much less pronounced (Wiencke and Bischof, 2012).

This means that the current classification system of algae primarily into microalgae and macroalgae is insufficient to describe the multitude of differences among existing species. For instance, microalgae are unicellular microorganisms (prokaryotic or eukaryotic), which are able to accumulate biomass by the photosynthetic process using sunlight, water and carbon dioxide. In this group alone, there are more than 200,000 microalgal species but only approximately 72,000 species have been classified (Guiry, 2012). These microplankton are the original photosynthetic agents which figured out how to harness food directly from sunlight, in a briny medium. From a few to a few hundred micrometers (μm) in size, microalgae individually are invisible to human vision, but collectively are not only dazzling in color but are major drivers in maintaining homeostasis for oxygen and other biospheric properties in aquatic environments.

On the other hand, macroalgae are eukaryotic photosynthesizing multicellular organisms that are somewhat more differentiated than microalgae, however less than plants. Widely known as seaweeds, these organisms have had central importance in some civilizations. Beyond the category of micro and macroalgae, a bevy of other aquatic vascular plants come in complexifying algae taxonomies. Hydrophyte variety is almost as dazzling as their terrestrial varieties. The term '*hydrophytes*' is commonly applied to the vascular plants, mosses, and larger forms of algae (in particular filamentous algae and the stoneworts – *Characeae*) which are visible to the naked eye; smaller algae and phytoplankton are therefore excluded. Aquatic plants are typically divided into four categories – emergent, submerged, floating leaves and free-floating. However, these distinctions are arbitrary and some species exhibit different growth forms according to environmental conditions, showing clear environmental plasticity.

When it comes to aquatic environments and ecosystems, microalgae are the dominant primary producers and the base of the food web together with phototropic cyanobacteria (collectively known as phytoplankton). However, other species are intimately connected and linked to phytoplankton, especially aquatic heterotrophic bacteria which interact with phototropic organisms. Phytoplankton–bacteria interactions are multifaceted and can create many different ecological relationships, from cooperative to competitive (Seymour et al., 2017). Co-occurrence of bacteria and algae can be traced back to billions of years ago. Their coexistence in similar ecosystems has promoted a plethora of useful and effective interactions among these two groups over evolutionary time scales. These phytoplankton-associated bacteria are capable of producing various biologically active secondary metabolites and play a central role in algal growth and metabolism (Variem and Kizhakkedath, 2021). Without any doubt, in some cases these relationships can also be considered within the framework of symbiosis. The physical interface where these interactions occur is the region surrounding an individual phytoplankton cell, a microcosm that has an enormous impact on the planetary macrocosm since it is responsible for more than 50% of the oxygen we breathe and is the sink for about 40% of CO_2, one of the main greenhouse gasses. Here, metabolites can be rapidly exchanged between the phytoplankton itself and the bacterial cells. This region is named the 'phycosphere' analogously to the same region surrounding a root in soil, colonized by thousands of different microorganisms, and its importance as a fundamental ecological interface in aquatic ecosystems has been recently confirmed (Seymour et al., 2017). The phycosphere can be described as a microscopic ecosystem with macroscopic impact.

Because of their myriad forms and interaction, algae as a category surpasses any taxonomic attempts to simplify or box in these ranging species of organisms, proliferating in size, shape, domain, and function. Water plants transform what we think of as a plant, but also, by bending the category 'plant,' help us better understand the fundamental symbiosis and multispeciality of terrestrial plants as well. Perhaps because of algal flexibility and responsiveness to context and environment, in late capitalism, they are also now a final frontier for instrumentalization.

3 Industrializing Algae

After the Second World War, a period of prosperity in the US went together with a race for new sources of energy. Since the turn of the 21st century, algae has become the focus of extractive research projects. Algae are expected to deliver wonderful promises in all kinds of fields: energy, materials, food and feed, pharmaceutics, cosmetics, and more recently ecosystem services and restoration. As new characters of the biotechnology enterprise, *Spirulina* is already on the market billed as a 'superfood', *Chlorella, Galdieria, Nannochloropsis* and other algal organisms are framed as and reduced to 'precious lipids' and 'proteins', both for the cosmetic and food industry. The algal bonanza promising an industrial *deus ex alga* for our scalability woes rings in a moment of biotechnological supremacy, which seems to know no bounds. When considering the advancements of biotechnology brought forward at the widest European research unit dedicated to microalgae, the lives of these organisms are the "green gold of the future" slated to fulfill all our human needs, as miniature slaves (Wolkers et al., 2011).

In these current schemes, global capital's renewed interest over algae is an instrumentalization of algae that permits apologetics for continued extractivism elsewhere. Just like in other bait-and-switch strategies such as carbon capture and storage, the promised metrics of algae are always higher than what they actually do. Yet, algae are not machines, but collective metabolisms.

The growing hypoxic zones of the oceans are as much an effect of algal dysbiosis as a reflection of the meat-grinder disposable culture globalized which has produced this bombardment. While the captains of industry used to proclaim "the solution to pollution is dilution!" – and what was the ocean but the biggest pollution sink of all? – by the time Rachel Carson started sharing her findings in the 1950s and early 1960s, aquatic ecosystems the world over were already in the process of slow-motion collapse.

One field of research attracting multibillion investments is the use of algae as feed for cattle. If cows were a country, they would be the third largest emitter of greenhouse gasses, after China and the US (Barthelmie, 2022), with the global numbers of livestock three times increased over the last 50 years, now representing about 60% of total mammalian biomass. Feeding algae to cows is a way to develop other sources of protein-rich feed and to lower production costs. It is also presented as innovation towards sustainability. For example, one team of researchers found that including seaweed in cow feed can reduce their methane emissions by 82%, but point to problems of scalability (Roque et al., 2021). Or, as Forbes writes: "Just a sprinkle of the SVD to existing livestock feed and poof: less methane from the cow" (Kart, 2020), where SVD refers to a feed product for cattle derived from the seaweed *Asparagopsis taxiformis*. A startup (Symbrosia, recently selected as part of the 2020 Solver class by MIT Solve) emphasizes the successful traits of *A. taxiformis* asserting that replacing just 0.4% of a cow's feed with the seaweed extract reduces the amount of methane the cow produces by more than 90%. This 'cowbucha' of *A. taxiformis*, consumed by generations of Hawaiians calling it *limu kohu* or *limu līpehe*, is an unlikely but much-hyped downstream solution for the upstream problem of overconsumption.

Critics of this pursuit have seen the problem as a kind of 'go on and don't change your lifestyle habits or the factory farm system of meat production, we'll just wave our magic algae wand over here and the problem will go away.' Overstating how much methane feeding seaweed to cows reduces can lead to faulty moral accounting, potentially allowing people to consume even more unsustainable meat, now that they can do so with less climate guilt. This creates a sort of Jevon's Paradox for meat, where the less damage we think it is causing, the more willing we are to overindulge. Again, rather than reduce the total number and intensity of cows under human control, the path of least corporate resistance has turned to kelp as a miracle drug to add in small percentages to cow feed. With headlines claiming that kelp additives can reduce cow methane effluent by up to 85%, who needs to eat less meat when we can simply supplement impoverished cow diets of corn, soy, and other genetically engineered staple crops with a bit of sea plant? After all, given the sea often lacks formal jurisdictions, there are few property rights in it, things move around, making it the perfect place for a first-mover enterprise to achieve a monopoly.

In the spirit of markets, since plant-based protein sources have consistently acquired greater importance as components of vegan and vegetarian diets becoming increasingly popular, but also to supplement eco-conscious animal-inclusive diets, algae present an intriguing possibility to supply global protein demand and minimize the existing protein gap among the socioeconomically disadvantaged. Some species of seaweed and microalgae are known to contain protein levels similar to – and often higher than – those of traditional sources, such as meat, eggs, soybeans, and dairy. Furthermore, seaweed and microalgae also have higher protein yields per unit area compared to terrestrial crops, such as soybeans, legumes, and wheat, while delivering essential micronutrients. For instance, *Spirulina* contains 180% more calcium than milk, 670% more protein than tofu, 3,100% more β-carotene than carrots, and 5,100% more iron than spinach (Burrows, 2021).

Cattle farming for meat and dairy production is not the only human activity where algae are currently exploited, considered as a mere tool to reach a problematic semblance of sustainability, transformed often into passive agents in this transition. Another example is the utilization of algae as an advanced technology to capture CO_2 through biosequestration.

Ocean afforestation is the latest ploy in playing energy accounting. While undoubtedly a good idea to stop bunker oil shipping, deepwater drilling, and other ways in which industrialism has colonized the seas through pollution, this is not what afforestation tycoons have in mind. In the lucrative carbon sequestration markets, storing carbon in seaweed would involve farming seaweed in massive quantities in order to reduce CO_2. Through the photosynthetic process, wild seaweeds incorporate dissolved inorganic carbon in the upper layers of the ocean into plant tissue. Floating kelp can suck up the carbon and then fall to the bottom of the ocean when they die, decomposing and locking in stored carbon for millions of years in the seabed. Existing natural seaweed ecosystems sequester roughly 173 million tons of carbon annually (Krause-Jensen and Duarte, 2016). As the race for marine carbon sinks heats up with our planet, researchers and corporations are looking for more ways to exploit seaweed for sequestration purposes. Atmospheric carbon dioxide removal is now viewed as the necessary Hail Mary measure to keep on polluting as usual but buy our indulgences through technofixes.

While farming sea kelp is a welcome low-tech bioremediation measure, it is often being done for instrumental reasons. Burying seaweed as biosequestration of carbon can easily become an excuse to continue polluting, the perfect substitution. Seaweed dying and sinking to the bottom of the ocean, stores about 10% of the total CO_2e emissions of our annual anthropogenic input from only automobiles (Krause-Jensen and Duarte, 2016). On their own, seaweed seem like an excellent carbon sequesterer. But we easily tend to forget that seaweed – and phytoplankton – are at the bottom of the food web, and are eaten by other organisms which themselves produce rather than consume CO_2. The logic of *deus ex alga* finds yet another exemplar in the strikingly opposite observations *in vitro* versus *in vivo* science provides, when seaweed ecosystems do not perform the trick we had hoped (or counted on) (Gallagher et al., 2022). Large-scale monoculture operations require a growing environment with as little complexity as possible, a "blank slate" with little to no biodiversity. The aim of this is to maximize efficiency, and create an *in vitro* environment *in vivo*. But life is complicated, and recursively effects and infects its environment, however intentionally sterile. Water too, has often been viewed as a *tabula rasa* to distill and infuse with different elements.

Moreover, such a scheme might have knock-on effects for other species that haven't been accounted for yet. For instance, reductions in phytoplankton production and trophic exchanges of energy that could adversely impact fisheries and marine mammals (Boyd et al., 2022; Burns 2022). At the same time, seaweed farming can also lead to the introduction of invasive species, carried by macroalgae into offshore ocean ecosystems.

For other reasons (of dissimulation), even ExxonMobil has been in on algae mania, investing in a biotech company to "develop next-generation biofuels from photosynthetic algae" (ExxonMobil, 2018). After funding MIT researchers and Synthetic Genomics, Inc (now Viridos), ExxonMobil saw algae biofuel as the next step after their push for corn ethanol, which risks emitting more CO_2 than regular old oil extraction over the whole production process (ExxonMobil 2018). Exxon's goal was to furnish 10,000 barrels of algae biofuels daily by 2025. To achieve a higher oil content, they genetically engineered an algae strain to boost oil production from 20 to 40 percent, publishing such results in prestigious academic journals such as *Nature Biotechnology* (Ajjawi et al., 2017). This magic trick of turning living organisms into biocrude oil (with a whole lot of inputs and energy to achieve high temperatures and pressures, begging the question of thermodynamic efficiency) is an easy transition from fossilized algae and other bodies. Part of our larger cultural obsession of raising life just to extract from it, the giving nature of algae has become a cure-all for the externalities of petrol industrialization. But will working more closely with them also address the internalities?

Algae oil was presented as the next "green gold" (Wolkers et al., 2011), a trillion-dollar market, both for supplying post-peak oil biocrude and as a carbon sink (we're told: "algae is up to 400 times more efficient than a tree at removing CO_2" (Lamm, 2019)), allowing companies to see a future devoid of addressing the ethical concerns of endless extraction of work from algae. Ironically, this frenzy of interest for algae oil is now declining. All big oil firms have stopped their investment in R&D activities around algae biofuel and even ExxonMobile announced in March 2023 that it withdrew from the race. With billions of dollars invested in this research field, algae proved to be an excellent marketing tool for those companies but obviously commercially disappointing, since they were not performing as expected.

If it's not for oil, the other benefits seen in algae are still an exciting prospect for tech companies around the globe. Part of peak instrumentalization, algae-as-objects become increasingly implicated in transnational economies.

Coccolithophore, a type of unicellular, eukaryotic phytoplankton, already eyed by scientists tantalizingly as a potential carbon sink (Smith et al., 2012), now has additional labor burdens placed on it by industry as it has been discovered that the alga absorbs carbon as it grows and turns it into calcium carbonate, indistinguishable from limestone. Like roughly 200 other species of marine phytoplankton, *coccolithophores* cover themselves with calcium carbonate shells called "coccospheres," presumably to protect against micro-zooplankton predation – their main predator. By protecting themselves with this shield of coccoliths (forming the coccosphere), microzooplankton pass them by and graze on more easy prey of other species of phytoplankton (Mayers et al., 2020). But since industry is hungry for cement, *coccolithophore*'s ability to biomineralize looks like the perfect mine to factory farm, replacing open pit limestone quarries. The World Economic Forum (WEF) tells us in one of their slick videos (World Economic Forum, 2022) that as cement is the second-most consumed substance on earth after water, responsible for 8% of total CO_2e emissions, we've just struck an ecological treasure trove. Instead of mining limestone from the earth, we'll now just mine "biogenic" limestone substitutes from algae. Perhaps most telling of WEF's suppositions, is their claim that by shifting over from limestone pit mining to massive algal mining through terrestrial algae farms, now we won't have to "disrupt nature." We won't have to reduce cement production or use, or reuse the cement we have, or find more regional building materials; instead, we can now mine what looks like a less polluting source of base material, according to a single-metric approach. Because sacrificing quadrillions of algae lives for our cement industry and necrocapitalism isn't a disruption. Algae are just unicellular water plants – they have no lives of their own that would cause us to blink in instrumentalizing them to the hilt.

Materials science professor Will Srubar has commented about his biogenic limestone initiative, "If nature can grow limestone, why can't we?" subsuming algal life into human artifice (Hempstead, 2022). Here, algae become nebulous nature, at the command of *homo faber*, and the aim of calcareous microalgae in providing protection for themselves becomes irrelevant. The material is what matters, not the purpose for it. Not required is any reexamination of concrete culture, the proliferation of paving the earth, and generating buildings not built to last. Instead, Srubar encourages that we just "plug and play" biogenic limestone into modern cement production processes (Simpkins, 2022). Biomimicry, which takes nature as inspiration to develop a design method, can be applied at the levels of an organism, a behavior, or an ecosystem. It is fruitful as long as it is dialogical and regenerative. Applying biomimicry methods at an ecosystemic level encourages holding a holistic vision of a specific problem and to reach a necessary balance. When it becomes a mercenary activity in order to break through the limits to growth, it fails to fulfill its biomimetic potential and becomes a form of biopiracy.

Cultivating marine algae on land-based farms could meet future nutritional demands from society and enhance environmental sustainability. In order to feed the 10 billion humans modeled to be on the planet by 2050, these land-based farms could produce way more than the 56 percent increase in food production needed (Agard, 2022). Since naturally 'carbon-eating' microorganisms are grown, this could mean that these farms will be carbon negative too. None of this is to say that we shouldn't eat algae and seaweed, and incorporate these water plants into our economies, climate plans, and other ways of creating more balance in our ecosystems and societies. We absolutely should. However, *how* we go about interacting with algae is key. If we wish to just use them for short-term gains, without taking the time to get to know them, it is unlikely that even their seemingly magic powers will do our species much good. Perhaps, by treating them so instrumentally, they may simply refuse to perform.

Algae are currently the main character in different approaches carried out by universities and labs in biotechnology, often partly sponsored by important players of the petrochemistry and agricultural industry they are positioned to replace (or continue by other means). Such a means of reproducing the same logic that produced the need to turn to algae lives as the latest sacrifice, instead of repositioning ourselves vis-à-vis algae, instead aims to sustain the unsustainable. This book instead explores what it would be like for us to get out from under our instrumentalizing imperative, asking what it would be like to become 'algae' (or algae-esque). Becoming algal entails a stripping away of the performativity of humanness, letting go of the standardization and homogenization of playing the role of the traumatized, civilized human, cut off from realizing our physical reliance on others. Algal being-in-solidarity with the life of our medium presents the task of acknowledging the for-whatness of all action, all work. To produce the very breath of life and metabolize without needing to kill evokes a model attainable by algae, but no human. Such beneficence can nonetheless serve as the horizon of human ethics in our making, extending towards inhabiting technologies of photosynthesizing rather than extracting.

4 Becoming Algae

This volume promiscuously mixes academic genres to return to the algal first-organism perspective: what it is like to be algae. Taking an ecological approach, a novel way of looking at critical plant studies in the binding medium of water, algal beings can transport us deeper into the collectivist origins of plants and their cycles. By attending to algal *Umwelten*, the figure-ground shift between environment and organism becomes that much more profound. In acknowledging the primacy of environment in determining being – and thus the preciousness of allowing organisms undisturbed to cultivate their own niches (especially when they are beneficial to others) – agency mixes up in a contrapuntal way behavior, action, and phenomenology.

The ambiguities in this book's title, *Being Algae: Transformations in Water, Plants*, arise from the discussion of water, plants, and water plants, as three separate but also inseparable aspects, manifesting the already uncertain position of algae within critical plant studies. Both macro, micro and beyond, algae and other water plants compose and are composed by the fertility of their medium. Water, like soil, is an active, alive medium, which gives birth to those organisms in its midst. Just as those agriculturalists who value the life of soil tend to regard plants with more respect, so too we wish that in offering a peek into the plant life of water, that we can better appreciate the micro and macro-organisms that compose its character.

5 Origins and Book Overview

This book elaborates on a series of workshops focused on the scientific as well as phenomenological anomalies of algae held in 2019 at Het Nieuwe Instituut, the Rotterdam-based Dutch National Center for Design, Architecture and Digital Culture. Directed by Johanna Weggelaar, who is conducting multidisciplinary projects focusing on algae, the workshops together explored the ethico-onto-epistemologies of algae alongside their biological, communicative, and artistic dimensions with an audience of artists, academics and students. With lectures and workshops by plant biologist Sergio Mugnai, environmental philosopher Yogi Hale Hendlin, and other academics and interpretive artists, the series' discussions rewarded our team with an archive that pushed us to give further academic voice to metabolize the emergent topics, especially around the recurring theme of the industrial exploitation of algae for geoengineering and bioreactor-based industrial production. Collecting a transdisciplinary selection of algae experts and interpreters in keeping with the diverse character of the original collaboration, the book is composed of thirteen chapters, organized according to three sections. We thank Claes Bech-Poulsen for permission to use the cover photo, *Tang 5*. We also are grateful to the Dynamics of Inclusive Prosperity Initiative and the University Library at Erasmus University Rotterdam for generously providing funding allowing this book to be Open Access.

The first section aims at introducing an algae-based perspective to the world of critical plant studies. Attempting answers to questions such as '*How can we encounter algae differently?*', the chapters here lean on us to reflect over renewed forms of morality and ethics when considered from a non-human, algal standpoint.

In the **first chapter** (*There's Something in the Water: Algae, Eliminativism, and Our Moral Obligation to Biological Beings*) M. Polo Camacho and Andrew Lopez question the ontology of algae starting from the premise that human beings are not the only bearers of value and remind us of the importance such claims holds to environmental ethics. Their chapter focuses on algal ecology to demonstrate algal systems' inherent processuality through an eliminativist approach to individuality. Having detailed how we can understand biological beings as processes, the authors conclude attempting to define the nature of moral obligation when directed at biological processes, considering for instance how pluralist organismality and processuality influence our consideration of the future.

In the **second chapter**, Jesse D. Peterson gives voice to certain species of algae encountered in the archipelago of Stockholm, reflecting about the way humans see them as mostly "harmful." Experimenting with "art-based data analysis" and multispecies "art of attentiveness," Peterson brings us into an underwater journey, delivering back on paper a speculative dialogue he has entertained with *Nodularia spumigena*, an algae species known for its toxic effects. The dialogue reveals the relational nature of the human-algae interactions responsible for the naturalization of some algae as "harmful." Entitled "*Seeking an Algal Perspective: Exploring 'Harmful' Algae through an Interview with* Nodularia spumigena," this chapter draws on recorded anecdotes, observations, and the thirty years old "Harmful Algae Newsletter," with the intention of taking seriously the need to explore algal historicity in order to explicate the ways in which maritime environments come to matter for humans and harmful microalgae. Throughout the conversation, the author discusses the relative benefits of attempting to understand human concerns from nonhuman perspectives, along with the ensuing contradictions that arise between human desires about algae and the oceans.

Having attempted to operationalize a responsible, relational mode of using algae that respects them as living beings, **chapter 3** *Contemplating Life, Death and Time Together with Diatoms*, continues the section. Here, Nina Lykke mobilizes her own experience of a sitting out dedicated to the corpo-affective contemplation of our relations to the more-than-human world as a prism to reflect on a planetary ethics of companionship. The text is written following a posthuman autophenomenographic methodology, bringing intuitions and biographical experiences at the center of her ethical speculations and supported by a vitalist materialist and immanence philosophical approach. The ethics of companionship proposed punctually troubles modern, colonial, extractivist onto-epistemological frameworks of biodiversity ignorance, while questioning the exceptionalizing of the human subject, supposedly presiding over a hierarchy of more or less "distant" and instrumentalizable others. Overall, the author makes a case for considering diatoms beyond widespread profit extraction activities, as wise ancestors, capable of teaching us lessons about life, death, and time.

Chapter 4 discusses the mounting deployment of photobioreactors (PBR) for algal work (to produce energy, food, etc.) in light of the ecological needs of algae. Decentering algae as single species through taking into account each algal species' multispecies mutualists forming the 'phycosphere' of algal habitat, the feasibility of sealed off laboratory sterility is questioned. The authors (Yogi Hendlin, Johanna Weggelaar, Natalia Derossi and Sergio Mugnai) suggest opening the typically monoculture and controlled environments of algal bioreactors to multi-species consortia both with other algae and with their bacteria symbionts. Attending to and reintroducing the microenvironment surrounding and interacting with microalgal cells potentiates the production of certain metabolites through interaction with cohabitating microorganisms. The chapter argues that through a better understanding of the phycosphere, more sophisticated forms of PBR incorporating algal-microbial consortia may be more productive (with less frequent inputs) and ethical (enriching the habitation of the target species) than single-species algal systems.

The **second section** explores aspects of human-algae relationships. Ancient relations or more recent histories are embodied in traditions, imaginaries and narratives. The material impact of these relations includes not only immediate physical interactions with the algal world, but also distant representational projections. Main inquiries here include: 'What have been and are various human communities' attitudes to algae? And how do the latter invite us to reconsider our own relationship to other organisms', 'How is research shaping our understanding of algae blooms'? These understandings of algae introduce plural models in the exploration of non-human alterities.

Chapter 5, *Algae in the Human World: Beauty and Taste Come First*, opens the section, its authors giving us a glimpse of how humans have interacted with algae throughout millennia. Focusing on two often overlooked facets linking us to algae, namely beauty and taste, the chapter describes the human journey in the company of algae from the earliest day of our species all the way until today, where algae are gaining increasing attention as a possible 'new' alternative food and material source. The text details algae's historical presence in museums and collections due to their beauty and kitchens and dinner tables as unique umami flavor providers. Overall, the Authors state that the history of humans' seaweeds inspire us to consider them as reminders, and guides to refresh our entangled life paths with seaweeds. For Ole G. Mouritsen and José Lucas Pérez Lloréns, one powerful way of reconnecting the algal organisms is paying closer attention to the way we sense them with our eyes and our palate.

Reinforcing the case advanced by chapter 5, the following chapter shows us once again that algae have been a valuable resource for crafts around the world. In **chapter 6** artist and designer Kathryn Larsen brings us inside her craft-making laboratory to see what it takes to work with algae. Particularly, taking the case of Danish "moss paint" and the Japanese "glue walls", the chapter (an investigation of algae's applications, inspired by indigenous and vernacular craft traditions) argues that by using algae material-based design as a resource and a tool for education, we can begin to build a cultural connection with algae once again in Western Europe. For Kathryn Larsen, beyond food, or worse, a rotting mass on a beach, we should see algae as potential craft material.

From Denmark and Japan, Mustafa Yavuz brings us to the Middle East, investigating diverse Arabic medicinal sources authored by medieval scholars, discussing their views on botanical taxonomy at that time. What role did algae play for medieval Islamic natural philosophers, physicians, or polymaths? In **chapter 7** (*Uses of and Considerations on Algae in Medieval Islamic Geography*) Yavuz opens a window into medieval uses of algae in the Islamic Civilization (11th to 13th century) through the examination of books of medieval medicine and pharmacy. Attempting to decipher the form of human-algae relationships in the Islamic Middle Ages, the author supplies an informative base in order to question the imaginaries and narratives of the past, but also providing epistemological and ontological basis allowing us to question our contemporary knowledge of and in relation to algae. Mustafa Yavuz illustrates that from poems in literature, to drugs in medicine, medieval people "naturally" incorporated the algae in their daily and social lives.

If in medieval Arabic culture algae were part of daily life, today our perceptions seem rather binary: natural blooms are often toxic, but in the laboratory we can produce precious compounds from them. Composed by Gustaaf Hallegraeff, **chapter 8** (*Microalgae and Human Affairs: Massive Increase in Knowledge Drives Changes in Perceptions of Good and Bad Blooms*) demonstrates how an increase in knowledge of microalgae has been driving ever changing perceptions of good and bad algal blooms, our recognition of the central role microalgae play on our planet, as well as for our human future. The chapter traces the role algal culturing techniques, electron microscopes, molecular genetics, satellite imagery, as well as shellfish and finfish farming have had on our understandings of algal microorganisms: the awe and wonder characterizing our understanding of the organisms gave way around the 1980 to negative perceptions of adverse impacts from harmful algal blooms, negative feelings which have then recently been replaced by a new type of biotechnological and bioengineering amazement, based on their photosynthetic and reproductive capacities and focused on their primordial role in evolution and present oxygen production. Arguing for the critical importance of such organisms for the future of our planet, the author claims it is perilous for our own human survival to ignore this 'critical creation'.

Responding to the fact that a disproportionate amount of applied scientific research concerning algae involves exploitative practices, focusing for instance on algae market potential within biofuels and bioplastic industries, **the third section** critically examines the uses and misuses of algae within human society. However, it does not only deal with the limits and promises of an algae based economy, it also lingers upon entanglements between algae commercialisation and fake stories of indigeneity or national security and labor exploitation. Next, it explores the potential of using them differently, considering for instance polycultures potentials and drawbacks, as well as their urbanistic potential to become a connector between the urban and the marine environment.

Making Marimo-algae its main character, in **chapter 9** Jon Pitt maps the role the charismatic organism had over an imagined indigeneity, beginning around Lake Akan and spreading throughout Japan (and ultimately the world at large). The author tries to consider what it means for the alga to have been embraced by the Lake Akan Ainu community in the name of preservation twice-over: of the endangered alga and of Ainu identity itself. Following this alga for Pitt means to superimpose through its narrative the historical, literal and folkloric elements of the Marimo involvement with the Lake people as well as with the dancing and singing practices linked to it, all the way to the contamination issues entrained by the network of Marimo's international distribution consequent to its popularity.

Melody Jue's **tenth chapter** (*"A Seaweed Goes to War": Agar as a Thermal Medium in C.K. Tseng's Research at the Scripps Institution of Oceanography (1943–1946)*) makes the reader travel into Agar's experiments and its entanglements with USA national security. Agar is an important seaweed product used to culture and identify bacteria in water supplies, milk, and medical supply contexts, one that has almost entirely been provided by Japan previous to WWII. Delving into the work of Chinese phycologist C.K. Tseng at the San Diego Scripps Institution of Oceanography, and particularly on the scientist's metaphor of a soldier-agar, the author shows how Tseng's wartime scientific writings tell a story about Agar as a thermal medium of labor in transpacific contexts. Analyzing the ways in which Tseng addressed both eastern and western thermal epistemes when describing Agar as a thermal medium, a culturing medium, and finally a medium of labor, the author brings forward a dual arguments: first, that the Japanese scientist overcomes 'thermal objectivity', a term referring to a temperature that would be independent from culture and perception; second, that Tseng's personification of Agar as a soldier is not only literary, but also a reflection of his own conditions of labor in the laboratory under the shadow of the war.

Within the framework of bio-integrated architecture and 'photosynthetic cities', Brenda Parker and Marcos Cruz's **chapter 11** (*Augmented Polycultures:*

Scaling up Algal Ecosystems and Design of a Biofouling Aesthetic) opens by acknowledging that the set-up of algae cultures for biotechnological purposes mostly comprise of enclosed systems and pure monocultures depending on complex maintenance apparatuses. Their text attempts a leap beyond monocultures and challenges the reader with a design for polyculture heterogeneity. The authors discuss the aesthetic drawbacks engendered by polycultures – namely biofouling and decay. The chapter shows that these conditions have been historically rejected because they reflect neglect and diseases, to then propose a design that accepts such material impacts as generative qualities. For Brenda Parker and Marcos Cruz, the quality of urban environments will depend and thrive on the augmentation of polycultures and a renewed ineffable aesthetics.

In **chapter 12** (*Phytofictions and Phytofication*) artist and designer Julia Lohmann explores kelp's potential as a sustainable material for making, while delivering her creative mode of production through the notions of *seaweedness, phytofictions* and *phytofication*. Seaweedness comprises the inherent properties of macroalgae that guide her creative process. Phytofictions in turn is a term capturing the co-imagined seaweed futures Julia Lohmann has invented with diverse publics throughout encounters at the Victoria and Albert Museum, London. Finally, the process of phytofication describes the ambition to become kelp, progressively taking place as we work with it. Can working with algae transform our actions and reflections towards a less destructive, perhaps even regenerative practice? This chapter attempts an answer to this question.

Lastly, **chapter 13** by Soo Jung Ryu and Cintia Organo Quintana is a call to planners, developers, designers, ecologists and citizens to grasp the opportunity to re-envision our current business-as-usual waterfront areas in light of a rise in sea levels. *Seaweed as the Denizens of the New Commons in the Anthropocene* seeks to investigate the unexplored potential of seaweed critical coastal ecosystem, pondering upon its potential to become a connector between the urban and the marine realm and establishing its ability to occupy the critical zone of transformation as a new form of commons. The authors' discussion stimulates recognition of the importance of engagement to the marine world through the lens of the seaweed, opening windows of transformative changes allowing us to fundamentally re-envision how coastal cities approach the sea.

Having assembled alliances on the topic of algae between literature, philosophy, biology, anthropology and art, we hope that the vignettes offered in this volume can support reflection on the possibilities for collaborative approaches and more-than-human diplomacies. Against a monolithic plant studies taxonomy, we plead for the algalization of humans, in attempts to appreciate our interspecies colocation with algae within the context of strategies for living and persisting in the current tempest of separation-induced climate change.

References

Agard, S. 2022. These land-based algae farms could feed 10 billion humans by 2050. October 11, 2022. https://interestingengineering.com/science/land-based-algae-farms.

Ajjawi, I., J. Verruto, M. Aqui, L.B. Soriaga, J. Coppersmith, K. Kwok, L. Peach, E. Orchard, R. Kalb, W. Xu, T.J. Carlson, K. Francis, K. Konigsfeld, J. Bartalis, A. Schultz, W. Lambert, A.S. Schwartz, R. Brown, E.R. Moellering et al. 2017. Lipid production in *Nannochloropsis gaditana* is doubled by decreasing expression of a single transcriptional regulator. Nature Biotechnology 35 (7): 647–52.

Barthelmie, R.J. (2022) Impact of dietary meat and animal products on GHG footprints: The UK and the US. Climate 10: 43.

Boyd, P.W., Bach, L.T., Hurd, C.L. et al. 2022. Potential negative effects of ocean afforestation on offshore ecosystems. Nat Ecol Evol 6, 675–683.

Braje, T.J., T.D. Dillehay, J.M. Erlandson, R.G. Klein, T.C. Rick. 2017. Finding the first Americans. Science 358 (6363): 592–94.

Burns, W. 2022. Can kelp help? The potential role of 'ocean afforestation'. Illuminem. July 13, 2022. https://illuminem.com/illuminemvoices/2bba1334-8bae-43f0-b543-73da972bb20b.

Burrows, D. 2021. Algae as human food. January 11, 2021. https://www.safefood.net/food-safety/news/algae.

Cheng, S., W. Xian, Y. Fu, B. Marin, J. Keller, T. Wu, W. Sun, X. Li, Y. Xu, Y. Zhang, S. Wittek, T. Reder, G. Günther, A. Gontcharov, S. Wang, L. Li, X. Liu, J. Wang, H. Yang, X. Xu, P.M. Delaux, B. Melkonian, G. Ka-Shu Wong, M. Melkonian. 2019. Genomes of subaerial Zygnematophyceae provide insights into land plant evolution. Cell, 179 (5) 1057–1067.

Darwin, C. 1832–1836. The Voyage of the Beagle, Volume III, Journal and Remarks, p. 305.

Erlandson, J.M., M.H. Graham, B.J. Bourque, D. Corbett, J.A. Estes, R.S. Steneck. 2007. The kelp highway hypothesis: marine ecology, the coastal migration theory, and the peopling of the Americas. Journal Island Coastal Archaeology 2 (2): 161–74.

ExxonMobil. 2018. Advanced biofuels and algae research. ExxonMobil Climate Solutions. September 17, 2018. https://corporate.exxonmobil.com:443/Climate-solutions/Advanced-biofuels/Advanced-biofuels-and-algae-research.

Gallagher, J.B., V. Shelamoff, C. Layton. 2022. Seaweed ecosystems may not mitigate CO_2 Emissions. ICES J Marine Sci 79 (3): 585–92. https://doi.org/10.1093/icesjms/fsac011.

Guiry, M.D. 2012. How many species of algae are there? J Phycol 48(5): 1057–63.

Hempstead, M. 2022. Algae-grown limestone could be the key to 'carbon negative' cement production. July 1, 2022. https://www.springwise.com/innovation/property-construction/algae-grown-limestone-for-cement/.

Kart, J. 2020. Hawaiian seaweed makes cows 90% less gassy – And that's good for climate change. November 21, 2020. https://www.forbes.com/sites/jeffkart/2020/11/21/hawaiian-seaweed-makes-cows-90-less-gassyand-thats-good-for-climate-change/?sh=712c08f25c4b.

Kaufmann, F.H. 1965. Early and modern upper taxon classification of plants. The American Biology Teacher 27 (10): 792–97.

Kharkongor, D., P. Ramanujam. 2014. Diversity and species composition of subaerial algal communities in forested areas of Meghalaya, India. Int. J. Biodiversity, Article ID 456202.

Krause-Jensen, D., C.M. Duarte. 2016. Substantial role of macroalgae in marine carbon sequestration. Nature Geoscience 9 (10): 737–42.

Lamm, B. 2019. Algae might be a secret weapon to combatting climate change. Quartz. October 1, 2019. https://qz.com/1718988/algae-might-be-a-secret-weapon-to-combatting-climate-change/.

Lawrence, A.M. 2022. Listening to plants: conversations between critical plant studies and vegetal geography. Progress in Human Geography 46(2), 629–651.

Mayers, K.M.J., A.J. Poulton, K. Bidle, K. Thamatrakoln, B. Schieler, S.L.C. Giering, S.R. Wells, G.A. Tarran, D. Mayor, M. Johnson, U. Riebesell, A. Larsen, A. Vardi, E.L. Harvey. 2020. The possession of coccoliths fails to deter microzooplankton grazers. Front. Mar. Sci. 7: 569896.

O'Connor, K. 2017. Seaweed: a global history. Reaktion Books Ltd, London, UK.

Ragan, M. 1998. On the delineation and higher-level classification of algae. Eur J Phycology 33: 1.

Roque, B.M., M. Venegas, R.D. Kinley, R. de Nys, T.L. Duarte, X. Yang, E. Kebreab. 2021. Red seaweed (*Asparagopsis taxiformis*) supplementation reduces enteric methane by over 80 percent in beef steers. PLoS ONE 16(3): e0247820.

Ruhfel, B.R., M.A. Gitzendanner, P.S. Soltis, J.G. Burleigh. 2014. From algae to angiosperms – inferring the phylogeny of green plants (Viridiplantae) from 360 plastid genomes. BMC Evol Biol 14: 23.

Seymour, J., S. Amin, J.B. Raina, M. Stocker. 2017. Zooming in on the phycosphere: the ecological interface for phytoplankton – bacteria relationships. Nat Microbiol 2, 17065.

Simpkins, K. 2022. Cities of the future may be built with algae-grown limestone. CU Boulder Today. June 23, 2022. https://www.colorado.edu/today/2022/06/23/cities-future-may-be-built-algae-grown-limestone.

Tetu, S.G., I. Sarker, V. Schrameyer, R. Pickford, L.D.H. Elbourne, L.R. Moore, I.T. Paulsen. 2019. Plastic leachates impair growth and oxygen production in *Prochlorococcus*, the ocean's most abundant photosynthetic bacteria. Commun Biol 2: 184.

The Seaweed Site. 2022. https://www.seaweed.ie/aquaculture/introduction.php#:~:text=Seaweed%20was%20first%20consumed%20in,it%20as%20a%20food%20source. Retrieved 28th September 2022.

Variem, S.S., V.K. Kizhakkedath. 2021. Phycosphere associated bacteria; a prospective source of bioactive compounds. Biologia 76: 1095–1098.

von Sachs, J. 1890. History of Botany (1530–1860). Translated by H.E.F. Garnsey. Oxford: Clarendon Press.

West, J.B. 2022. The strange history of atmospheric oxygen. Physiol Rep 10(6): e15214.

Wiencke, C., K. Bischof (eds). 2012. Seaweed biology. Novel insights into ecophysiology, ecology and utilization. Ecological Studies Vol 219, Springer Publishers.

Wolkers, H., M.J. Barbosa, D.M.M. Kleinegris, R. Bosma, R.H. Wijffels, P.F. Harmsen. 2011. Microalgae: the green gold of the future?: large-scale sustainable cultivation of microalgae for the production of bulk commodities. Wageningen UR – Food & Biobased Research. https://edepot.wur.nl/170781.

World Economic Forum. 2022. Carbon-negative cement made from algae in world first. https://www.weforum.org/videos/carbon-neutral-cement-made-from-algae-in-world-first.

CHAPTER 1

There's Something in the Water: Algae, Eliminativism, and Our Moral Obligations to Biological Beings

Andrew Lopez and M. Polo Camacho

Abstract

This paper seeks to bring two disparate areas of research into conversation with each other: 1) philosophy of biology on conceptions of organismality, and 2) environmental ethics on the determination of bearers of value. We believe that environmental ethics, as a field invested in the value of beings beyond the human, can benefit from work focused on determining organismality. In this paper, we take algal systems as our engine of thought and argue for an eliminativist position that organisms as individuals do not exist; instead, we propose that organisms are constellations of processes. Conceptions of organismality, however, do capture something about the world, and so our eliminativism gives way to a realist eliminative pluralism about organismality; in other words, we argue that accepting different conceptions of organismality does not lead to relativism, because different conceptions of organismality capture different features of life, broadly understood. This eliminative pluralism can ultimately aid us in ethically attending to the more-than-human world around us.

Keywords

ethics – organisms – environmental ethics – algae – organismality – individuation – process

1 Introduction

A central question in the philosophy of biology is how heterogenous organisms[1] emerge from seemingly homogenous parts (Minelli & Pradeu 2014). Indeed,

1 We here treat 'organism' as synonymous with 'biological individual.' We wish to note, however, that some biologists and philosophers make a conceptual distinction between

© ANDREW LOPEZ AND M. POLO CAMACHO, 2024 | DOI:10.1163/9789004683310_003
This is an open access chapter distributed under the terms of the CC BY-NC 4.0 license.

when looking past the world of "medium-sized dry goods"—the world of tables, chairs, coffee mugs, and leather-bound books—the world *appears* to contain distinct biological individuals: rabbits, turtles, and octopuses. What's more, these "biological individuals" are *apparently* bound by cells containing genes, a genetic code, strings of A's, C's, T's and G's. Indeed, much work has been done in philosophy to make sense of this relationship (Pradeu 2014, Love 2014).

Note, however, that this question takes for granted what it means for an organism to count as a biological individual. Common sense dictates, for example, that moles are biological individuals, distinct from tigers, which in turn are distinct from elephants and beetles. The idea, roughly, is that the world contains a wide array of organisms, which may differ not only in terms of their taxa, but also within taxa. My pet axolotl differs from the caracara Mexican eagle at the Los Angeles Zoo and Botanical Gardens in terms of their species category, but my pet axolotl is also distinct from other axolotl amphibians in terms of their individuality.

These discussions matter, as they have implications for environmental-ethical decision-making. This is because what ultimately counts as a biological being matters when considering what is valuable, and, thus, of moral significance. Traditional approaches in ethics, for instance, have been criticized for being too human-centered, too focused on what human beings owe each other, and not focused enough on what human beings owe the environment, and the non-human organisms therein (Routley 1973, Goodpaster 1978). Indeed, some have argued—from an ethical perspective—that non-human organisms bear value, and are therefore worthy of moral consideration (Rolston 1994, Callicott 1984). The idea, generally speaking, is that human beings are not the only bearers of value: non-human biological individuals bear value also, and this is significant when thinking about our moral obligation to the environment.

So, not only do philosophical discussions in developmental biology take for granted what an organism as a biological individual is—and how it differs from other biological systems or entities—the same goes for debates in environmental ethics. Without some sense of how to delimit organisms as biological individuals, we cannot understand how biological individuality can inform our understanding of moral significance.

With a specific focus on algal ecological systems, this chapter makes explicit what it means to have moral obligations to biological beings. We begin by

---

'organism' and 'biological individual,' with the latter meaning that the entity in question is subject to natural selection.

entertaining eliminativist[2] approaches to biological individuality, which we define in the sections below (Merricks 2003). These approaches hold that biological individuals do not exist *strictly speaking;* what exists, rather, are biological parts operating in concert. We argue that the eliminativist approach, though promising, leaves us with an open question: if biological beings are merely biological parts operating in concert, then what is the nature of the parts and their relations? We address this question by proposing that biological beings exist as biological processes. What's more, these processes can be understood through different accounts of organismality.[3] Specifically, we propose that algal ecological systems, like harmful algal blooms—which comprise a variable biological network of chemicals, species, and organisms—are but constellations of biological processes. After highlighting some of the theoretical benefits of this approach, we end by noting that if we hold moral obligations to biological beings, then we hold moral obligations to biological processes.

2 Algal Ecologies & Biological Individuality

Much has been said in the philosophy of biology and theoretical biology about algae. The science of algae has figured in discussions about signalling systems and perception (Ganson 2017), cultural naming practices and their implications for the epistemology and ontology of algae (Kendig 2020), as well as the evolution of algal systems (Hanschen 2017). The effects of algal research on research methodologies in the biological sciences have also been explored (Nickelsen 2017). Algal-environment interactions (Odling-Smee et al. 2013) provide further support for niche-construction—a debated evolutionary process in the philosophy of biology, whereby organisms modify their environments to support their adaptive capacities. Finally, certain microalgae are also very sensitive to their environments (Bacova et al. 2020), so much so that the environment induces heritable traits and characteristics indicative of epigenetic mechanisms.

2 By 'eliminativism,' we mean a philosophical approach that seeks to eliminate a term or concept from our conceptual toolbox, rather than revising it or providing an account of it with the use of more basic concepts. For example, think of philosophers who argue that we should eliminate concepts like 'belief' in favor of empirically-supported concepts from psychology or neuroscience (Churchland 1981).

3 By 'organismality,' we mean the qualities or properties that make something an organism. As hinted here and argued for later, what qualities or properties make something an organism will depend on context and our aims.

This chapter shows how a proper understanding of algae informs our moral obligations to the environment. A recurring theme in environmental ethics is how we hold moral obligations to both human beings *and* non-human organisms (Leopold 1970, Routley 1973, Goodpaster 1978). Indeed, from an ethical perspective, some commentators have maintained that non-human organisms hold value independently of human valuers, and have argued that organisms are worthy of moral consideration (Rolston 1994, Callicott 1984). The issue with many of these discussions is that they take for granted what it means to be an organism, or—what we refer to hereafter as—a biological being. Even if we accept the claim that biological beings are worthy of moral consideration, this invites the question: what *is* a biological being?

Discussions about biological individuality are pervasive in analytic philosophy of biology. Ellen Clarke (2013) has recently advocated for an account of biological individuality that appeals to the concept of multiple realizability, or the idea that the existence of some type of entity can be realized in different ways or with different materials.[4] According to Clarke, biological beings are composed of mechanisms which may be multiply realized, meaning that both mechanisms can be achieved in a variety of ways.[5] We will not discuss Clarke's view here further, as the main point is simply that there are philosophically rich discussions about the nature of biological individuality.[6]

4 Consider the artifact of a chair as an example of multiple realizability. The existence of chair types can be realized in many ways. Chairs can be made of wood, metal or plastic. What's more, chairs can have multiple legs of varying sizes. Debates over whether a computer or artificial intelligence can 'think' in the same way we do is another popular instance of this, as it focuses on whether minds are multiply realizable.

5 Specifically, she holds that a biological individual consists of two mechanisms: a policing mechanism and a demarcating mechanism. The former is "any mechanism that inhibits the capacity of an object to undergo within-object selection," while the latter is "any mechanism that increases or maintains the capacity of an object to undergo between-object selection" (2013, p. 421, 424).

6 Our own take on Clarke's account is that her account fails to capture the biological individuality of algae. This is because algae reproduces sexually (via two distinct gametes) or asexually (via cell division), thus complicating the demarcating mechanism. *Cyanophyceae* algae reproduce almost completely asexually, suggesting there is no mechanism increasing or maintaining the capacity for algae to undergo between-organism selection. Moreover, Clarke's policing principle seems counterintuitive, especially when we consider facts about organisms. Note that, for Clarke, policing mechanisms inhibit the capacity for an organism to undergo within-object selection, but if such is the case, what to do with chimeric organisms, i.e. organisms with two distinct genetic codes? Many organisms, such as cats, plants, and algae, may contain two sets of DNA, but this poses no issues in terms of within-object selection. Clarke also assumes that biological individuality is extensionally equivalent with organismality; it should be noted, however, that the two concepts may differ in fundamental ways, as noted earlier. We do not explore this topic further here.

Here, we consider and evaluate an eliminativist account of biological individuality. However, before we delve into this, it's worth discussing some of the characteristics of algae that uniquely mark its place in the life sciences and the biological world, generally. Firstly, algae hold a special place in evolutionary history, as evidenced in the following quote from Samir Okasha (2019):

> If you look at a slug on a strawberry plant in your garden, for example, it takes a considerable leap of imagination to accept that you, the slug, and the plant all share a common ancestor if we trace back far enough. And yet evolutionary theory tells us that this is true—the ancestor of all plant and animal life was a single-celled protist (similar to algae) estimated to have lived some 1.6 billion years ago. (2019, pp. 22–23)

Though algae can range from single-celled organisms (diatoms) to multicellular organisms (red algae), algae share similarities with beings like slime molds and amoebas. They are all protists. Protists are eukaryotic organisms that neither belong to the kingdom of plants, animals, or fungi. The term "protista" is thought to have originated with Ernst Haeckel (1866), who used the term to designate microscopic organisms and unicellular organisms that evade macroscopic classification.[7]

Algal organisms are also unique in that they bear a special *causal* relationship to their ecological surroundings. To fully appreciate this causal relationship, consider the manner in which algae factors into ecological assessments of water bodies. According to R. Jan Stevenson, algal systems have been used for about a hundred years in scientific contexts to provide ecological assessments. In some cases, assessments are generated using various methods that track causal relationships between algal systems and other biological systems in water bodies. As Stevenson (2014) notes, these "stressors" (i.e. factors that exert causal influence on algae) may be studied using stressor identification, causal pathway analysis, or cause-and-effect analysis, to name a few (p. 452). Harmful algal blooms (or HABS), cases where algae grow exponentially and produce toxins that harm other organisms, uniquely illustrate such causal relationships. In 2011, Lake Erie had one of its greatest HABS in history. The harmful effects were caused by agricultural practices and meteorological conditions, and in particular the use of fertilizer that contains dissolved reactive phosphorus, combined with rainy weather that caused a high intake of nutrients and phosphorus into the lake (Michalak et al., 2013). These considerations highlight a significant fact about algae—they are organisms whose properties are causally linked to other factors, in this case phosphorus and nutrient levels.

7 Indeed, there's been some debate about the term (Corliss 1988).

In short: algae ultimately evade kingdom classification and are inextricably linked with other organisms and factors.

It's now worth considering how these facts about algae bear on philosophical discussions about biological individuality. To be clear, the question at stake here is not about individuality generally but rather about *biological* individuality; some work has been done on this question (Love & Brigandt 2017).

2.1 *Algae & Eliminativism about Biological Individuality*

Given the causal interdependency and interconnectedness of algae, along with the asexual characteristics of some algae, we begin by proposing an eliminativist account of biological individuality supported by Trenton Merricks's Overdetermination Argument.[8] The eliminativist account has been proposed by others (Haber 2013, Okasha 2011), but we hope to grant the approach some metaphysical weight.[9] These accounts are considered "eliminativist" because they propose the elimination of objects (or organisms!) in favor of the parts that comprise them. There are no biological individuals in the same way there are no baseball bats, tractors, or books, strictly speaking. What exists, rather, are parts organized bat-wise, tractor-wise, book-wise, or even salamander-wise.

Merricks's argument for the elimination of everyday objects may be applied to cases of biological individuality. Before applying this argument, however, it's worth illustrating Merrick's position with the following case: in the ninth inning, Salvador hits a homerun against a visiting baseball team. The ball shatters a window in the stadium. The first premise, roughly, is that if the baseball exists—as something over and above the parts that comprise it—then it would overdetermine the shattering of the window. This is because the shattering would be caused by both the leather particles and cork particles *as well as* the baseball as a whole. The next premise is that the shattering of the window was not overdetermined in this manner. The shattering was not caused by *both* the baseball and particles in the baseball. Therefore, the baseball does not exist, strictly speaking.[10]

Merricks's argument may be used to support the claim that biological individuals do not exist. To illustrate this, consider HABS. HABS are made up of biological parts—such as species, organisms, cellular machinery and chemical entities therein. These parts taken together cause harmful effects, which include the production of toxins, eutrophication, and the creation of dead

8 It should be noted that Merricks deploys the argument differently than we do here. Merricks's argument applies to ordinary objects, whereas we apply it to biological individuals.

9 See Love and Brigandt (2017).

10 See Jaegwon Kim (1998) for more on the Exclusion Argument.

zones in water bodies. Given all this, we might say that the various parts that make up HABs cause harmful effects in the same way that the parts of the baseball bring about the shattering of the window. In the case of HABs, these effects are caused by a wide array of cellular and chemical factors operating in concert, which include cellular and extracellular activity and nutrient levels in the algal system. In the case of the baseball, the particles in the baseball—cork, rubber, etc.—operate in concert to shatter the window. The baseball (conceived here as an object over and above the particles that constitute the baseball) is irrelevant to the shattering of the window. By analogy, the HAB—conceived as a separate object from those chemical and biological factors mentioned above—is causally irrelevant to the harmful effects associated with algal blooms. The shattering of the window is not caused by *both* the baseball and the particles that make up the baseball, in the same way that the harmful effects caused by HABs are not brought about by the HABs and the properties and factors that make up the HABs. Thus, the harmful ecological effects mentioned here are caused by the wide array of cellular and chemical factors operating in concert, nothing else. The conclusion here is that HABs do not exist over and above the biological and chemical parts that comprise them. If we apply the above argument to the biological world as a whole, the conclusion is much stronger. The conclusion would be that no biological beings (e.g., rabbits, ducks, otters) exist *strictly speaking*. Rather, what exists are biological parts organized rabbit-wise, duck-wise, or otter-wise.

One may object, of course, that the conclusion above is premature because HABs *are not themselves individual organisms in the first place*. Given this, there is nothing to eliminate. This is because, as noted above, HABs are made up of a variety of factors, which include various species and the individual organisms therein. In other words, the conclusion simply doesn't follow, because HABs were never individual organisms in the first place! We maintain that the Overdetermination Argument runs *even if we concede this point*. Let us consider the diatom *Pseudo-nitzschia australis*, which is found in HABs. This microalgae found in HABs produce toxins such as domoic acid. This toxin is produced, caused by the unique cellular machinery found in the diatom. It is *not* the case that this toxin is produced by the diatom *and* the cellular machinery therein, as this would imply overdetermination. Given this, what exists isn't the diatom, but the unique cellular parts of the diatom operating in concert.

2.2 *Open Question*

The aim in this section was to make the case that biological beings exist only as biological parts operating together. However, this invites the question: if biological beings, at bottom, are simply parts operating in concert, then how are these parts best understood? In other words, what is the nature of these parts and their relationship to each other? In the section below, we offer an answer to this question that draws from process-oriented approaches in the philosophy of biology.

3 Biological Beings as Processes

Having argued for the elimination of biological beings, we make two proposals in the rest of the chapter: 1.) taking an eliminative-pluralist approach to biological beings, and 2.) taking all different kinds of biological beings as fundamentally kinds of processes.

Though process-oriented approaches have received some attention within analytic philosophy of biology (Nicholson and Dupré 2018), we take inspiration for our eliminative-pluralist approach from Marc Ereshefsky (1992), who has argued that we should "[e]liminate the term 'species' and replace it with a plurality of more accurate terms" (1992, p. 681). Ereshefsky's aim is to do away with the idea that we should use one overarching species concept and replace it with multiple, specialized concepts of species that may be incompatible with each other, but correctly reflect important features of the tree of life. 'Biospecies,' 'ecospecies,' and 'phylospecies,' for example, emphasize reproduction, ecological niche, and lineage respectively, and all accurately identify features of the tree of life, even if these different definitions of species are in various cases incompatible with each other.

We deploy Ereshefsky's unique eliminative pluralism, but apply it to the concept of organism.[11] Specifically, we argue that we should make use of different and potentially incompatible concepts of 'organism' to think about biological beings, whether these be based on immunological models (Pradeu 2010), integrated functional metabolic units (Godfrey-Smith 2013), bounded individuality (Torres & Trainor 2008), contextualized dynamic assemblages exhibiting high cooperation and low conflict (Díaz-Muñoz et al., 2016), or other grounds. Taking this kind of pluralist approach enables us to intelligibly discuss organisms of different kinds and of different scales, whether it's beavers, hydrozoa, or Gaia itself (Doolittle 2020). This is of course not to say that beavers, hydrozoa, and Gaia are all the same kind of organism. Different accounts of organismality

11 Here we may be understood to be doing something similar to Mariscal & Doolittle (2020), who propose an eliminative pluralism concerning 'life,' though they also propose treating all of terrestrial life beginning with the last universal common ancestor (LUCA) as an individual. We are amenable to this move, but ultimately have a different goal in mind.

will clearly delimit differently. This is additionally made all the more plausible if we take care to distinguish between an entity that persists and an entity that reproduces. For instance, worker ants count as organisms on many lay and scientific accounts of organismality, but they do not individually reproduce; rather, the colony does. Most accounts would also consider the last human being an organism, even if it meant the species is doomed to extinction due to a lack of mating partners. Keeping this distinction in mind is useful, as the lack of reproduction and being a member of a population of one have traditionally been major criticisms of the proposal that Gaia is an organism. If we separate our concept of an organism from our idea of what is subject to selection, we can avoid these criticisms.[12]

While our approach may seem drastic, it is not new; some biologists have already argued that "[m]ultiple organism concepts can usefully coexist," though also noting that we should be clear about which concept we deploy and why (Pepper & Herron 2008, p. 626). This pluralism quickly highlights its usefulness when considering the particular kinds of issues encountered within environmental ethics concerning value theory. For instance, a focus on species (whether as a general concept or the more specialized senses elaborated by Ereshefsky) or ecosystem stability and function within environmental ethics has led to fierce criticisms from theorists concerned with the moral worth of individuals (Calarco 2017, p. 3); the focus on a lineage or population or ecological niche at the expense of individual members runs the risk of enacting what Tom Regan has condemned as 'environmental fascism' (1983, p. 361-2). On the other hand, sentiocentric accounts (accounts which prioritize conscious or sentient beings) such as Regan's that have focused only on 'subjects of a life,' or the capacity to experience pleasure and pain emphasized by utilitarian accounts, fail to account for the value other kinds of individuals may have (plants being the most common example, as well as our engine of thought in this essay, algae). Phytocentric accounts simply swap in plants for sentient beings, inheriting the limitations and problems of sentiocentric accounts. It is not clear where algae fall, given the difficulty in classifying them, but doubtless many would have criticisms of a phycocentric account that placed algae at the heart of moral consideration!

Still, ecocentric, sentiocentric, phytocentric accounts, etc., seem to capture *something* about the moral landscape, albeit only segments of it. The same follows for different accounts of organismality. Indeed, it is unlikely that we can get away from certain common sense intuitions about organisms, including the idea that some entities are more obviously organisms than others. As

12 For an attempt to 'Darwinize' Gaia, however, see Doolittle (2017).

Subrena Smith (2017) notes, "it seems inevitable that one's point of departure will be paradigmatic cases and that one's initial assumptions about what organismality amounts to will be based on them" (2017, p. 5). Smith draws the conclusion that where we draw the line for organismality—or as we suggest, what conception of organismality is deployed—will to some extent depend on our explanatory aims. Given this, we mustn't delude ourselves into thinking that we can conclude "what organisms really are in a way that is wholly independent of one's explanatory interests" (2017, p. 5). However, the flexibility of our approach is meant not just for explanatory interests, but for our ethical goals as well; different conceptions of organismality serve as lenses for thinking about the more-than-human world and assessing cases and situations that draw our gaze (Sherwin 1999).

We argue that organisms of all kinds are ultimately manifolds of processes. By a 'process,' we mean a homogenous set of occurrences, wherein the parts are linked together through specific causal chains, or by a functional unity. These sets of occurrences are inherently indefinite: while an otter may finish its swim in the sea, 'swimming' as a process is homogenous and indefinite, for it has no necessary temporal bounds. And though the example used here is of an organism (an otter) engaging in a process, the otter is itself a system of processes: organisms are long-term systems, a process that remains stable or seemingly unchanging long enough so as to seem static, individual (in other words, a being). Toying with the timescale or the passage of time allows for intuitively grasping this point: hitting fast-forward on a human life makes a human being look more like a process than a thing or substance; slowing down our perception of a dust-devil might make it seem like a thing or substance rather than an ephemeral process. An organism, then, is a metastable[13] system, one that has emerged from the relations between other processes and become its own individuated process at a certain level of description irreducible to

13 By 'metastable,' we here help ourselves to an ecological understanding of the term. Imagine a ball (a system) rolling around on a table with various different troughs (attractors) filled with dyes that the ball could roll into and around. Shake the table and the ball will roll in a trough for a while, and eventually resettle. Shake the table enough and the ball may roll into another trough, changing its properties by rolling around in a different dye and eventually settling in this new trough. Ecosystems undergo transitions between different troughs, or dynamic equilibria; while we can focus on an ecosystem's ability to return to equilibrium within one trough (engineering resilience), ecologists also study an ecosystem's ability to persist within one trough among others, given that the troughs or state of equilibria are themselves dynamic and changing (ecological resilience) (Holling 1996). That is to say, not only does the ball move around the table in and between troughs, but the table and its troughs change over time as well. With this said, we propose that organisms as systems of processes are like ecosystems.

another.[14] Of course, we tend to see these metastable processes as individual beings, and as mentioned earlier, it is likely that we cannot avoid seeing some forms of organization as organisms and as beings. But a process view that describes these seemingly static entities as themselves processes engaged in relations with other processes is compatible with the claim that this is not necessarily how they appear to us in experience. The human being is the result of a complex interaction between various dynamic bodily systems (nervous, digestive, endocrine, muscular, cardiovascular, etc.), though appears to us as a being and not a macro-process. The fact that a human being appears to us as a being *is not in conflict* with the claim that it is identical to this complex interplay of systems that is itself a process, and one that is entangled in co-constitutive relations with other human beings (who themselves are processes). And these processual systems are engaged in a complex web of relations with processes at 'higher' (social, cultural, political, national, etc.) and 'lower' (chemical, biomechanical, physiological, etc.) levels of description. This embeddedness in larger systems and constitution through hierarchical organization extends to all organisms, with no easy answer for which level is correctly determined to be an individual. Matthew Herron (2017) argues that for colonial volvocine algae, "there are three kinds of biological units that can contend for at least a degree of individuality: the cells, the colonies, and the clones" (2017, p. 69). He concludes that asking which is the individual is the wrong question. Rather, the question should be concerned with identifying the amount of individuality present at each of these three different levels of hierarchical organization.

The elaboration of a process metaphysics—via the work of Gilles Deleuze and Félix Guattari—may be useful for cashing out its relevance for ethics. Similar to certain thinkers in the Western philosophical tradition—most notably Benedict Spinoza—Deleuze and Guattari's philosophy is a philosophy of immanence. That is, the account they provide does not posit any transcendent entities or structures; there is only this world. But unlike the metaphysical system proposed by Spinoza (or others), their world is not parsed into one Substance and modes of Substance, with the latter driven to persevere in their nature by conative drive, in a struggle against other modes to avoid annihilation. Processual systems do not first possess a nature and then express this nature. Rather, their account holds that the nature of processes is developed in their interactions with other processes. Given this, we here tread familiar ground in the Western canon, in that we hold to there being an intimate connection between *what an entity is* and *what an entity does*; in essence, this

14 For discussion of this irreducibility as exemplified with physiological and ecological models of lynx-hare predation, see (Dupré 1993, p. 114-6).

implies an intimate connection between ontology and ethics (ethics understood broadly to encompass ethology and behavior). The kind of being we are informs what we do.[15] For example: if the human being is a rational animal, then our rationality, our capacity for reason, plays some role in what we do. The process view, however, focuses on what an entity does, and since a process need not be determined (though perhaps delimited), then what an entity does ultimately has some level of flexibility in its expression. We emphasize 'some,' since entities are not maximally malleable; it is of course trivial to point out that an organism cannot withstand all forms of change or perturbation without running the risk of collapse or annihilation. Additionally, the functional unity of organism-as-processes militates against an overly constructivist[16] view of what kinds of relations processes may enter into with other processes.[17] Though that being said, there is still much room for a diversity of assemblages and relations. For instance, four species of algae have been found as epibionts (organisms that live on the skin or surface of other organisms) of sea turtles (Majewska et al. 2021).

---

15 More strongly, the kind of being we are informs what we ought to do and how we ought to be.

16 The emphasis on what an entity can become sometimes leads to the mistake of treating the entity as an ideal or social construct free from causal influences. To give an example from critical animal studies: the call by some theorists to transform our relationships with the more-than-human world includes changing our current connections to animals from relationships of domination to friendship, as in the case of coming to see other domesticated animals (beyond culturally-specific companion species) in a different light (e.g., Americans coming to see cows as more than just property, dairy providers, or meat and exploring friendships with them). Domesticated animals are amenable to this because of their species histories that are intertwined with ours. The same is not so for many wild animals. As Matthew Calarco notes in his discussion of Joe Hutto's interactions with mule deer, "we should be clear that Hutto understands he is not working with beings who can be shaped in any manner he might wish. Mule deer, like human beings, are animals with a robust genetic inheritance and longstanding ecological relations that are neither easily displaced nor transformed. As such, Hutto does not see himself as working against these factical constraints and inheritances but rather within and through them in order to see what other kinds of possibilities might be found therein" (2018, p. 49-50).

17 Processual systems are affective in structure, meaning that they are characterized by their capacity to affect and be affected. James Gibson's concept of affordance can explicate this point: an affordance is that which an entity or an environment furnishes or provides unto another entity or environment, for better or worse (2015, p. 119). For example: a chair can afford a person an object to sit on, to sleep, to step onto, to sell, to throw through a window, to use as kindling, etc. The number and kind of relations that entities can potentially enter with each other results in a combinatorial explosion of possibilities too numerous to index and which remain indeterminate prior to actualization (though our previous comments on flexibility and collapse still apply).

Processes are in a complex interplay with other processes; some processes stretch out into the future in an attempt to perpetuate themselves and subordinate other processes, while other kinds of processes aim at some resolution, achievement, or event. For example, organisms that engage in niche construction shape their environments to make them more conducive to their survival. Algae, for instance, "change levels of atmospheric redox states and influence energy and matter flows by modifying nutrient cycles" (Odling-Smee et al. 2013, p. 5). Additionally, organisms can respond to the niche construction of other species through their own practices of niche construction, exhibiting a complex weaving of dynamic life-processes. Post & Palkovacs (2009) note how a complex interplay between algae and guppies influences the niche construction of the other. In areas with higher guppy predation, guppies are smaller and reach maturity at an earlier age, but they also release more nitrogen and phosphorus into their environment through excretion, which promotes algae growth. Algae, in turn, exert influence on guppy evolution: algae produce carotenoids (yellow, orange, and red pigments) within their environment, and male guppy color is sensitive to their presence (2009, p. 1634-5).

Organisms, as processes, have to regulate their interactions with transformative forces and relations, and minimally maintain their organizational unity. Engaging with transformative forces and processes can lead to the annihilation of a processual system. The kinds of processes a system engages with—and the extent or rate of engagement—play an important role in informing our practical engagement with the world, and hence are important for thinking ethically through and about algal systems and HABS.

Our proposal, then, has implications for the nature and function of HABS. On our account, HABS are biological beings, and biological beings are, at bottom, biological processes.[18] HABS are also made up of other biological processes, which include species, organisms, chemical properties, etc. These processes, given our accounts, are co-constituting. That is to say, HABS are processes that constitute other processes, such as the diatom *Pseudo-nitzschia australis*. Thus, HABS do not exist *as substances*. Rather, what exists are properties—species, organisms and chemicals—operating in concert to engender certain effects. Given all this, our answer to the open question posed in Section 2.2, is that the parts comprising algal ecological systems, and their relations, are best understood as biological processes.

---

18 In this, we share affinities with the 'organicism' of biologists like Ludwig von Bertalanffy, John Scott Haldane, Edward Stuart Russell, Conrad Hal Waddington, Paul Alfred Weiss, and Joseph Henry Woodger.

4 Our Moral Obligations to Biological Processes

Thus far, we have detailed the elimination of a standard conception of organismality in favor of a pluralist conception, which, in turn, is grounded in the idea that all kinds of organisms are processes. While we have taken on some metaphysical commitments, we have not yet addressed what our approach ultimately implies for environmental ethics, or the claim that we have moral obligations towards biological beings. After all, thinking of organisms as processes—rather than static entities—is perfectly compatible with destabilizing or destroying them. On its own, this metaphysical commitment is value-neutral: for instance, our perspective allows us to talk about the resilience of a particular processual system or cyclical process without touching upon any of the reasons for why we might think a system or cycle is bad, as is the case with a eutrophic lake (Desjardins et al. 2015). Eutrophic lakes are rich in nutrients that can foster algal blooms, and they can absorb or handle significant perturbations and remain in a stable state and, hence, be considered a resilient system. But this resilience would not satisfy the process-minded environmental ethicist since this is a lake devoid of animal life due to the lack of oxygen as a result of algal life-cycle processes.

If we are to consider the idea that organisms are bearers of value, that organisms are ultimately processual systems, and that different conceptions of organismality enable us to engage with different phenomena as organisms, then we need a sense of what the value they bear consists in and how to best respect, preserve, and foster this value. Thomas Birch defines moral consideration as such: "To give moral consideration to X is to consider X (to attend to, to look at, to think about, where appropriate to sympathize or empathize with X, etc.) with the goal of discovering what, if any, direct ethical obligations one has to X" (Birch 1993, p. 315). Birch notes that this definition rehabilitates a typically disregarded meaning of consideration: when the ethically engaged individual holds something in consideration, it is understood as holding it in regard, or perceiving its value. Birch does not dismiss this understanding, but calls us to also emphasize consideration as attentive contemplation (1993, p. 327). This brings on the important task of understanding the long-term dynamics of algae within an ecosystem. As Theodore Smayda notes, "[t]he salient criterion to use in defining whether a 'harmful' species is in bloom and the distinctive feature of such blooms lie not in the level of abundance, but whether its occurrence has harmful consequences" (1997, p. 1135). It is here that environmental ethicists can fruitfully engage with what we can learn about algae by critically analyzing what these harmful consequences are and whether they constitute a form of wrong. For instance, the presence of blooms and whether they are deemed 'harmful' may depend on what the bloom threatens; what may be at stake are human economic interests, and not necessarily the ecosystem, as it is possible to investigate how blooms impact the nutritional and metabolic dynamics of the ecosystem and whether this impact falls within the natural long-term variance seen within this ecosystem (1997, p. 1134). An environmental ethic, in aiming to respect the value of the more-than-human world, will be wary of deeming a bloom harmful solely on the basis that it does harm to human economic interests in aquacultures.

In attending to processes, we are also called to understand their inherent temporal nature; biological processes extend out into the future, in ways that can—and cannot—be anticipated. Considering processual systems requires paying attention to the factors that enable processes to perpetuate themselves, to renew cyclically, to form new constellations of processes, and to understand what perturbs them, disrupts them, or causally determines their development. The aim here is not to uncritically foster any process. While novelty, creativity, and expression are integral to our processual metaphysics, it does not yet imply that these should be enabled or fostered in any instance. Cancers are an example of a deviation of a processual system that gives rise to novelty and creativity (e.g. teratomas are tumors that develop muscle, bone, teeth, hair, etc.), but cancers also destabilize the constellation of processes of which they are part, and very often lead to their dissolution. Fostering HABs as processes that are part of a world in flux can lead to the proliferation of various species of algae, while at the same time killing, poisoning, or suffocating other aquatic organisms. We should not aim to uncritically valorize or eschew stability or change; measured engagement with both is necessary for survival as well as for becoming otherwise. As Deleuze and Guattari note regarding ethical and political practices of becoming, "[y]ou have to keep enough of the organism for it to reform each dawn; and you have to keep small supplies of signifiance and subjectification, if only to turn them against their own systems when the circumstances demand it" (1987, p. 160).

Nor is the aim to *always* foster processes. If an environmental ethic is to take, as its starting point, the claim that we have moral obligations to biological beings, it must also contend with death. Organisms as processual systems die, often at the hands of other organisms. The moral obligations that we have toward processual systems are contextual and dynamic. As a result, we will find ourselves in situations that call upon our common-sense understanding of organismality, in which case we must recognize our moral commitments to persons, flora, and fauna. At other times, however, our moral obligations must expand far beyond this common-sense notion, and thus be concerned with the flourishing of larger, diffuse organisms, such as Gaia, and the constellation of processes that constitute it. Ford Doolittle and Andrew Inkpen (2018) have recently developed ITSNTS[19] theory "as a way of conceptualizing evolution by natural selection (ENS) that allows processes as well as things to be 'units of selection,' selected for persistence and re-produced but not reproducing" (2018, p. 1). While our aim here is different, the metaphor of a process being a song iterated by a chorus of singers allows for a way to conceptualize a concern with both the macro-process and the micro-processes that realize the macro-process. Attending to both the song and the singer(s), and the expression of both should be the concern of an environmental ethics, one that is facilitated by our eliminative-pluralist approach to organismality. Our proposal, then, implies that if we have moral obligations to biological beings (which include algal systems), they're ultimately moral obligations *toward biological processes.*

5 Conclusion

We have sought to bring two areas of research into conversation with each other: philosophy of biology on conceptions of organismality, and environmental ethics on the determination of bearers of value. We believe that environmental ethics, as a field invested in the value of beings beyond the human, can draw from philosophical discussions about organismality. Our position is that organisms as individuals do not exist; instead, we propose that organisms are constellations of processes. Broadly, these constellations are situated in webs of relations in a manner familiar to environmental ethics. Instead of a wholesale rejection of organismality, our eliminativism gives way to a realist pluralism about organismality. In other words, accepting different conceptions of organismality does not lead to a relativism about what counts as an organism, because different conceptions of organismality capture different features of life, broadly understood. What's more, this eliminative pluralism can ultimately aid us in ethically attending to the more-than-human world around us.[20]

19 "It's The Song, Not The Singer," pronounced "It's nuts."

20 What conceptions of organismality we deploy will depend on the practical issues that present themselves to us as one of many constellations of processes invested in the world. However, we have not said much on what this attending may look like in any concrete situation.

We have stated that we endorse the Western philosophical tradition's claim of an intimate connection between what an entity is and what an entity does. However, we have eschewed a substance-based view in favor of a process-based view: we have turned away from the view that an entity exists and then expresses its nature and embraced the view that an entity's nature and what it does are co-constitutive. Ontology and ethics dynamically influence and develop each other, so to speak. But this extends past our common-sense notions of bounded individuality, which demarcates us from other biological beings. Our practices not only develop our nature, but the nature of others, and their practices in turn shape us. As processes bound up in webs of relations with each other, we find that "[b]ecoming is always double," as Deleuze & Guattari are often quoted (1994, p. 109).

This capacity to affect and be affected means that we must be attentive to the quality of the relations between processes (as stated earlier, seeing other beings as processes is perfectly compatible with any number of actions). Other processes cannot be shaped into whatever we want, just as we cannot be transformed in any way whatsoever by their effects on us without risking annihilation. Understanding these relations with the world and internalizing this understanding as an *ethos* for our co-development is key for being able to accurately attend to the needs and potentials of other biological processes, mirroring Vine Deloria Jr.'s description of 'American Indian metaphysics' as "the realization that the world, and all its possible experiences, constituted a social reality, a fabric of life in which everything had the possibility of intimate knowing relationships because, ultimately, everything was related" (2001, p. 2).

Our engine of thought in this paper has been algae and algal blooms. Blooms, particularly harmful ones, serve as valuable examples for understanding this perspective and some of the key considerations drawn from ecological concerns about HABs. Understanding algal blooms and their effects requires us to attend to the qualities of the process (its concentration, rate, location, constitution, etc.) at various levels of description, which aids us in determining what our practical steps ought to be. What algae and algal blooms are, and what they can and ought to be, will be understood in terms of what their relations are to the other processes they affect, including ourselves. This may seem obvious. After all, we specifically singled out *harmful* algal blooms, which involves an evaluative component (recall our mention of eutrophic lakes as stable and resilient). But calling our attention to this helps us think more clearly about how our pluralistic approach to organisms as processes interacts with other normatively-laden concepts in biology such as sustainability or conservation, and how environmental ethics can make important contributions to thinking about and aiming toward a dynamic world.

References

Bacova, R., Kolackova, M., Klejdus, B., Vojtech, A., & Huska, D. (2020). Epigenetic mechanisms leading to genetic flexibility during abiotic stress responses in microalgae: A review. *Algal Research*, *50*, 101999.

Birch, T. H. (1993). Moral Considerability and Universal Consideration. *Environmental Ethics*, *15*(4), 313–332. https://doi.org/10.5840/enviroethics19931544

Calarco, M. (2017). Beyond the Management of Pe(s)ts. In J. Stanescu & K. Cummings (Eds.), *The Ethics and Rhetoric of Invasion Ecology* (pp. 1–16). Lexington Books.

Calarco, M. (2018). The Three Ethologies. In D. Ohrem & M. Calarco (Eds.), *Exploring Animal Encounters: Philosophical, Cultural, and Historical Perspectives* (pp. 45–62). Palgrave Macmillan.

Callicott, J. B. (1984). Non-Anthropocentric Value Theory and Environmental Ethics. *American Philosophical Quarterly*, *21*(4), 299–309.

Churchland, P. (1981). Eliminative Materialism and the Propositional Attitudes. *The Journal of Philosophy*, *78*(2), 67–90.

Clarke, E. (2013). The Multiple Realizability of Biological Individuals. *The Journal of Philosophy*, *110*(8), 413–435.

Corliss, J. O. (1998). Haeckel's Kingdom Protista and Current Concepts in Systematic Protistology. *Stapfia*, *56*, 85–104.

Deleuze, G., & Guattari, F. (1987). *A Thousand Plateaus: Capitalism & Schizophrenia* (B. Massumi, Trans.). University of Minnesota Press. (Original work published 1980)

Deleuze, G., & Guattari, F. (1994). *What is Philosophy?* (H. Tomlinson & G. Burchell, Trans.). Columbia University Press. (Original work published 1991)

Deloria, V., Jr. (2001). American Indian Metaphysics. In V. Deloria Jr. & D. Wildcat (Eds.), *Power and Place: Indian Education in America* (pp. 1–6). Fulcrum Resources.

Desjardins, E., Barker, G., Lindo, Z., Dieleman, C., & Dussault, A. C. (2015). Promoting Resilience. *The Quarterly Review of Biology*, *90*(2), 147–165. https://doi.org/10.1086/681439

Díaz-Muñoz, S. L., Boddy, A. M., Dantas, G., Waters, C. M., & Bronstein, J. L. (2016). Contextual organismality: Beyond pattern to process in the emergence of organisms. *Evolution*, *70*(12), 2669–2677.

Doolittle, W. F. (2017). Darwinizing Gaia. *Journal of Theoretical Biology*, *434*, 11–19. https://doi.org/10.1016/j.jtbi.2017.02.015

Doolittle, W. F. (2020, December 3). Is the Earth an organism? *Aeon*. https://aeon.co/essays/the-gaia-hypothesis-reimagined-by-one-of-its-key-sceptics

Doolittle, W. F., & Inkpen, S. A. (2018). Processes and patterns of interaction as units of selection: An introduction to ITSNTS thinking. *Proceedings of the National Academy of Sciences*, *115*(16), 4006–4014. https://doi.org/10.1073/pnas.1722232115

Dupré, J. (1993). *The Disorder of Things: Metaphysical Foundations of the Disunity of Science*. Harvard University Press.

Ereshefsky, M. (1992). Eliminative Pluralism. *Philosophy of Science*, *59*(4), 671–690. https://doi.org/10.1086/289701

Ganson, T. (2018). The Senses as Signalling Systems. *Australasian Journal of Philosophy*, *96*(3), 519–531.

Gibson, J. (2015). *The Ecological Approach to Visual Perception*. Psychology Press. (Original work published 1979)

Godfrey-Smith, P. (2013). Darwinian Individuals. In F. Bouchard & P. Huneman (Eds.), *From Groups to Individuals: Evolution and Emerging Individuality* (pp. 17–36). MIT Press.

Goodpaster, K. E. (1978). On being morally considerable. *Journal of Philosophy*, *75*(6), 308–325.

Haber, M. (2013). Colonies Are Individuals: Revisiting the Superorganism Revival. In F. Bouchard & P. Huneman (Eds.), *From Groups to Individuals: Evolution and Emerging Individuality* (pp. 195–217). MIT Press.

Haeckel, E. (1866). *Generelle morphologie der organismen. Allgemeine grundzüge der organischen formen-wissenschaft, mechanisch begründet durch die von Charles Darwin reformirte descendenztheorie.*

Hanschen, E. R., Davison, D. R., Grochau-Wright, Z. I., & Michod, R. E. (2017). Evolution of Individuality: A Case Study in the Volvocine Green Algae. *Philosophy, Theory, and Practice in Biology*, *9*(3). https://doi.org/10.3998/ptb.6959004.0009.003

Herron, M. D. (2017). Cells, Colonies, and Clones: Individuality in the Volvocine Algae. In S. Lidgard & L. K. Nyhart (Eds.), Biological Individuality: Integrating Scientific, Philosophical, and Historical Perspectives (pp. 63–83). University of Chicago Press.

Holling, C. S. (1996). Engineering Resilience versus Ecological Resilience. In P. C. Schulze (Ed.), *Engineering Within Ecological Constraints* (pp. 31–44). National Academy Press.

Kendig, C. (2020). Ontology and values anchor indigenous and grey nomenclatures: A case study in lichen naming practices among the Samí, Sherpa, Scots, and Okanagan. *Studies in History and Philosophy of Science Part C: Studies in History and Philosophy of Biological and Biomedical Sciences*, *84*, 101340.

Kim, J. (1998). *Mind in a Physical World: An Essay on the Mind-body Problem and Mental Causation*. MIT Press.

Leopold, A. (1949). *A Sand County Almanac*. Oxford University Press.

Love, A. C. (2014). The erotetic organization of developmental biology. In A. Minelli & T. Pradeu (Eds.), *Towards a Theory of Development* (pp. 33–55). Oxford University Press.

Love, A. C., & Brigandt, I. (2017). Philosophical Dimensions of Individuality. In S. Lidgard & L. K. Nyhart (Eds.), *Biological Individuality: Integrating Scientific, Philosophical, and Historical Perspectives* (pp. 318–348). University of Chicago Press.

Majewska, R., Ashworth, M. P., Bosak, S., Goosen, W. E., Nolte, C., Filek, K., Van de Vijver, B., Taylor, J. C., Manning, S. R., & Nel, R. (2021). On Sea Turtle-associated *Craspedostauros* (Bacillariophyta), with Description of Three Novel Species. *Journal of Phycology*, *57*(1), 199–218. https://doi.org/10.1111/jpy.13086

Mariscal, C., & Doolittle, W. F. (2020). Life and life only: A radical alternative to life definitionism. *Synthese*, *197*(7), 2975–2989. https://doi.org/10.1007/s11229-018-1852-2

Merricks, T. (2003). *Objects and Persons*. Clarendon Press.

Michalak, A. M., Anderson, E. J., Beletsky, D., Boland, S., Bosch, N. S., Bridgeman, T. B., Chaffin, J. D., Cho, K., Confesor, R., Daloğlu, I., DePinto, J. V., Evans, M. A., Fahnenstiel, G. L., He, L., Ho, J. C., Jenkins, L., Johengen, T. H., Kuo, K. C., LaPorte, E., ... Zagorski, M. A. (2013). Record-setting algal bloom in Lake Erie caused by agricultural and meteorological trends consistent with expected future conditions. *Proceedings of the National Academy of Sciences*, *110*(16), 6448–6452. https://doi.org/10.1073/pnas.1216006110

Minelli, A., & Pradeu, T. (Eds.). (2014). *Towards a Theory of Development*. Oxford University Press.

Nicholson, D. J., & Dupré, J. (Eds.). (2018). *Everything Flows: Towards a Processual Philosophy of Biology*. Oxford University Press.

Nickelsen, K. (2017). The organism strikes back: Chlorella algae and their impact on photosynthesis research, 1920s–1960s. *History and Philosophy of the Life Sciences*, *39*, Article 9. https://doi.org/10.1007/s40656-017-0137-2

Odling-Smee, J., Erwin, D. H., Palkovacs, E. P., Feldman, M. W., & Laland, K. N. (2013). Niche Construction Theory: A Practical Guide for Ecologists. *The Quarterly Review of Biology*, *88*(1), 3–28. https://doi.org/10.1086/669266

Okasha, S. (2011). Biological Ontology and Hierarchical Organization: A Defense of Rank Freedom. In B. Calcott & K. Sterelny (Eds.), *The Major Transitions in Evolution Revisited* (pp. 53–63). MIT Press.

Okasha, S. (2019). *Philosophy of Biology: A Very Short Introduction*. Oxford University Press.

Pepper, J. W., & Herron, M. D. (2008). Does Biology Need an Organism Concept? Biological Reviews, 83(4), 621–627.

Post, D. M., & Palkovacs, E. P. (2009). Eco-evolutionary feedbacks in community and ecosystem ecology: Interactions between the ecological theatre and the evolutionary play. Philosophical Transactions of the Royal Society B: Biological Sciences, 364(1523), 1629–1640.

Pradeu, T. (2010). What is An Organism? An Immunological Answer. *History and Philosophy of the Life Sciences*, *32*(2/3), 247–268.

Pradeu, T. (2014). Regenerating theories in developmental biology. In A. Minelli & T. Pradeu (Eds.), *Towards a Theory of Development* (pp. 15–32). Oxford University Press.

Regan, T. (1983). *The Case for Animal Rights*. University of California Press.

Rolston, H., III. (1994). Value in Nature and the Nature of Value. In R. Attfield & A. Belsey (Eds.), *Philosophy and the Natural Environment: Royal Institute of Philosophy Supplement: 36* (pp. 13–30). Cambridge University Press.

Routley, R. (1973). Is there a need for a new, an environmental ethic? In *Proceedings of the XVth World Congress of Philosophy* (Vol. 1, pp. 205–210). Sofia Press.

Sherwin, S. (1999). Foundations, Frameworks, Lenses: The Role of Theories in Bioethics. *Bioethics, 13*(3/4), 198–205.

Smayda, T. J. (1997). What is a bloom? A commentary. Limnology and Oceanography, 42(5, part 2), 1132–1136.

Smith, S. E. (2017). Organisms as Persisters. *Philosophy, Theory, and Practice in Biology, 9*(14). https://doi.org/10.3998/ptb.6959004.0009.014

Stevenson, J. (2014). Ecological assessments with algae: A review and synthesis. *Journal of Phycology, 50*(3), 437–461.

Torres, J.-L., & Trainor, L. (2008). On organism: Environment buffers and their ecological significance. *Biology & Philosophy, 23*(3), 403–416. https://doi.org/10.1007/s10539-007-9107-5

CHAPTER 2

Seeking an Algal Perspective: Exploring "Harmful" Algae through an Interview with *Nodularia spumigena*

Jesse D. Peterson

Abstract

As agricultural runoff, wastewater, and other forms of pollution have increased the number and range of microalgae in aquatic environments during the last century, "harmful" algae have become a being of political and scientific interest. Taking seriously algal historicity and temporalities, this chapter asks in what ways do the oceans matter to "harmful" microalgae. It addresses the conundrum of trying to take on an algal perspective and using this perspective as a lens for understanding different ways that maritime environments come to matter for humans and harmful microalgae. Based on recorded anecdotes, observations, and approximately thirty years of reporting in the "Harmful Algae Newsletter" (HAN), I develop and interview the speculative algal character, *Nodularia spumigena*, which offers an alternative reading of these sources. Such speculative imagining underscores the relative benefits for attempting to understand human concerns from nonhuman perspectives, the relational interactions between humans and algae that permit the naturalization of some algae as "harmful," and the ensuing contradictions that arise between human desires about algae and the oceans. Seeking an algal perspective leads to more-than human ethical reflections which can assist in engendering reconciliation of human interests and activities in accordance with other beings.

Keywords

algae – microalgae – harmful algal bloom – speculative fabulation – interview – more-than-human – multispecies studies – oceans – anthropomorphism – toxicity

As agricultural runoff, wastewater, and other forms of pollution have increased the number and range of microalgae in aquatic environments during the last century, "harmful" algae have become a being of political and scientific interest.

© JESSE D. PETERSON, 2024 | DOI:10.1163/9789004683310_004
This is an open access chapter distributed under the terms of the CC BY-NC 4.0 license.

As a form of "stranger species" or "alien," algae and other microbial forms represent ambiguous, ambivalent beings whose alterity creates "heterotopias, baroque topologies that mix the dystopian and utopian, the hospitable and the hostile" (Helmreich 2009, 28). For instance, human encounters with algae draw individuals and their communities into political engagements, which offer opportunities for different forms of response to the environmental issues that come alive when algae appear in ways that perturb people (Helmreich 2009; Waterton and Tsouvalis 2015; Waterton 2017). Moreover, cultural values and beliefs as well as material scientific practices have contributed to different notions and perceptions of the toxic dinoflagellate *Pfiesteria piscisida* (Paolisso and Chambers 2001; Schrader 2010). In "The Time of Slime: Anthropocentrism in Harmful Algal Research," feminist philosopher Astrid Schrader looks at environmental science on harmful algal blooms and points out that "the human subject remains the only historical actor, while the microbes are assumed to continue to act as they have always done, existing in a fixed relationship to their environment, which they presumably cannot change" (2012, 87). Going against this assumption, Schrader argues that society must consider algal histories and temporalities in order to grapple with questions such as "to whom the ecological transformations of the oceans matter" (2012, 89). Realizing that ecological transformations of the oceans matter to more than just people, we can ask ourselves in what ways do oceans matter to "harmful" microalgae.

Taking seriously the need for exploring algal historicity in order to explicate different ways through which maritime environments come to matter for humans and harmful microalgae, this chapter draws on recorded anecdotes, observations, and approximately thirty years of reporting in the "Harmful Algae Newsletter" (HAN). Methodologically, I frame and augment my analysis of these records by employing multispecies "arts of attentiveness," which encompasses "getting to know another in their intimate particularity" and "learning ... to cultivate worlds of mutual flourishing" (van Dooren et al. 2016, 17). In addition, this analysis is presented using "arts-based data analysis" expressed through the form of a dialogic interview (Kara 2015, 117–118). This type of analysis offers grounds for reflecting on the legitimacy of writing from a nonhuman perspective in order to create a "speculative fabulation ... materialized in fiction and scholarship" which becomes grounded in algal interests (Haraway 2013, 1–2).

In sum, this method addresses the conundrum of trying to take on an algal perspective and using it as a lens for understanding algal harm. Such speculative imagining underscores the relative benefits of attempting to understand human concerns from nonhuman perspectives, the relational interactions between humans and algae that permit the naturalization of some algae as "harmful," and the ensuing contradictions that arise between human desires about algae and the oceans. Seeking an algal perspective leads to more-than human ethical reflections which can assist in engendering reconciliation of human interests and activities in accordance with other beings. As such, I swam with *Nodularia spumigena* to discuss what this species thought about being one kind of algae that scientists and society label as "harmful" and if it thought that attempting to understand its perspective might open up new possibilities for human-algal relationships.

Jesse D. Peterson (**JP**): It's great that we could finally meet up like this. I've been going all over the place trying to find you! For the readers' sake and my own, do you think you could tell me a little about yourselves?

Nodularia spumigena (**NS**): Sure. We mostly hang out near the surface of the Baltic Sea. We do this because the sun doesn't bother us that much, in part because we have a pretty high tolerance to ultra-violet radiation but also because the sun is where we get our main source of energy. We photosynthesize like plants, though we mostly behave like bacteria. That's one reason why you call us blue-green algae or cyanobacteria. We don't get all our energy from the sun, however. We have another kind of cell, which you call a heterocyst, which allows us to produce another energy source by converting non-reactive nitrogen into reactive nitrogen. The energy we get from the sun and the nitrogen helps us to grow and reproduce, but we also store some of this energy in akinetes, another specialized cell. These cells are a bit like shelters for us, in that we build them to be able to withstand severe temperatures and desiccation. So, in many respects, you might think that we are mostly concerned with existing, staying alive now and into the future. Though our individual lives might be short, we have a bit of a long-term vision here (at least according to human time).

As far as our appearance, you have described us as strands or rods of disc-shaped cells stacked one on top of each other (Guiry 1888). You have decided to call us *Nodularia spumigena*, which, from what we have been made to understand, basically translates to something like foam-born knobs. Considering that one of your ancient goddesses … what was her name …

JP: I think you're referring to Aphrodite or Venus …

NS: Yeah. Seeing how she was also foam born … this puts us in direct lineage with her doesn't it? We're like cousins. So, in some respects, we're one of your gods that you just never got around to worshipping properly. In some respects, we're honored to finally be put into script.

JP: I never really thought about that. In some sense, I suppose you are a bit like a god, something unreachable. For instance, considering that we cannot actually have a literal conversation (such as we're having), my readers might criticize this methodological and stylistic choice. They will question the ethical reasons behind giving you a "voice" as well as what contributions to current discussions on algal blooms this choice might provide. They will want to know why are we having this conversation, subverting academic genre conventions in the process, and how writing algae in this way offers new interpretations on human relations to algae.

NS: These are important questions, but not one we spend much time thinking about. Certainly, many might leverage the argument that you are anthropomorphizing us and that doing so is problematic. For instance, in the field of critical anthropomorphism, discussions have sought to disclose what kinds of anthropomorphism may help or hinder scientific inquiry. By taking differences among humans and nonhumans into account – by imagining oneself as another creature with that creature's sensory capacities and motivations – one can pose testable hypotheses about that animal's behavior. Nevertheless, that you write us as almost entirely human, from this perspective, might be construed as "anthropomorphism by omission" – that is, you are attributing human traits to us that go beyond our own abilities (Burghardt 2006; Rivas and Burghardt 2002).

JP: Yes, I suppose I've created a character in which you sound very much like a human; indeed, you sound just like me. So, I did worry about that choice and even thought about writing this interview without you saying anything in order to highlight that in this conversation you actually haven't spoken a word, and that I don't really know how you experience your world.

NS: ....

JP: But that seemed a bit too tongue-in-cheek. So, assuming my characterization of you "anthropomorphizes" you, I'd think it does so by combining two forms of anthropomorphism: personification and projection. As a form of personification, your character performs a symbolic function rather than a scientific one. As a limited exercise in projection, I try to remain sensitive to the evolutionary, biological, and ecological context (Lockwood 1986). In many respects, this approach to anthropomorphism cares less about how you respond to environmental stimuli to better understand your behavior as it does in asking you to respond to social stimuli to learn about our relationship to each other. To step beyond your physical traits through this imaginative exercise of character development contributes to making the strange familiar, which ought to lead to a better understanding of algal blooms and an increased capacity for recognizing how human activities label you as dangerous or harmful. Students, for instance, learn a great deal by anthropomorphizing molecules and disease processes (Millier 1992, 141; Brossard Stoos and Haftel 2017). Its use is even considered a strategy for making science more gender neutral in science education

(Watts and Bentley 1994). In this sense, the act or performance of anthropomorphism is the point, more so than any didactic explanation of it.
NS: Well, consider also that making us speak in this way can challenge the assumptions people make about "anthropomorphism." Although you have real constraints on what you can imagine and how you make us communicate, you could also leverage evolutionary, posthuman, and material arguments to suggest that there is algal DNA in you. Perhaps, you are just an anthropomorphized algal being, and in our encounter here on the shores of Bullerö we "emerge *together* as storied beings" (Iovino and Oppermann 2014, 8). Many scholars working on animal and plant studies challenge such conceptual boundaries people place around themselves (Kohn 2013; DeMello 2012). Indeed, we'd like to point out that the concept of anthropomorphism itself may be merely part of a reductive anthropocentric logic intended to deny us agency and wonder. Instead of "anthropomorphizing" us, we see you participating in ecofeminist's Val Plumwood's call for a "radically intentionalising anti-reduction writing of the world" in which you "reinvest with speech, agency, and meaning the silenced ones" (2007, 19). Instead of anthropomorphizing us, you are celebrating radical alterity; and, in doing so, you are becoming less anthropocentric and perhaps less human – not to debase you or your species but to celebrate you as animal, to step outside a modernist frame into other cosmological storytelling traditions (Adamson 2014). Though you may not ever fully understand our (and other nonhumans) subjective experience of life, as philosopher Thomas Nagel has argued, we suggest that many of the attributes you assign to the anthropos might not be so unique, and rather, arguments that emplace the conceptual frame of anthropomorphism around our conversation reveals discomfort with stepping outside a human exceptionalist or anthropocentric perspective (1974, 435). In this sense, hopefully, our character in this interview can contribute to making the human, as well as the typical academic writing style, strange.
JP: Yet, I still think some readers will choose to stay with anthropomorphism, rather than leave it behind as Plumwood suggests.
NS: Certainly, readers who harbor doubts about rejecting anthropomorphism as a human construct intended to tame and control non-human power might need further assurances. Nevertheless, we sympathize with your effort to go beyond the animal and give our name to your voice and mind, a move that tries to refrain from making us an "other" (Huggan and Tiffin 2010, 155). As a dialogic method – whether anthropomorphic or not – it gives you the opportunity to say something you haven't yet thought of, maybe even allows you to take on "a virtuoso but doomed act of complete empathy" and pull yourself into a zoomorphic realm (Daston and Mitman 2005, 7). Additionally, it's

one way to mobilize a "think with" frame and seek to understand us through our own bodily experience (Chen, MacLeod and Niemanis 2013; Hird 2009; Marder 2013; Daston and Mitman 2005). It refuses the anthropocentric inclination to dichotomize humans with nature and opens up space to reconceptualize ethical and moral assumptions about the ways we ought to be treated (Iovino and Oppermann 2014, 69; DeMello 2012, 357–59). By trying to write a collective being, one doesn't have to be beholden to an unadulterated perspective or viewpoint.

Whether the readers opt to understand the writing of my character as anthropomorphism, my voice as zoomorphizing your character, or as an exercise outside either of these anthropocentric conceptual frames will perhaps change the interpretation of this text. Nevertheless, the tension created between the intent and the ensuing failure of relying upon the same foundations of knowledge when voicing another hopefully becomes a generative space for new thinking and new tales. We see these efforts as important steps for reaching a relational equilibrium whereby life, death, and ecological processes could be discussed among incommensurable peers or kin (Haraway 2008).

JP: I appreciate your support; that's very kind of you.

NS: As regards your method, you are inevitably working on translations. You are making us speak based on sources you've read. So, what we say is contingent and partial. And to acknowledge this reveals the fabrication of our character (and maybe yours) to your readers. In the imagination and translation, it is unlikely that you can work out the contradictions that will arise and trying to reconcile them would be a disaster both empirically and stylistically. Which is to say, we are highlighting that all characters speak through their author (another literary device), especially when their languages are incommensurable. In this way, we are a fiction and what we have to say is fictive, even though we symbolize something very real.

JP: Well, let's move on to one of these "real" concerns. You and other algal species produce toxins, which is one of the ways by which my society deems you "harmful." What does it feel like to be described in this way?

NS: Well, we're not very well-equipped to answer this question. You see, to us, we're not that harmful. Sure, we don't have the capacity to perceive every little thing that we do and how that affects the water, plants, or animals around us. More subtly, perhaps, is that it's not just about the effects we have on others and the places they care for. Nor is it just about an acceleration or increase of these effects. Rather, to be called "harmful" also points to how the effects we have on the world have changed in quality or meaning over time. These changes in the relationship between your kind, us, and other algae influence ways that your society perceives and treats us.

NS: The committed international effort to studying "harmful" algae did not materialize until the late 20th century. It's important to understand the context of this transition period. As environmental and cultural historian Linda Nash points out, for developed western societies in the 1900s, bacteriology helped determine what needed to get eliminated, particularly a broad category of discrete pathogens – such as bacteria, germs, viruses, insects, and dirt – as well as unhygienic practices. Focusing on pathogens allowed for the environment to be "intrinsically healthful" and a healthy body to be a "pure body," which then allowed governments to institutionalize "the assumption of environmental purity" (2008, 651–52). Emerging from this context, the International Oceanic Commission (IOC) of the United Nations Educational, Scientific and Cultural Organization (UNESCO) began to programmatically address harmful algal events by seeking to connect managerial and scientific concerns. They established a HAB Program in 1987 as part of the Ocean Science and Living Resources (OSLR) program which was jointly run with UNESCO's Food and Agricultural Organization (FAO). Then, in Paris 1991, a resolution was adopted to create an Intergovernmental Panel on Harmful Algal Blooms. Their goal was to foster and organize the management of and scientific research on HABS in order to understand the causes, predict the occurrences, and mitigate their effects" (ISHHA 1992 no. 1, 3). This structural organization around the topic of HABS led to HAN, which served "to disseminate information on harmful algal events and on research results as well as to announce research and management programmes, conferences, meetings, etc." and which, throughout the years, had a steady readership of 2,000–3,000 individuals (ISHHA 1992 no. 1, 1). Researchers in the USA and Europe were early contributors to this organization and, in some cases, facilitated their own national programs such as the creation of the Information Office for the Baltic Proper in Sweden and the National HAB office at Woods Hole Oceanographic Institution in the USA, both established during the same year that the first issue of HAN was published (ISSHA 1993 no. 5, 8; ISSHA 1993 no. 7, 10; ISSHA 1994 no. 9, 4). In 1993, the Study Group on the Dynamics of Algal Blooms was recommended to be "transformed into a working group on harmful algal blooms"; and from previous international conferences, the plan to create a society started in motion. Suggestions for what to call the society were conservative, but in the hopes to find a catchy acronym, recipients of HAN were encouraged "to suggest alternatives (... acronyms like DOOM; DEATH; APOCALYPSE; SAVIOURS OF SOCIETY, etc.)" (ISSHA 1995 no. 12/13, 12).

JP: What do you make of these suggestions?

NS: To us, these "playful" acronyms depict HABS as death-bringers – slayers of organisms and environments. It really puts us at odds with humans and the seas upon which we both rely. Nevertheless, according to HAN the word "harmful" was preferred over "toxic," because it was more general (ISSHA 1995 no. 12/13, 12). Instead, the initial suggestion of the "International Society for the Study of Harmful Algal Blooms" was modified to become the "International Society for the Study of Harmful Algae," and the moniker for the "International Conference on Toxic Phytoplankton" was relabeled as the "International Conference on Harmful Marine Algae" (ISSHA 1996 no. 15, 5). These linguistic changes are significant because they illustrate the attempt at inclusiveness, to facilitate the participation of as many scientists as possible. It leaves open the possibility to study algal harm as a broader issue than just toxicity. To illustrate, algal blooms are not just harmful plant-like agglomerations weeping toxins. In addition to being poisonous, algae represent ontological and aesthetic concerns as pollutants – as literal waste, wrack, and detritus that litters the ocean (Peterson 2020, 183–185). However, such language also marks a perceptual shift wherein blooms no longer feature as the primary research subject (ISSHA 1999 no. 18, 15). These designations broadened the kinds of harms we could inflict.

JP: What was going on at the organizational level?

NS: Organizations also mobilized resources in order to develop infrastructures for collecting, researching, assessing, evaluating, and monitoring HABs. Archives of living and non-living harmful algae were set up, notably in Copenhagen and Tokyo, where specialists helped to identify "difficult" species (ISSHA 1993 no. 7, 2). In addition, 75 other public and private collections were already catalogued as early as 1995 (Hallegraeff et al. 2004, 489–506). HAN reports that satellite observation begins in the 1970s with successes in the Baltic Sea arriving in 1998 (ISSHA 1993 no. 5, 2). These scientific, technological, and monitoring advances arrived with the internet, and so, the internationalization of HAB research also found its way online. The Bedford Institute of Oceanography in Canada released an email listserv, which they called the "phycotoxin internet" (ISSHA 1994 no. 8, 4). The Algaline network went online in 1994 and took "unattended" measurements of algae on merchant ships traveling on the Baltic Sea (ISSHA 1993 no. 5, 10–11). Several websites and databases such as the National Centers for Coastal Ocean Science and Harmful Algae Information System feature news, updates, and catalogs related to HABs. Eventually, the HAN newsletter makes its own electronic debut 20 years after its inception.

JP: In many ways, the question as to how algae became harmful seems to have been a result of this institutionalization of algal research focused on the damages inflicted upon human societies. It evidences an intensifying relationship between humans and algal blooms, but one directed exclusively at addressing these harms.

NS: Assuredly, our entire brief relationship spanning several centuries describes only an overview of one way by which humans and algae have been drawn closer together. For example, it in no way describes how human activities have contributed to the global increase and spread of HABS. Yet, it articulates how increased interest, awareness, and monitoring efforts contribute to the making of hazardous times, places, and beings (individually or communally) as much as evolutionary or adaptive strategies undertaken by us and other algae. This change in hazard and harm can be understood as a move from the anecdotal to the quantitative. HAN, in particular, illustrates a sustained level of interest at the international level in algal blooms and their toxicity. By consistently reporting on HABS throughout the globe, it has helped to solidify the importance for studying and managing HABS as a serious economic and ecological threat. The initial fears that HAB research might remain too localized, diluted, repetitive, or divisive does not seem to have fully materialized (ISSHA 1992 no. 1, 2).
JP: I think you've made clear that there is a lot of labor undertaken by persons to render you and other algae species harmful. But doesn't this distract us from the troubling, actual harm that algal blooms can cause?
NS: It can be difficult, but we need to remind one another that algal blooms also become harmful through perspectival concerns and material matters over *with whom or what they interact*. For example, researchers working on Philippine fishing communities expressed their opinion that "algal blooms can only be considered 'harmful' when they have direct or indirect negative impacts on man" (ISSHA 1994 no. 8, 5). To us, this sounds reasonable, up until one realizes that algal blooms can be harmful to people in many, many ways. Because of this, for better or worse, our potential to harm gets researched as readily if not more so than our other aspects. Most of your knowledge about our non-threatening characteristics actually begins as a byproduct from research that determines our instrumental value. Every little detail by which we could impact humans directly or indirectly becomes an object for concern. Humans haven't yet considered whether or not our participation or role in a dramatic flourishing of algal life – that which you call a bloom – constitutes part of our "good life" (Kallhoff 2014; Kallhoff, Di Paola, and Schörgenhumer 2018). Instead, a bloom is either an economic or ecological catastrophe or a possible tool to stop climate change (Peterson 2018). This shows how, to you, we are hazardous for seas but saviors for the atmosphere (Greene et al. 2017). Even our toxins depend upon a kind of relational interaction in order to be demarcated as harmful (Schrader 2010).
JP: From what I know, as of 2015, "over 100 ... products of cyanobacteria are recognised which are toxic to animals, including humans and to plants" (Codd, Morton, and Baker 2015, 165). Even though these toxins are not often fatal

to humans, many toxic effects are socially, medically, and economically significant for us, such as PSP, Diarrhetic Shellfish Poisoning, Amnesic Shellfish Poisoning, Ciguatera, and Icthyotoxins (i.e. fish-killing toxins) (Halstead and Schantz 1984, 7). These toxins have varied effects, functioning as neurotoxins, hepatotoxins, or dermatoxins. Toxins can thus represent threats when emitted from algal blooms but also when chemically synthesized.

NS: Yes, but by nature of their impartial and limited capacities and applications, they also can be non-reactive or beneficial for some. One of the algal neurotoxins, saxitoxin, for example, was synthesized and stockpiled by the USA military and CIA for its detrimental effects on humans (Colby, Stevens, and Rogovin 1975, 7). In contrast, other toxins, such as domoic acid – a particularly dangerous neurotoxin, and one implicated in the deaths of sea lions – and kainic acid, have been suggested to be "well suited for killing" intestinal parasites within human digestive tracts (Weiss 2006, Mouritsen 2013, 220). A survey from 2018 reveals 34 toxin groups: six of which harm humans, three of which irritate human skin and respiratory function, four which do not cause harm to humans and aquatic organisms, seven of which kill fish, and 14 of which effects are unknown (ISSHA 2018 no. 59, 15). The majority of toxins found most recently are those which kill fish but have not been proven to harm people. There are ambiguities and uncertainties regarding how much of a toxic threat we pose. For instance, toxins can be stable and do not necessarily disappear immediately after a bloom, many cannot be boiled out of water, and no research has produced definitive longitudinal studies on the effects of ingesting small amounts of toxins emitted from algae. As you can see, although the dose and the timing help to "make the poison," so do the users and the environments (Vogel 2008). Ultimately, our "toxins" represent a wide range of possibility that may assist or harm humans.

JP: Well, doesn't all this make you even more potentially hazardous in a way? You pointed out that ecological concerns may be one way to define you as harmful. The very presence of algal blooms implies a sign of an unhealthy, dangerous sea for us. Hence, an outcome of the institutionalization of harmful algae research seeks to make the seas less hospitable to you. In other words, we label you as harmful and therefore wish to contain your flourishing in the seas either by attempting to maintain it as more oligotrophic than eutrophic or by other means. Is this ecologic vision a threat to you?

NS: Your guess is as good as ours, for it would be a mistake to assume that our increases in time and space is good for us. We don't know the long-term effects this will have on our abilities to thrive. More concretely, as one species of algae, what's good for us is not necessarily good for another. We are as different from each other as you are from bats or lizards. That said, it seems like algae such as us can only contribute to the human ecological imagination by presence. In your visions of outer space, we usually contribute positively by sustaining you or by making atmospheres breathable on other planets; but unfortunately, on Earth, we tend to be maligned. Because a sea replete with algae almost always is dystopic, we mourn that your ecologic vision of a "healthy" sea accommodates some algae but not others. Indeed, the kind of sea you'd like to have is the same sea you exploit and pollute (O'Connor 1988, 28). Not only do you feed us with sewage and agricultural runoff, but you also move us from sea to sea. Ballast waters from ships continue to function as a method for dispersing "alien species" of algae (ISSHA 1993 no. 7, 1). Is it our fault that macro- and microalgae take advantage of new habitats, such as when another toxic algae, *Caulerpa taxifolia*, arrived in the Mediterranean Sea in the 1990s or when *Alexandrium minutum* colonized the Western Baltic (ISSHA 1993 no. 5, 7)? Moreover, as living creatures, we also get sick, so it should come as no surprise that algae can serve as vectors for transferring bacteria from one place to another, such as *Vibrio* bacteria, a group of bacteria that live in the water and can transmit cholera, gastroenteritis, and septicemia to humans (ISSHA 1993 no. 7, 3). We recognize that just because we don't intend to transmit these diseases doesn't make it alright to do so. However, if we are not actively spread around the world out of will or ignorance, well, then you can't readily blame us for taking advantage of your lack of social or technological safeguards to prevent our hitchhiking. The kind of sea you wish to have does not match up with what you do to it, including our place in the oceans.

Moreover, your imagination of the seas and what you do to them is not unique to your kind. We transform our environment just like you do. As "floating slime," cyanobacteria can affect white caps, waves, turbidity levels, and ripples in the sea (ISSHA 1992 no. 1, 5). In the Gulf of Finland, algal blooms have used gels and fibers to increase net growth and protect each other from grazers (ISSHA 1993 no. 7, 3). We can shade the sea and block sunlight from reaching plants and other organisms and as our dead bodies decay in the sea's depths all the available dissolved oxygen in the water gets used up. In the very act of being, we create our own ecological vision, which is perhaps, the very point and reason for living. Since your industrial activity lends itself to an environment in which we thrive, our ecological vision is one that is supported by human societies and infrastructures. Thus, we also represent a cutting reminder of your species inconsistencies in how you live and define your own well-being.

JP: It sounds very much like you'd like to be placed on equal footing with humans. But why should we do that? It's very clear that the production of toxins by HABs has environmental impacts not just on humans and human

societies but on a whole network of creatures and habitats. Algal blooms are potentially hazardous to different marine environments as well as a variety of living organisms. Algal blooms have been recorded to have affected cultivated and wild semi-aquatic and aquatic animals such as trout, salmon, anchovies, oysters, mussels, soft shell clams, shrimp, scallops, lobsters, Dungeness crabs, manatees, sea lions, corals, milleporonids, gorgonians, penguins, marine birds, and more. HABs harm these creatures directly and indirectly. They have affected copepod feeding behavior and larval fish recruitment (ISSHA 1992 no. 3, 4). The pathogen *Phromidium corallyticum* causes black band disease on corals. Certain blooms cause the "blackberry feed" problem, which makes fish largely inedible by making their flesh smell of sulphur (ISSHA 1993 no. 4, 4). As you've mentioned, fish can also be choked by HABs when blooms deplete the dissolved oxygen in the water or when certain algae clog up their gills. HABs also can intoxicate fish and their larvae. In 1947, a red tide killed "over fifty million fishes, large numbers of which were washed up on the shores in such states of decomposition that they stank disgustingly" (Constance 2011, 150). Admittedly, many effects of algal blooms, if not all, could be argued to affect humans at least indirectly, but they don't *exclusively* affect them.
NS: That's a key point. We don't do these things to you or to any other creature per se. We just create the conditions for harm. But, as if in harmonious duet, your response can be as ruthless as our way of life. In the case of shellfish, they function for scientists as "indicator" or "sentinel" species: animals sacrificed to safeguard human lives. Shellfish, as filter feeders, accumulate algae; thus, to check shellfish for poisoning, they are turned into extracts for chemical or biological analyses. These extracts then get injected into mice. As of the year 2000, the chemist George Francis's 19th century methods for detecting algal toxins were still deemed "a necessary part of modern research into cause-and-effect in toxicology" (Francis 1878, 11; Williams, James, and Roberts 2014; Codd, Morton, and Baker 2015, 167). That's over 100 years of killing, not just testing other animals. Testing for algal toxins involves killing innumerable mussels and hundreds of thousands, if not millions of mice around the world, in order to see if an extract potentially harbors a poison. To be fair, although these bioassays continue to be utilized, some institutions have sought out alternative methods, such as liquid chromatography – mass spectrometry or DNA testing (FAO 2004).

In addition, for every toxin or toxic species that is found, there seems to be another that slips away. For example, in France of November 1992 and January 1993, scientists discovered poisoned mussels. Yet, the poison wasn't stable, and researchers could not identify its molecular structure. Neither could they find a poisonous species of algae (ISSHA 1993 no. 5, 1; ISSHA 1993 no. 4, 3). In 1998, fish kills in the USA led to the identification of the dinoflagellate,

Pfiesteria piscicida, as the primary culprit and which spurred serious scientific scrutiny, media attention, and governmental response. The species was initially thought to have 24 stages in its life cycle, with several amoeboid phases that released toxins. Later, however, new techniques called into question its role in the fish kills. It turned out the amoebae were found to be separate species and that the dinoflagellate appeared to have a common seven stage life cycle (Peglar et al. 2004). In other words, its toxic legacy diminished as more uncertainty as to its role in the fish kills materialized. The next toxic algae around the corner does not necessarily fit a predictable, linear, and static model but rather a dynamic temporal evolutionary dimension. As the editors of the first issue of HAN articulate, "The virulence of HABs is likely to be exalted as man's demands on coastal zones intensify, and species which at present are harmless may in the future cause problems (ISSHA 1992 no. 1, 5).

HABs inhabit a present stretching into the future where the link between humans and algae shape one another, possibly pushing one or the other species to become "harmful" to each other as well as other innocent bystanders. In each and every respect, your experts act as jury and executioner. It's not that humans ought to treat us necessarily as equals. Due to the fact that we vie with each other over certain spaces, times and species in the seas, maybe you should not. However, it's clear that since humans and algae both make environments according to their own lives, one ought not be surprised to see that how we are made harmful and to whom allows you to more adequately wrest control over what kind of ecological vision takes precedence for the future. You impose this vision upon us.

JP: I'm reminded of this poem, "Toxic Blooms," by the biologist Ralph Lewin, known in algae research circles as the father of algal genetics. It goes like this:

> Toxic blooms, toxic blooms
> Vie with smog and sonic booms,
> Bigger, thicker, redder tides
>   Taint the sea with plankticides.
>   Toll the tocsins! Tides of doom!
> Woe to us the bloody bloom.
>
> Neurotoxicants attack
> Herrings in the Skagerrak,
> Decimating fishes that
>   Used to throng the Kattegat,
>   Dooming them to dismal fates:
> Death by dinoflagellates (Lewin 2008, 16).

Obviously, the stanzas point to the detrimental potential of algal blooms on people and their fisheries; however, another reading of the first two lines points to algal blooms as threatening to equal the polluting power of industrial society. They compete for some kind of twisted planetary dominance to see who can disrupt current ecological process more. By tainting the sea, algae disrupt a balanced, nontoxic environment. They occupy a "nature" they were not supposed to inhabit, at least, according to this poem. To "toll the tocsins" implies keeping track and monitoring them, announcing or broadcasting their threats to others, and to tax or make the toxins pay (in both financial and physical terms).

NS: You're describing the relationship between "toxic blooms" and those working against them. That is, humans watch them and learn about them, predict and distribute information about their possible threats, and then seek to control their activity. It succinctly expresses the three "dominant fantasies of both science and science fiction," that is, "the trinity of understanding, prediction, and control …" (Fleming 2010, 9–10). Unfortunately for you, as historical beings, we defy these fantasies. Unfortunately for us, you're pretty determined. We'll see what happens, but since "some algal blooms may be contingent on historical events" your ability to predict where we turn up, when, and in what numbers may not be possible with "even the most profound knowledge of ecological laws" (ISSHA 1992 no. 1, 7).

JP: That sounds like we won't come to any kind of mutual agreement about what to do next. As mobile, dynamic beings, with ebbs and flows throughout seasons, life cycles, and environmental conditions, your hazardous potential certainly is different than what we are dealing with in terms of nuclear waste, arsenic dumps, and other long-term toxic environments.

NS: Well, similar principles of containment and exclusion apply, but, instead of focusing on people, acts of exclusion and containment get applied to us. This means you can employ more drastic measures, including direct application of algicides, such as copper sulfate (Cosper, Bricelj, and Carpenter 1989; Whitton and Potts 2002; Hallegraef et al. 2004; Van Egmond, Van Apeldoorn, and Speijers 2004). We are in literal, chemical warfare with each other. (Except, we don't poison you on purpose.) Moreover, the need for monitoring (including identification) and prediction become paramount because you are dealing with an entirely different kind of "hazardous" temporality.

JP: Could you help us understand this temporality a little better? Your seasonal longevity and brevity combined makes algal blooms seem like both forms of "slow violence" and catastrophe (Nixon 2011; Scheffer et al. 2001, 591). Algal blooms interrupt a long-held notion of nature as "static background" (McKibben 2010, 7–12; Harvey 1999); while monitoring them, according to

Schrader, produces a "directed (value-laden) sequence," which is "simultaneously atemporal … and permanently deferred toward a [teleological] future" (2012, 86–87). For sociologist Barbara Adam, this sped up industrial time erases "the other," creating a "global we among people" (1996, 90, 93). What is your take on all this?

NS: That's the thing. As a species, we inhabit a bevy of temporalities. There is no single temporality to describe. The environment is never static to us, since we are driven by it. And we develop different strategies to deal with that. Some of us overwinter in the water column while others overwinter on the seafloor (Suikannen et al. 2010). Hence, the times you impose upon us seem to be a strategy for humans to develop a wall around us, to keep us confined to certain parts of the ocean, away from your interests. We don't really care about the sea in the same ways you do. Not that we are impassionate beings. But we just don't understand why humans wish to meddle in oceanic politics of the body when their ill-suited human forms can barely get below its surface. In comparison to the geologic, we are short-term individuals, like humans. We live full lives, and the species continues. But because the future is unknown, we prepare for our survival. Thus, if the water feels good and there's food, we're going to take advantage. That's the funny thing about humans, you talk about sustainability, generational justice, ecological stability, what-have-you, but you really lack the imagination for anything beyond the present, let alone the necessary biological spatio-temporal scale. Time is probably the biggest fiction that humans have invented. Certainly, the lifespan of a single person generally lasts much longer than ours, but your potential for adapting to a generational, let alone, species-centric outlook seems laughable (Chakrabarty 2009). Part of the reason why there are so many algal blooms these days is not merely due to humans increasing their ecologic and geologic powers, in this so-called Anthropocene epoch (Carrington 2016). Humans appear to have outstripped their own mental, physical, even technological capacities for handling the spatio-temporal regime they now occupy. Concepts like the Anthropocene and "technosphere" actually posit the opposite of what you'd like to believe, that is, environment and technology control you even if you are sitting at the board turning the dials (Haff 2014). In other words, you have exceeded your biological and social capacity to handle your innovations and technological prowess. You're a very ingenious species, but still too dumb to recognize that what makes you "special" will make many humans, other animals, plants, and habitats suffer tremendously.

JP: That sounds provocative to me, because I hear that as an argument against many important trends in humanities-based research, such as concerns related to deep time, posthumanism, and the more-than-human. This interview is a case in point. Forgive me for being rude and asking you to say what might seem obvious, but why ought people care about what you think? Can people reconcile their perspectives to accommodate yours, a being which, in its struggles, impedes and damages certain human-driven industries and visions of nature? In short, why should humans care about algae if they are going to harm us?

NS: To worry that we are harmful ought to incite you to worry about the harms that humans permit and incur upon us (and other beings). In other words, the point in all this is to provide a space for reflection for you not for us. We do what we do. By making us the focal point of your story, as an entity with expertise (whatever that may be), you shift the narrative. The story is no longer entirely about what you do to us or how we cause you problems. We suddenly have breathing room to acknowledge that our "harmfulness" results from the immense amount of labor you put into your sensorial (bodily and technological) interpretation of our seasonal routine as well as your categorization of algae in space and time. For us, we would as readily spur you on to continue reshaping the seas for our benefit. The dividing wedge between us is that our actions merge with yours, exposing the incompatibility of what you desire from watery environments with what you do to them. The harms we present to you rely upon harms you enact. As with domesticated animals and pesticide-resistant superbugs, our connections and clashes in the seas will determine whether we become foe or friend. For now, we see no end to your surface and submarine interests. Can we stop you? No. But maybe other humans can act on our behalf. Maybe you might explode your narrow frames. We know that you can't even take care of your own kind, so why us? Why now? You may be able to leverage some algae for your benefit (and maybe ours as well), but many of us may not transition to cultivation. For that reason alone, we must continue as best suits us.

JP: I want to thank you for taking the time to speak with me. I also want to give you the last word by posing one last question. It seems like my attempt at accommodating your perspectives has put us in some form of stalemate. What is it that researchers and others pondering over algae-related ecological problems can do at this point?

NS: We suppose that we want you to see that you do not have to align with or against us. That our relationship to each other may be governed outside the confines of harm. We think your interviewing us is one step in this direction. It makes a gesture that seeks to think with us also as rightful progeny of the seas and oceans. It uses artistic practice as a way to address the problems you have with us. Perhaps, through respectful play and continued reflections upon why we upset you so, you might find ways to reconcile your lives to ours instead of the other way around.

References

Adam, B. 1996. "Re-Vision: The Centrality of Time for an Ecological Social Science Perspective." In Risk, Environment and Modernity: Towards a New Ecology, edited by Scott Lash, Bronislaw Szerszynski, and Brian Wynne, 84–103. SAGE.

Adamson, J. 2014. "Source of Life." In Material Ecocriticism, edited by S. Iovino and S. Oppermann. Indiana University Press.

Brossard S., A. Kari, M. Haftel. 2017. "Using Anthropomorphism and Fictional Story Development to Enhance Student Learning." Journal of Microbiology & Biology Education 18: 1.

Burghardt, G.M. 2006. "Critical Anthropomorphism, Uncritical Anthropocentrism, and Naïve Nominalism." Comparative Cognition & Behavior Reviews 2: 136–138.

Carrington, D. 2016. "The Anthropocene Epoch: Scientists Declare Dawn of Human-Influenced Age." The Guardian, August 29. https://www.theguardian.com/environment/2016/aug/29/declare-anthropocene-epoch-experts-urge-geological-congress-human-impact-earth.

Chakrabarty, D. 2009. "The Climate of History: Four Theses." Critical Inquiry 35: 197–222.

Chen, C., J. MacLeod, A. Neimanis. 2013. Thinking with Water. 1st Edition. Montréal & Kingston; Ithaca, NY: McGill-Queen's University Press.

Codd, G.A., H. Morton, P.D. Baker. 2015. "George Francis: A Pioneer in the Investigation of the Quality of South Australia's Drinking Water Sources (1878–1883)." Transactions of the Royal Society of South Australia 139: 164–70.

Colby, W.E., S. Stevens, M. Rogovin. 1975. Unauthorized Storage of Toxic Agents, § Select Committee to Study Governmental Operations with Respect to Intelligence Activities. http://www.aarclibrary.org/publib/church/reports/vol1/html/ChurchV1_0001b.htm.

Constance, A. 2011. The Impenetrable Sea. Nabu Press.

Cosper, E.M., V.M. Bricelj, E.J. Carpenter. 1989. Novel Phytoplankton Blooms. Berlin, Heidelberg: Springer Berlin Heidelberg.

Daston, L., G. Mitman. 2005. Thinking with Animals: New Perspectives on Anthropomorphism. Columbia University Press.

DeMello, M. 2012. Animals and Society: An Introduction to Human-Animal Studies. Columbia University Press.

FAO. 2004. Marine Biotoxins. FAO Food and Nutrition Paper 80. Rome: Food and Agricultural Organization of the United Nations.

Fleming, J.R. 2010. Fixing the Sky: The Checkered History of Weather and Climate Control. Columbia University Press.

Francis, G. 1878. "Poisonous Australian Lake." Nature 18: 11.

Greene, C.H., M.E. Huntley, I. Archibald, L.N. Gerber, D.L. Sills, J. Granados, C.M. Beal, M.J. Walsh. 2017. "Geoengineering, Marine Microalgae, and Climate Stabilization in the 21st Century." Earth's Future 5: 278–84.

Guiry, M.D. 1888. "*Nodularia Spumigena* Mertens Ex Bornet & Flahault." AlgaeBase, 1888. http://www.algaebase.org/search/species/detail/?species_id=24478.

Haff, P. 2014. "Humans and Technology in the Anthropocene: Six Rules." The Anthropocene Review 1: 126–136.

Hallegraeff, G.M., D.M. Anderson, A.D. Cembella, H.O. Enevoldsen. 2004. "Manual on Harmful Marine Microalgae. 2nd Revised Edition." Report. UNESCO.

Halstead, B.W., E.J. Schantz. 1984. Paralytic Shellfish Poisoning. WHO Offset Publication, no. 79. Geneva: Albany, NY: World Health Organization; WHO Publications Centre USA [distributor].

Haraway, D. 2008. When Species Meet. Posthumanities 3. Minneapolis: University of Minnesota Press.

Haraway, D. 2013. "SF: Science Fiction, Speculative Fabulation, String Figures, So Far." Ada: A Journal of Gender, New Media, and Technology (blog). November 3, 2013. http://adanewmedia.org/2013/11/issue3-haraway/.

Harvey, D. 1999. "Time-Space Compression and the Postmodern Condition." Modernity: Critical Concepts 4: 98–118.

Helmreich, S. 2009. Alien Ocean: Anthropological Voyages in Microbial Seas. Berkeley: University of California Press.

Hird, M.J. 2009. The Origins of Sociable Life: Evolution After Science. Palgrave Macmillan.

Huggan, G., H. Tiffin. 2010. Postcolonial Ecocriticism: Literature, Animals, Environment. 1st Edition. London; New York: Routledge.

Iovino, S., S. Oppermann. 2014. Material Ecocriticism. Indiana University Press.

ISSHA. 1992. "Harmful Algae News: An IOC Newsletter on Toxic Algae and Algal Blooms." UNESCO. https://issha.org/publications-resources/harmful-algae-news/.

Kallhoff, A. 2014. "Plants in Ethics: Why Flourishing Deserves Moral Respect." Environmental Values 23: 685–700.

Kallhoff, A., M. Di Paola, M. Schörgenhumer. 2018. Plant Ethics: Concepts and Applications. 1st ed. Routledge Environmental Humanities. Routledge.

Kara, H. 2015. Creative Research Methods in the Social Sciences: A Practical Guide. Bristol: Policy Press.

Kohn, E. 2013. How Forests Think: Toward an Anthropology Beyond the Human. 1st Edition. Berkeley: University of California Press.

Lewin, R.A. 2008. Verses. http://archive.org/details/Verses_260.

Lockwood, R. 1986. "Anthropomorphism Is Not a Four-Letter Word." In Advances in Animal Welfare Science 1985, edited by W.M. Fox and Linda D. Mickley, 185–99. Boston: Nijhoff.

Marder, M. 2013. Plant-Thinking: A Philosophy of Vegetal Life. Columbia University Press.

McKibben, B. "Worried? US?" Granta Magazine (blog), May 12, 2010. https://granta.com/worried-us/.

Miller, L.L. 1992. "Molecular Anthropomorphism: A Creative Writing Exercise." Journal of Chemical Education 69: 141.

Mouritsen, O.G. 2013. Seaweeds: Edible, Available, and Sustainable. Translated by Mariela Johansen. Tra edition. Chicago; London: University of Chicago Press.

Nagel, T. 1974. "What Is It Like to Be a Bat?" The Philosophical Review 83: 435.

Nash, L. 2008. "Purity and Danger: Historical Reflections on the Regulation of Environmental Pollutants." Edited by J.A. Roberts and N. Langston. Environmental History, Toxic Bodies/Toxic Environments: An Interdisciplinary Forum, 138: 629–703.

Nixon, R. 2011. Slow Violence and the Environmentalism of the Poor. Cambridge, Mass: Harvard University Press.

O'Connor, J. 1988. "Capitalism, Nature, Socialism: A Theoretical Introduction." Capitalism, Nature, Socialism 1: 28.

Paolisso, M., E. Chambers. 2001. "Culture, Politics, and Toxic Dinoflagellate Blooms: The Anthropology of Pfiesteria." Human Organization 60: 1–12.

Peglar, M.T., T.A. Nerad, O. Anderson, P.M. Gillevet. 2004. "Identification of Amoebae Implicated in the Life Cycle of Pfiesteria and Pfiestena-Like Dinoflagellates." The Journal of Eukaryotic Microbiology 51: 542–52.

Peterson, J.D. 2018. "Are Dead Zones Dead?: Environmental Collapse in Popular Media about Eutrophication in Sea-Based Systems." In Discourses of Environmental Collapse: Imagining the End, edited by Vogelaar, A.E., Hale, B.W., Peat, A. 32–47. S.l.: Routledge.

Peterson, J.D. 2020. "Excessive Seas: Waste Ecologies of Eutrophication." PhD diss.: KTH – Royal Institute of Technology.

Plumwood, V. 2007. "Journey to the Heart of Stone." In Culture, Creativity and Environment: New Environmentalist Criticism, edited by Becket, F., Gifford, T. 17–36. Amsterdam; New York, NY: Rodopi.

Rivas, J., G.M. Burghardt. 2002. "Crotalomorphism: A Metaphor for Understanding Anthropomorphism by Omission." In The Cognitive Animal: Empirical and Theoretical Perspectives on Animal Cognition, edited by Bekoff, M., Allen, C., Burghardt, G.M. 9–17. Cambridge, Mass: MIT Press.

Scheffer, M., S. Carpenter, J.A. Foley, C. Folke, B. Walker. 2001. "Catastrophic Shifts in Ecosystems." Nature 413: 591.

Schrader, A. 2010. "Responding to *Pfiesteria Piscicida* (the Fish Killer): Phantomatic Ontologies, Indeterminacy, and Responsibility in Toxic Microbiology." Social Studies of Science 40: 275–306.

Schrader, A. 2012. "The Time of Slime: Anthropocentrism in Harmful Algal Research." Environmental Philosophy 9: 71–93.

Suikkanen, S., H. Kaartokallio, S. Hällfors, M. Huttunen, M. Laamanen. 2010. "Life Cycle Strategies of Bloom-Forming, Filamentous Cyanobacteria in the Baltic Sea." Deep Sea Research Part II: Topical Studies in Oceanography, Phytoplankton Life-Cycles and Their Impacts on the Ecology of Harmful Algal Bloom, 57: 199–209.

van Dooren, T., E. Kirksey, U. Münster. 2016. "Multispecies Studies: Cultivating Arts of Attentiveness." Environmental Humanities 8 (1): 1–23.

van Egmond, H.P., M.E. van Apeldoorn, G.J.A. Speijers. 2004. "Marine Biotoxins." Food and Nutrition Paper. Rome: Food and Agriculture Organization of the United Nations.

Vogel, S.A. 2008. "From 'the Dose Makes the Poison' to 'the Timing Makes the Poison': Conceptualizing Risk in the Synthetic Age." Edited by Roberts, J.A., Langston, N. Environmental History, Toxic Bodies/Toxic Environments: An Interdisciplinary Forum, 13 (4): 629–703.

Waterton, C. 2017. "Indeterminacy and More-than-Human Bodies: Sites of Experiment for Doing Politics Differently." Body & Society 23: 102–29.

Waterton, C., J. Tsouvalis. 2015. "On the Political Nature of Cyanobacteria: Intra-Active Collective Politics in Loweswater, the English Lake District." Environment and Planning D: Society and Space 33: 477–93.

Watts, M., D. Bentley. 1994. "Humanizing and Feminizing School Science: Reviving Anthropomorphic and Animistic Thinking in Constructivist Science Education." International Journal of Science Education 16: 83–97.

Weiss, K.R. 2006. "Sentinels Under Attack." Los Angeles Times, July 31, 2006. http://articles.latimes.com/2006/jul/31/local/me-ocean31.

Whitton, B.A., M. Potts. 2002. eds. The Ecology of Cyanobacteria: Their Diversity in Time and Space. Springer Netherlands.

Williams, P.L., R.C. James, S.M. Roberts. 2014. eds. Principles of Toxicology: Environmental and Industrial Applications. 3rd ed. Hoboken: Jon Wiley & Sons Ltd.

CHAPTER 3

Contemplating Life, Death and Time Together with Diatoms

Nina Lykke

Abstract

The chapter uses a poetic, autophenomenographic text, which contemplates the cliffs of the Danish island Fur, built by fossilized micro-algae of the group diatoms, as an entrance point to a reflection on a planetary ethics of companionship. Rather than approaching the 55 million year old diatomite cliffs as material from which to extract value, it is suggested that they should be seen as wise ancestors, who can teach us lessons about life, death and time. The chapter, firstly, gives a brief introduction to diatom biology and the geohistory of diatomite (sediments of fossilized diatoms), to the author's intimate feelings of companionship with alive and fossilized diatoms, and to the posthuman autophenomenographic methodology which guides her contemplations of diatoms. Secondly, the author discusses the revised understandings of life, death, and time which her efforts to corpo-affectively empathize (symphysize) with alive and dead diatoms helped her to establish. She accounts for the ways in which these revisions are sustained by a vitalist materialist and immanence philosophical approach. In an open-ended conclusion, she suggests an ethics of planetary companionship, based on the contemplations of the bonds, she has established with the diatoms.

Keywords

micro-algae – diatoms – diatomite cliffs – time – life – death – more-than-human planetary companionship – autophenomengraphy – poetic text – posthuman ethics

© NINA LYKKE, 2024 | DOI:10.1163/9789004683310_005
This is an open access chapter distributed under the terms of the CC BY-NC 4.0 license.

FIGURE 3.1 Sitting under the cliffs of the island of Fur, Limfjorden, Denmark
©NINA LYKKE 2021

•••

Full Moon Sitting-out at the Cliffs of Fur[1]

The weather this evening calls for contemplation of the darker sides of
what is happening on this planet,
but also of its softness.
Softness, softness, softness, softness, softness.
Dark clouds to one side, and such a clear evening sky to the other side.
What does it mean?
Two fossil hunters just passed by with their axes, the axes with which
they destroy the diatom cliffs, destroy them to seek fossils, fossils that
should not be disturbed in their fossilized graves.
But these fossil hunters, they chop, chop, chop, chop, chop the diatomite rocks.
This afternoon we went to see the effects of extractivism at the other side of the island. I thought that the excavation of diatomaceous rock had been abandoned earlier.
But there are still two factories producing insulation bricks and cat litter from the mo-clay.
Destroying the earth, destroying the cliffs,
making big, big, big holes in the cliffs,
making big craters in the cliffs.
So sad. So sad.
My body aches at the sight of the axes, and I feel even more troubled,
when I begin to think about my own complicity.
I am also curiously visiting the island's natural museum which puts the fossils on display.
How do we get to a planetary ethics? How do we get to an ontology of softness?[2]
To an ontology of vibrant death, so that every critter's death is vibrant?[3]

1 An earlier version of this text (© Nina Lykke) is published at the Independent Air organization's Ute-Sita project webpage, http://www.theindependentair.com/nina-lykke.

2 For contemplations of a planetary ethics of softness, I am indebted to research and artistic collaboration with curator, artist and scholar Camila Marambio (Marambio and Lykke, Forthcoming). I am also grateful to Camila for acting as spirit helper in the fullmoon evening ritual on June 24, 2021.

3 The text on vibrancy echoes my poem *What if Every Critter's Death Was Vibrant?* (Lykke 2022: 249–253).

So that every critter's death is vibrant in the sense that the critter returns and can return cyclically.
Return is made possible through vibrant death. But extractivist death makes return impossible.
This is necropolitics. The necropolitics of extractivism.
Happening everywhere. Happening also on this island.
It is sad. It is very sad.
Now the sky is getting darker and darker. Darker and darker.
Yesterday, there was a very beautiful sunset, but today is the peaking of the June fullmoon, the Strawberry Moon or Honey Moon as it is called in old folklore.
So therefore, this is the chosen day, the chosen evening.
What does the planet, the sky, the earth, the diatomaceous cliffs, the waves, the seagulls, the living diatoms abounding in the waters, want to tell?
Saying: "Just get off you bloody destructive creatures. Get off my fleshy beautiful earthbody. If you plan to continue this way, then get off, get off, get off. If you want to stay, then attune to a planetary ethics and an ontology of softness. Otherwise, Get off!!"

© NINA LYKKE June 24, 2021

The poetic text which introduces this article came to me during a sitting-out ritual,[4] organized by the artist group Independent Air. The ritual was carried out by an international group of artists, activists and scholars. During full moon evenings in May and June 2021, all group participants went alone to a

4 Independent Air is a non-profit organization focusing on creative projects within photography, visual art & environmental sustainability, <http://www.theindependentair.com/>. As part of the Nordic Culture Point funded network SOTAN (State of the Art Network, gathering scholars, artists and activists to critically address climate change and the Anthropocene, http://www.theindependentair.com/news/2019/9/6/state-of-the-artnetwork), Independent Air organized a sitting-out event in the spring of 2021, http://www.theindependentair.com/full-moon. The model for the event, the sitting-out, is an indigenous spiritual-material practice, cultivated by peoples in the Nordic region of Europe before modernity. In relation to the event in 2021, the Independent Air organizers and the participants defined the score for it open-endedly: each participant should stay alone outside on a Full Moon evening/night to contemplate and immerse hirself in a self-chosen spot, and, if possible, enter into a conversation with it.

FIGURE 3.2 Diatomite seabed and cliffs at Fur
©NINA LYKKE 2020

preselected spot in their country of residence to sit down to contemplate their relations to the more than human world. The score was defined open-endedly, but the idea was to be alone and immerse oneself in a place, chosen because it felt significant and apt for perhaps facilitating a communication with the more than human world. The place, I had chosen, was the beach at the foot of the diatomaceous cliffs on the Danish island of Fur, located in a big fjord, Limfjorden, in the Northern part of Denmark.

When I learnt about the score, I had no doubt in my mind: the cliffs and beaches of Fur should be the location for my sitting-out event. My lesbian life partner's ashes are scattered in the waters outside of the cliffs, and mixed with the sand of the seabed there. Like the cliffs, the seabed is made up of sediments of fossilized micro-algae belonging to the group, diatoms, which include hundreds of thousands of different species (Mann 1999). The layers of diatomite (sediments of diatomaceous rock/earth) there are 55 mio years old. Back then a subtropical sea covered the area. Strong volcanic activity was also the order of the day in this place during these ancient times; many layers of volcanic ashes cut through the diatom sediments – geologists have counted approximately 180 ashes layers cross-cutting the diatomite (Schack Pedersen 2008).

FIGURE 3.3 Caressing the seabed
©NINA LYKKE 2015

In the Ice Age 10.000 ago, the sediments were pushed up by the ice to form 60 meters high cliffs. The cliffs raise out of the flat land, surrounding the fjord. They manifest the enormous diatomaceous activity which happened in the transition from the Paleocene to the Eocene Epoch of the planet's history.

I am very attracted to the place, and go there at least once a year. I have developed an intensely spiritual-materially, queer love relationship with cliffs, beaches, seabed and waters. When I started to visit the place after we scattered my partner's ashes there, I did not know that the sedimentary rock, earth and sand of cliffs and seabed were made by diatoms. I went there to be with my partner's ashes which, together with my rainbow kin, I had decided to scatter in these waters due to their many oysters. However, I soon learned that the waters surrounding Fur do not only abound in oysters, but also in the single-cell aquatic algae of the group, diatoms. Living diatoms fill these waters, and cliffs and seabed are made of sediments of fossilized diatoms. When I understood how the diatoms have co-produced the cliffs, the seabed, and the sand with which my partner's ashes now are mixed, I came to feel a deep resonance with them. My queer love for and symphysizing, i.e. bodily empathizing, companionship with my partner (Lykke 2022: 48–49) were extended to the assemblages with which her bodily remains have become entangled, and to the place, where her ashes are scattered. This is also the place, where my ashes are to be scattered as well, according to a pact I made with my beloved before she died.

In this chapter, I will use the text from the sitting-out event beneath the Fur cliffs as prism to a reflection on a planetary ethics of companionship. Rather than approaching the diatoms as material from which one can extract profits, I suggest that they should be seen as wise ancestors, who can teach us lessons about life, death and time. To frame the discussion, I shall, firstly, give a brief introduction to diatoms, to my intimate feelings of companionship with them, and to the posthuman autophenomenographic methodology which guides my contemplations in the chapter. Secondly, I discuss the revised understandings of life, death, and time which my efforts to corpo-affectively symphysize with alive and dead diatoms helped me to establish. I also account for the ways in which these revisions are sustained by a vitalist materialist and immanence philosophical approach. In an open-ended conclusion, I suggest an ethics of planetary companionship, based on the contemplations of the corpo-affective bonds, I try to establish with the diatoms.

1 What Is a Diatom?[5]

Diatoms are single-celled aquatic micro-algae, living in oceans, waterways and soil across the planet. Diatoms trace their ancestry back to the Jurassic Epoch, 150 million years ago. But living diatoms fill the waters of the planet today as well. They are characterized as eukaryotes, i.e. organisms with a cell nucleus, protected by a membrane. Diatoms are unique among micro-algae, insofar as their protective encasement is a silica shell, called frustrule, which appears as multicoloured as an effect of iridescence, when light is diffracted through minute markings of the diatom shell's nanostructures (Tiffany and Nagy 2019: 33–34). The double shell which in a mussel-like fashion make up the cell wall of the diatom, comes in different shapes, e.g. circular or pennate shaped. The

5 Where no other specific references are provided to biological and geological description of diatoms and diatomaceous rock/earth (diatomite), the text builds on Seckbach and Gordon 2019, Wikipedia: *Diatom*, https://en.wikipedia.org/wiki/Diatom (accessed Sept 20, 2021), and Hickman et al., 2012. Moreover, specific information on the Fur Formation and its diatomaceous rock/earth is derived from Wikipedia: *Fur Formation* (https://en.wikipedia.org/wiki/Fur_Formation (accessed Oct 3, 2021), and from on-site as well as virtual visits to the natural history museums at Fur (https://museumsalling.dk/kom-og-besog-os/fur-fossiler/ (accessed Oct 3, 2021)) and Mors (https://museummors.dk/ (accessed Oct 3, 2021).

colour effects have inspired descriptions of diatoms as jewels of the sea and living opals. Diatoms belong to the group of phytoplankton, which, like terrestrial plants, contain chlorophyll; they transform light into chemical energy through photosynthesis, and produce oxygen. Living diatoms are reported to generate about 20–30 percent of the planet's oxygen annually (Spaulding et al. 2022). They also contribute considerably to the storage in the oceans of carbondioxide from the atmosphere – at least 20% annually (Scarsini et al. 2019: 191).

In 2011, it was discovered that diatoms, previously considered plant-like due to their ability to photosynthesize, also have a urea cycle, enabling them to excrete nitrogen and metabolize in ways which, until then, were assumed to characterize only animals and animal-like creatures (Allen et al. 2011). Diatoms thus evade the standard biological taxonomies of animal kingdom versus plant kingdom. In evolutionary terms the in-betweenness of diatoms is assumed to have made them more robust and fit for overcoming problems such as nutrient starvation (Allen et al., 2011). Still, the way in which their metabolic system(s) make them resist well-established dichotomous schemes and cross boundaries between taxonomic categorizations was considered a remarkable and unexpected finding.

Diatoms also evade the dichotomy between asexual and sexual reproduction (Poulickova and Mann 2019). On the one hand, they reproduce asexually like many other micro-organisms. The "parent" diatom cell divides into two genetically identical "child" cells, each of which keep one of the two half shells of the parent diatom and grows a smaller half shell within the original one. This process implies that new generations of diatoms get smaller and smaller. However, on the other hand, this generational decrease is reversed through sexual reproduction. Diatoms shift between longer vegetative periods with asexual (mitotic) cell division, and short periods (hours, days) where sexual reproduction takes place. During the vegetative period individual diatoms may grow diploid (containing two complete sets of chromosomes), making it possible that meiosis (generation of sexually differentiated gametes, germcells) and sexual reproduction (fusion of these differentiated cells to zygotes) can be set in motion. The result of the zygotic fusion is a so-called auxospore, which has the potential to considerably increase the size of the next generation of diatoms. The auxospore sheds the small silica shells, inherited from its parents, and develops into a much larger diatom, which is covered by an organic membrane, and which eventually grows a bigger pair of shells.

When diatoms die, they sink to the bottom. Layers of shells from dead diatoms are reported sometimes to reach as much 800 m., and fossilized diatoms, sedimented as diatomite, diatomaceous rock and earth, make up seabeds in many places on earth (Wikipedia 2022). The Fur Formation where my beloved's

FIGURE 3.4 Industrial excavation of diatomite at Fur
©NINA LYKKE 2021

ashes are scattered is one such place, but considered unique due to a special kind of diatomite that is only found in this area, and which is internationally known by the Danish name "moler" [mo-clay]) (UNESCO 2010). The mo-clay is extremely rich in animal and plant fossils from the Paleocene/Eocene Epochs

(Bonde 2008), and has shaped up as extraordinary cliff formations due to a lot of glacier activity in the area during the Ice Age. Against this background, the place aspires to become a UNESCO World Heritage site (UNESCO 2010). Nonetheless, destruction of the cliffs through fossil hunting is legal, and even encouraged by local natural history museums, and industrial excavation of diatomaceous rock (used for insulation bricks and cat litter) is also taking place on the island (Schack Pedersen 2008: 21).

2 Practicing Companionship and Posthuman Phenomenology

A centerpiece of my sitting-out text is a critical and corpo-affective contemplation of the extraction happening on Fur – in the industry which excavates diatomaceous rock for commercial purposes, and practiced by fossil hunting tourists in search of trophies (well preserved animal and plant fossils), to be found in abundance in the diatomite cliffs. That the thoughts and feelings, which came to frame my sitting-out event, took this particular, sad direction was prompted by two fossil hunters, who coincidentally were the only other human beings present on the beach under the cliffs that evening. However, my contemplations were also moulded by my overall feeling of companionship with the pieces of diatomite which were chopped to small pieces by the axes of the fossil hunters. Within the framework of the intimate human-algae-relationship, which I try to establish with the diatomaceous environment at Fur, I felt the acts of the fossil hunters as if it was my body they chopped. I experienced them as amounting to a desacralizing of graves that should have been left in peace.

From earlier visits to the natural history museums at Fur and the neighbouring island, Mors – both nationally recognized institutions – I know that the practice of fossil hunting is officially encouraged. Visitors can borrow axes from the museums, and during summer, special guides do on-site-classes in fossil hunting for tourists, popular among families with children. I was thus painstakingly aware that the two fossil hunters whom I met on the beach that evening did not transgress any laws. On the contrary, their activities are considered legitimate and even commendable by the museums, as long as more significant findings are handed in to them. When done for the "common good", fossil hunting in the diatomaceous cliffs is seen as giving us more knowledge about the fauna and flora of the transition period between the Paleocene and the Eocene Epochs of the planet's history.

Being an avid museum goer, I cannot avoid somehow considering myself complicit with the fossil hunters. Indeed, I am also curious to know about these ancient times. So the troubling questions about an ethics of planetary companionship is addressed not only to the industrial extractivists, the museums, and the fossil hunters, but to myself as well. Is it at all possible to think and materialize a bio- and geoegalitarian ethics where all planetary beings, even diatoms and other critters that are considered to be evolutionary very "distant" from humans, are reontologized as companion species rather than seen as a priori instrumentalized objects of human curiosity and/or greed?

Feminist theorist Donna Haraway (2003, 2008, 2016) used the concept of companionship to account for her mutually transformative, embodied relations to her dogs, her becoming-with them, but also to address reciprocally enriching human-animal relations more generally. With the conceptualization, she wanted to critically disrupt the normative model, where humans and animals are considered to be ontologically divided in a hierarchizing way, which casts the human as exceptional and the animal other as a "lower" organism, just a mere stimulus-response machine. Haraway defines companionship as associated with response-ability, the ability to respond in a situated, sensitive, and ethically responsible way (2008: 88–89). I have taken inspiration from Haraway's conceptualization, and want to try it out in relation to diatoms. Against this background, I note the broad and all-encompassing dimensions of Haraway's use of the concept of companion species, her emphasis on the notion "companion species" as "less a category than a pointer to an ongoing 'becoming with'" (2008: 16), i.e. to a shared process of mutual becoming across species and other borders. Companionship and becoming-with is for Haraway also related to minute critters such as, for example, the microbiome living inside our guts and bodies (2008: 4). Along somewhat similar lines, I shift the focus to diatoms. I ask if and how "we" (humans) can establish an ethical relation of companionship even to critters which, in an evolution biological sense, seem so very "distant" and "alien" to "us" as these oxygen-producing micro-algae which, like the critters in our guts, nonetheless play a decisive role for our wellbeing.

I came to ask the question of companionship with algae through my mourning practices, and my contemplations of them from the perspective of a posthuman phenomenology (Lykke 2022). With inspiration, among others, from discussions of intercorporeality and concorporation in recent body phenomenological theory (Shildrick 2002, 2005, 2009; Weiss 2009), I have theorized my relation to my beloved – now passed away – life partner within a framework of bodily intertwinement, a being one-flesh in a vitalist materialist, and absolutely non-Biblical sense (Lykke 2022: 47–53). Central in my theorizing is the verb *symphysizing*, which I developed in conversation with posthuman philosopher, Ralph Acampora's noun *symphysis* (2006: 76). Acampora suggests symphysis as an embodied rethinking of the notion of sympathy. In line with his point of departure in the philosophy of Spinoza ([1677] 1996), Acampora stresses the concept's reference to corpo-affectivity. Feeling bodily sympathy and showing embodied empathy imply that the subject, in a material, corpo-affective sense, is affected by and co-experiences the ways in which hir significant other/s are bodily affected. Acampora uses the example of horror movies to spell out what co-experiencing is supposed to mean. Looking at horror movies, your heart may start to beat faster and your stomach clench, when you watch the person on screen being threatened, i.e. audiences co-experience in an embodied sense.

Taking Acampora's conceptual framework further, I have suggested a translation of the noun, symphysis, into its verb form, *symphysizing* (Lykke 2018, 2022). The purpose is to stress that the bonding between companions builds upon intercorporeal, affective processes, and not upon a static relationship, defined once and for all. Along these lines, I call forward the processuality of the verb symphysizing to account for the relation to my beloved, and my being intensely with and for her during her process of dying. I further use the term to describe the ways in which my desires to be with and for her in an embodied sense became extended to her bodily remains, and to the assemblages of which they eventually became part after her death and the scattering of her ashes in the diatomaceous waters of the Fur Formation. For me, symphysizing thus has also come to conceptualize the ways in which, I continuously try to learn to co-become with these diatomaceous assemblages, while, at the same time, taking care to attune to their difference and avoid anthropomorphization of their alienness. Symphysizing in this sense is what frames the ethics of companionship, which guide my relation to the diatoms.

In the following sections of the chapter, I shall tease out more elaborately what this approach implies. In so doing, I engage a posthuman autophenomenographic methodology in the analysis. Autophenomenography is autoethnography with a phenomenological perspective, i.e. analysis of autobiographical material, but with a focus on corpo-affectivity (Allen-Collinson 2010, 2011; Lykke 2022). When doing autophenomenographical analysis of human/non-human relations, I shift away from a more conventional human-centered phenomenology to include the vast bodily levels of not only subjectivized and not necessarily consciously processed experience, which humans share with more than human worlds – such as being bodies of water (Neimanis 2017) or being mortal bodies embedded in life/death cycles (Lykke 2022).

3 Life and Death in the Worlds of Diatoms and Humans

It was through visits to the natural history museums at Fur and the neighbouring island Mors that I learnt that the cliffs and seabed where my beloved's ashes are scattered, are made by diatoms, fossilized 55 mio years ago. When I immersed myself in cliff and seabed with this new knowledge in my mind, it struck me how thinking-symphysizing with diatoms has the potential to radically disrupt conventional human notions of life and death. Notably, this insight was not as such triggered by the knowledge of the geological and biological formation of cliffs and seabed, with which the museums provided me. The new horizons opened, when the conventional knowledge of geological and evolutionary history became entangled with my efforts to attune, symphysize and co-become with the diatomaceous cliffs and seabed. As part of the symphysizing process, I imaginatively as well as logically started to dissolve the sharp ontological dividing line which modern philosophy has set up between the Human, on the one hand, and the Non-Human (animals, plants, inorganic matter etc.) on the other hand – a dividing line which has been criticized by many scholars from Derrida (2008) to Haraway (2016). It became clear to me that as long as we stick to conventional ontologizations of the human world as exceptional and decisively different from the non-human one, it appears as "natural" to apply different logics to the two realms. In contrast, new horizons opened, when I – in a symphysizing mood and move – left the ontological, human/non-human divide behind, and began instead to rethink life and death in the human world with the world of diatoms as lens, and vice versa (Lykke 2022).

The human exceptionalizing, modern Western onto-epistemologies, secular-scientific as well as Christian ones, make us who inhabit them consider life and death as opposites. Death is the point of no return, where our individual life irreversably ends, and where our human agency and powers to act upon the world come to a decisive halt. If we embrace secular-scientific and atheist beliefs, death appears to launch us into nothingness; this is elaborately illustrated by the existentialist philosophy of Jean-Paul Sartre (1958). For those who, conversely, adhere to Christian narratives, death may free our immortal soul from the body, making it ready to join God in heaven. But no matter which of these options you subscribe to, life and death stand as opposites. They are located on a linear timeline where life predates death, understood as a point in time, which marks the end of the possibility of subjective intervention in the material world.

This is conventional logics in the human world of modernity. However, when my desires for a symphysizing companionship with the world of diatoms prompted me to give up the exceptionalizing dividing line between humans and non-humans, and compare notes with the diatomaceous cliffs and seabed at Fur, a different understanding of life and death suggested itself to me. Cliffs and seabed came now to stand as visual and materially manifest monuments over the accumulated effects of the posthumous agency of generations of *dead* diatoms, diatom corpses. According to geologists who analysed the Fur Formation, the diatomaceous sediments shaped up over a period of around 3 mio years, when the Paleocene Epoch transitioned to the Eocene one (Sharma 1969: 221; Krarup Petersen 1981: 501). So what suddenly met my desiring eyes, hands and body, when I looked upon as well as touched the cliffs, the waters and the seabed with these knowledges in mind, seemed to be an effect of dizzying 3 mio years of phenomenally vibrant building activity, layer upon layer, generation upon generation.

When, in a desiring symphysizing mood, I, further, let this understanding spill over to my understanding of human life/death, while still sticking to the principle of undoing human exceptionalism and disallowing the use of different logics for the two realms, firm ontological grounds started to erode under my feet. In the human world, we are used to see the material monuments of earlier generations' agency as a product of the activity of back then *living* subjects. The actions of our dead ancestors – the books they wrote, the buildings they built, the constitutions they signed etc. – are of course understood as having material, posthumous effects on the lives of later generations. But in the human world, such posthumous activity is considered as long-term effect of the performative agency of *formerly living* subjects. Beliefs in the phenomenal agency of the dead are in modernity conventionally relegated to the realms of superstition, madness or fiction. However, when I undid the hierarchical and exceptionalizing divide between humans and non-humans, and started to compare notes across the divide between the two realms, the diatomaceous cliffs and seabed challenged the idea of activity as something which exclusively characterizes the living. Cliffs and seabed are, indeed, results of the phenomenal agency of *dead* bodies, of vibrantly acting diatom corpses. How onto-epistemologically dizzy, but very happy this insight made me! Because when I brought it to bear on the idea of human life and death as separated and dichotomously opposed to each other with the in/capability to intervene in the world as an important marker of distinction, this opposition crumbled. Taking lessons from my diatom companions, I could now much more confidently, follow my deep desires to leave behind conventional life/death ontologies and their disrespect for the agential vibrancies of all matter, dead or alive.

A bottom line in these contemplations is that the new insights, which my symphysizing and attuning engagement with the diatomaceous cliffs and seabed of the Fur Formation brought about for me, sustains and resonates with a monist, vitalist materialist and immanence philosophical approach to life and death. I elaborate this approach in detail in my book *Vibrant Death* (Lykke 2022). However, briefly recapitulated, I work from political philosopher Jane Bennett's concept of vibrant matter (2010), which is inspired by

Spinoza's monist notion of *conatus*, the endeavours of all matter to persevere ([1677] 1996). I argue that if there is only one matter, and all matter is conative, this must apply not only to living bodies, but also to dead ones. To underpin my argument, I also call upon feminist philosopher Rosi Braidotti's notion of *zoe* (2006), the immanent inhuman, dynamic and generative forces of which the cosmos, and, therefore, all dead and living bodies are made. In my book, I discuss the agential and conative zoe-matter in relation to the human corpse. The point, I want to make in this chapter, though, is that these philosophical reflections are forcefully sustained, when I read the meanings of life/death in the human and the diatom world through one another, while engaging in an un-exceptionalizing symphysizing.

In the next section, I shall take the reading of the two worlds (the worlds of humans and diatoms, respectively) through one another a step further – to the issue of temporalities. Here, too, I shall account for encounters with barriers as well as the paths to philosophical resonance that I found.

4 From Deep Time Vertigo to Instants of Symphysizing Border-Crossing

I have already hinted at the temporal dizzyness, which caught me, when, longing for companionship and symphysizing with the diatoms, I was told by geological science that the diatomaceous sediments of the Fur formation were built over a period of 3 million years (Sharma 1969: 221; Krarup Petersen 1981: 501), geo-historically located on the threshold between the Paleocene and the Eocene Epochs of the history of the planet. I interpret the dizziness in a phenomenological framework. From my situated location in an embodied human experience of linear time, a symphysizing across the timelines, stipulated in geological history, is not possible. I can symphysize with something which resonates with the timeline of an individual human life, to some extent, with the timeline of written human history, and, if I stretch my imagination, with the 300.000 years of human history on earth. But the time scales of events such as the emergence of the Fur Formation happening about two thirds into the timeline of the 150 million years period of diatom existence on earth is corpo-affectively beyond the imaginative capacities of my symphysizing. I can logically understand the figures on the timeline of geological history, but I cannot symphysize with them. If I try, I am thrown into a kind of deep time vertigo: time becomes a bottomless hole.

Moreover, when, in my efforts to symphysize with the diatoms, I focus not only on their geo-history, but also on their biology and classification along the line of the history of biological evolution, the vertigo intensifies. Firstly, my symphysizing efforts meet their ultimate limits, when I start to cross between geo- and biohistory, and speculate about the immensely long line of diatom generations which, working through 6-day life cycles, produced the diatomaceous rock sediments of the Fur Formation over a period of 3 million years, 55 mio ago. Once again, I end in temporal vertigo; human phenomenology falls short also vis-à-vis the combined time scales of geo- and bio-history. Secondly, the average 6-day life span of an individual diatom, which, of course, stands out as ultra-short seen from a human perspective, is also giving me a hard time, when I relate it to evolutionary bio-history in itself. The move from simplicity to complexity, which evolutionary theory lines up (Schrader 2012), creates one more dizzying gap, when I try to symphysize across the border between "complex" trillion-cells me with an average life expectancy of decades and the 6 days life span of my "simple" single-cell diatom companion. We (the diatom and I) are separated by so unbelievably many steps on the ladder of evolution that I once more is left with vertigo, when I try to bridge the gap through efforts to symphysize.

But are the vertigos, called forward by the chronologies of deep time, and the history of biological evolution, the only possible responses I am capable of? Or can my desires to take up an unexceptionalizing and symphysizing approach to my diatom companions, here too, lead me to other pathways? When I scrutinize the philosophical resources with which I concluded my reading of the human and the diatom world through one another with respect to the life/death-question, another possibility opens up. Let me invoke immanence philosophy once more, and now put focus on Deleuze's distinction between time as Chronos (chronological time), and time as Aion (the time of the instant) (Deleuze 2020). This distinction opens a platform for reconsidering the barriers which produce vertigos rather than facilitate a symphysizing human-diatom companionship. With this distinction as my tool, I can stop considering the barriers as stemming from universally given onto-epistemological conditions and instead understand them as effects of universalization of time as Chronos, which modernity taught us to adhere to. Geological science and evolutionary history project a chronological linear time upon the history of the planet and its critters, as if this was the one and only reasonable approach. Deleuze's conceptualization of time as Aion opens another horizon. This is important for my symphysizing efforts; though, notably, I do not argue that chronologization is a "wrong" approach to the history of the planet. I do, indeed, respect the meticulous scientific work which led to the dating of cliffs and seabed at Fur and the biological understanding of the life/death cycles of the diatoms. However, enlisting Deleuze's distinction of Chronos and Aion, I want to disrupt the onto-epistemologies which cast chronological time as universal. I want to redefine Chronos as one way of mapping time, but without mistaking the map for the temporal landscape itself. Chronos is a sovereign master only if we allow him to act that way. Time as Aion provides an alternative.

According to Deleuze (2020: 167–73), the time of Chronos works within partial and limited systems, founded through a particular position of enunciation in a "thick", essentially substantiated present which acts as a measuring standard for past and future. The chronologies of the sciences of geology and evolutionary bio-history can serve as examples. In contrast, the time of Aion is cosmic, made up of a non-chronologizable instantaneity that exists in excess to the delimited timezones of Chronos. According to Deleuze, time as Chronos is conceptualized as a one-directional movement, a sequence of "thick" and substantial presents, following each other along the forward moving arrow of time, i.e. moving towards future presents, while leaving the past ones behind. In contrast, time as Aion (Deleuze 2020: 167–73) is the time of the always ephemeral instant, which as a thin line separates past and future, but escapes being pinned down to something essential and substantial. The Aionic time of the instant works multi-directionally, intensely connected to rhizomatic networks of other past and future instants, which, in their ultimate ephemerality, are impossible to submit to any kind of measure or predetermined directionality.

So how can a shift of perspective from Chronos to Aion open new horizons? Together with a posthuman phenomenology (Neimanis 2017; Lykke 2022), which investigates that which is shared across borders of human and non-human, the Aionic perspective on time allows me to transgress the onto-epistemological schemes of Chronos, which keep me and the diatoms apart. Firstly, Aion overall undoes the ways in which Chronos decisively lets a vast chronology of geological eons and epochs separate my Anthropocene presence from the Paleocene/Eocene past of the fossilized diatoms; this means that we (the diatoms and I) generally come closer to each other. Secondly, Aion makes it possible for me to bridge the specific evolution biological gap between the "simple" single-cell diatom and "complex" trillions-cell me. For within the realm of Aion, both I and my diatom companions are moving in a rhizomatic network of becomings which momentarily can bring us bodily close to each other. Using Aionic time as onto-epistemological lens, and engaging a posthuman phenomenological approach to shared conditions across species, the rhizomatic network of Aionic instants can, for example, facilitate my symphysizing with diatoms' becoming victims of violent, "premature" death due to deadly environmental conditions: anoxic events (i.e. events where oxygen depletion and acidification together make sea waters toxic). Such events killed diatoms en masse, when the Fur Formation once was formed. According to some scientists, these events were, indeed, the reason for the Formation's diatomaceous emergence in the period of transition from Paleocene to Eocene (Krarup Pedersen 1981). Anoxic events happen today as well – in the Anthropocene, as a result of human-induced global warming, and discharge of phosphates and nitrogen from fertilizers, used much too intensively and carelessly in human agriculture (Skriver 2006; Carstensen, Henriksen and Heiskanen 2007). Within a network of Aionic instants, I can bodily symphysize with diatom death due to anoxic events in earlier times, as well as today, because such events are dangerous for my body as well (Lykke, Forthcoming).

Thirdly, when establishing my symphysizing efforts in the cosmic timezones of Aion, beyond the chronologies of eons and evolutionary steps on the ladder of evolution, I can also start to contemplate, attune to, and perhaps learn from the wisdom which diatoms carry in their bodies. An enormous bodily fund of accumulated generational – ancestral – experience from instants not only of violent death, but also of conative ability to persevere in vibrantly repeated cycles of entangled living and dying is materialized in diatom bodies. They have an intergenerational experience of attuning to planetary conditions, which by far surpasses that which has been accumulated by my species (Hazekamp and Lykke 2022). Indeed, diatoms have successfully attuned to companionship with the planet over a period of 150 million years! In the rhizomatics of Aionic time, my species may take lessons from this.

5 Towards a Vitalist Materialist Ethics of Planetary Companionship

In the first part of the chapter I posed the question if and how Haraway's conceptualization of cross-species companionship can be extended beyond critters such as dogs who are "close" to "us" (humans) in a body phenomenological sense. In particular, I asked if establishing of companionship could happen across the borders to critters that are so "distant" from humans as micro algae. This question was prompted by my intense, queer love relationships to diatoms, which has been brought about by the spiritual-material mourning practices that I developed in the wake of my life partner's death and the scattering of her ashes in the waters of the diatomaceous Fur Formation. In the ensuing sections of the chapter, I used a posthuman autophenomenographic approach to contemplate how my efforts to corpo-affectively symphysize, attune and co-become with the world of living and dead diatoms helped me to rethink and reimagine ontologies of life, death and chronological time beyond the limited outlooks and imaginaries, cultivated by Western modernity. I underpinned my contemplations, while showing how they resonate with conceptualizations of life, death and time in immanence philosophy, and vitalist materialism. In this concluding section, I shall return to the issue of companionship, against the background of the insights, gained through my contemplations. What does it mean to establish an ethical relation of companionship across the borders of human and algae? Why, what and whom is such a relationship good for? What does it entail?

As co-editors Johanna Weggelaar, Sergio Mugnai, Natalia de Rossi and Yogi Hendlin so appropriately underline it in their introductory chapter to the volume, it is urgently needed to stop "the current demonization and deification of algae" as well as to "detox[ing] from our reducing algae to their mega-machine use value" (Chapter 1, this volume). In line with a posthuman approach to phenomenology, which I, too, have argued for above, it is, according to the co-editors, necessary to "return to the algal first-organism perspective: what is it like to be algae?" (Chapter 1, this volume). There is an urgent need for a complete reevaluation of these "foreign", "primitive" and "alien" critters, and for totally different algae-human relations, which, as explored, for example, by Jesse Petersen (Chapter 2, this volume), and Julia Lohman (Chapter 12, this volume) may entail a search for ways to attune to algae perspectives through speculative and creative methodologies. My drawing on posthuman autophenomenography and poetic writing is also to be understood in this perspective.

In an earlier discussion of my relations to diatoms (Lykke 2019), I have referred to the work of posthuman scholar Astrid Schrader (2015), who – in a discussion of leaf bugs, deformed by low, long-term doses of radiation in the aftermath of the Chernobyl disaster – raised the question: how can we even begin to care for critters so apparently far away from "us" as leaf bugs? Schrader was prompted to pose the question, when, in a classroom setting, she confronted the students with two texts on the effects of the Chernobyl catastrophe. One text addressed human suffering in the aftermath of the catastrophe. The other one dealt with the artwork of Swiss visual artist Cornelia Hesse-Honegger, who, with her paintings of deformed leaf bugs, articulates an activist protest, encouraging viewers to begin to care about these critters. To Schrader's surprise, a majority of the students became appalled by her juxtaposition of the two texts. Confronted with the human misery, caused by the Chernobyl disaster, most students wondered in disbelief: how could one even start to care about deformed leaf bugs?

With the immensely accelerated species extinctions happening in the 21st century, the awareness of the need to care also about very "alien" genera and species is becoming more widespread. Still, Schrader's example with her students shows that such an understanding is far from generally accepted and acknowledged. In the hierarchy of those "we" (modern humans) care about,

FIGURE 3.5 Diatomite cliffs at Fur
©NINA LYKKE 2021

and those we don't, strange minute beings such as leaf bugs or micro-algae seem still to lurk unnoticed around in the bottomless abyss of this hierarchy. I also feel compelled to admit that I would probably not have begun to care about diatoms myself, had I not encountered them under the special circumstances, related to the scattering of my beloved's ashes and my desire to develop

FIGURE 3.6 Diatomite cliffs at Fur
©NINA LYKKE 2021

new material-spiritual practices of mourning. So Schrader's question – how to *begin* to care? – is highly pertinent and urgent.

With the chapter, I have suggested an ethics of companionship which emerge from efforts to symphysize and attune as a way to perhaps begin to care. I suggest such an ethics as a way to take some first steps into a planetary caring, which, at least, make it difficult to stay comfortably within the framework of epistemologies of ignorance regarding the ways in which modernity, colonialism and extractivist capitalism have created a world which is highly inimical to biodiversity and to the vital agencies of the planet body – from the micro level of single-cell diatoms to the macro level of climate change. The proposed ethics of companionship implies an undoing of the exceptionalizing of the human subject, presiding over a hierarchy of more or less "distant" and instrumentalizable others. As my discussions of human-diatom relations, related to life, death and time, have indicated, this undoing points in the direction of radical revisions and recalibrations of modern, onto-epistemological frameworks. However, notably, the ethics of companionship for which I argue, is not a normative ethics which stipulates universal moral values. What I propose is a vitalist materialist, critically affirmative ethics, unfolding through pluriversal commitments to response-able practices of open-ended symphysizing attuning to, learning from and co-becoming with the more than human world.

References

Acampora, R.R. 2006. *Corporal Compassion: Animal Ethics and Philosophy of Body*. Pittsburgh, PA: Pittsburgh University Press.

Allen, A.E., C.L. Dupont, M. Obornik, A. Horak, A. Nunes-Nesi, J.P. McCrow, H. Zheng, D.A. Johnson, H. Hu, A.R. Fernie, C. Bowler. 2011. "Evolution and metabolic significance of the urea cycle in photosynthetic diatoms." *Nature* 473 (7346): 203–207.

Allen-Collinson, J. 2010. "Running Embodiment, Power and Vulnerability: Notes towards a Feminist Phenomenology of Female Running". In: E. Kennedy and P. Markula (eds.) *Women and Exercise: The Body, Health and Consumerism*: 280–98, London: Routledge.

Allen-Collinson, J. 2011. "Intention and Epoché in Tension: Autophenomenography, Bracketing and a Novel Approach to Researching Sporting Embodiment". *Qualitative Research in Sport, Exercise and Health* 3 (1): 48–62.

Bennett, J. 2010. *Vibrant matter: A political ecology of things*. Durham, NC, and London: Duke University Press.

Bonde, N. 2008. "Osteoglossomorphs of the marine Lower Eocene of Denmark – with remarks on other Eocene taxa and their importance for palebiogeography." Cavin, L., Longbottom, A., and Richter, M. (eds), *Fishes and the Break-up of Pangaea*. Geological Society, London, Special Publications 295: 253–310.

Braidotti, R. 2006. *Transpositions: On nomadic ethics*. Cambridge: Polity.

Carstensen, J., P. Henriksen, A.S. Heiskanen. 2007. "Summer algal blooms in shallow estuaries: Definition, mechanism and link to eutrophication." *Limnology and Oceanography* 52(1): 370–384.

Deleuze, G. 2020. *Logic of Sense*. trans. C.V. Boundas, M. Lester and C.J. Stivale, London: Bloomsbury.

Deleuze, G., Guattari, F. 1988. *A thousand plateaus: Capitalism and schizophrenia*. Transl. B. Massumi. New York and London: Continuum.

Derrida, J. 2008. *The animal that therefore I am (More to follow)*. Transl. D. Wills. New York: Fordham University Press.

Haraway, D. 2003. *The Companion Species Manifesto. Dogs, People, and Significant Otherness*. Chicago Il: Prickly Paradigm Press.

Haraway, D. 2008. *When species meet*. Minneapolis: University of Minnesota Press.

Haraway, D. 2016. *Staying with the Trouble: Making Kin in the Chthulucene*. Durham, NC and London: Duke University Press.

Hazekamp, R. and Lykke, N. 2022. "Ancestral Conviviality. How I fell in love with queer critters". *Forum+* Vol 29, Issue 3, pp. 30–36.

Hickman, C.P., L.S. Roberts, S.L. Keen, A. Larson, D.J. Eisenhour. 2012. *Animal Diversity*. 6th ed. New York: McGraw-Hill.

Krarup Pedersen, G. 1981. "Anoxic events during sedimentation of a Palaeogene diatomite in Denmark." *Sedimentology* 28 (4): 487–504.

Lykke, N. 2018. "When death cuts apart: On affective difference, compassionate companionship and lesbian widowhood." In: Juvonen, T., Kohlemainen, M. (eds): *Affective Inequalities in Intimate Relationships*. Routledge, New York, London: 109–125.

Lykke, N. 2019. "Co-Becoming with Diatoms. Between Posthuman Mourning and Wonder in Algae Research." *Catalyst. Feminism, Theory, Technoscience* 5 (2).

Lykke, N. 2022. *Vibrant Death. A Posthuman Phenomenology of Mourning*. London: Bloomsbury.

Lykke, N. (Forthcoming). "Contemplating Suffocation Through the Lens of Shared Cross-species Vulnerabilities". In: M. Gorska and M. Trakilovic (eds): *Social and Political Suffocations*, London, New York: Routledge.

Lykke, N. and Marambio, C. (Forthcoming). *Sandcastles – A Queerfemme Proposition on Cancer Ecologies*. Manuscript in preparation.

Mann, D.G. 1999. "The Species Concept in Diatoms". *Phycologia* 38 (6): 437–495.

Neimanis, A. 2017. *Bodies of Water: Posthuman Feminist Phenomenology*. London and New York, NY: Bloomsbury Academic.

Poulickova, A. and Mann, D.G. 2019. "Diatom Sexual Reproduction and Life Cycles". In: J. Seckbach and R. Gordon (Eds.) *Diatoms: Fundamentals and Applications*. Hoboken NJ: Wiley, and Beverly: Scrivener: 245–273.

Sartre, J.P. 1958. *Being and Nothingness: An Essay on Phenomenological Ontology*. trans. Hazel E. Barnes, London and New York, NY: Routledge.

Scarsini, Matteo, Marchand, Justine, Manoylov, Kalina M., and Schoefs, Benoit. 2019. "Photosynthesis in Diatoms". In: J. Seckbach and R. Gordon. Eds. *Diatoms: Fundamentals and Applications*. Hoboken NJ: Wiley, and Beverly: Scrivener: 191–211.

Schack Pedersen, S.A. 2008. "Paleogene diatomite deposits in Denmark: Geological investigations and applied aspects." *Geological Survey of Denmark and Greenland Bulletin* 15: 21–24.

Schrader, A. 2012. "The Time of Slime: Anthropocentrism in harmful algal research." *Environmental Philosophy* 9 (1): 71–94.

Schrader, A. 2015. "Abyssal intimacies and temporalities of care: How (not) to care about deformed leaf bugs in the aftermath of Chernobyl." *Social Studies of Science* 45 (5): 665–690.

Seckbach, J. and Gordon, R. (Eds). 2019. *Diatoms: Fundamentals and Applications.* Hoboken NJ: Wiley, and Beverly: Scrivener.

Sharma, V.P. 1969. "Early Tertiary Field Reversals Recorded in Volcanic Ash Layers of Northern Denmark." *Bulletin of the Geological Society of Denmark* 19: 218–223.

Shildrick, M. 2002. *Embodying the Monster: Encounters with the Vulnerable Self.* London: Sage.

Shildrick, M. 2005. "Transgressing the Law with Foucault and Derrida: Some Reflections on Anomalous Embodiment". *Critical Quarterly* 47 (3): 30–45.

Shildrick, M. 2009. *Dangerous Discourses of Disability, Subjectivity and Sexuality.* London: Palgrave Macmillan.

Skriver, J. 2006. *Limfjorden.* Nordjyllands Amt, Viborg Amt, Ringkjøbing Amt, Århus Amt.

Spaulding, Sarah A., Potapova, Marina G., Bishop, Ian W., Lee, Sylvia S., Gasperak, Tim S., and Jovanoska, Elena. 2022. *What are Diatoms?* Diatoms.org: supporting taxonomists, connecting communities. *Diatom Research* 36 (4), 2021: 291–304. doi:10.1080/0269249X.2021.2006790, and https://diatoms.org/what-are-diatoms, (accessed June 5, 2022).

Spinoza, B. De [1677] 1996. *Ethics.* London: Penguin.

Tiffany, M.A., and Nagy, S. 2019. "The Beauty of Diatoms". In: J. Seckbach and R. Gordon. Eds. *Diatoms: Fundamentals and Applications.* Hoboken NJ: Wiley, and Beverly: Scrivener: 33–43.

UNESCO. 2010. *Moler Landscapes of the Liim Fiord.* https://whc.unesco.org/en/tentativelists/5474/ (accessed Oct 3, 2021).

Weiss, G. 2009. "Intertwined Identities: Challenges to Bodily Autonomy". In: van de Vall, R., Zwijnenberg, R. (eds): *The Body Within: Art, Medicine and Visualization*: 173–186, Leiden: Brill.

Wikipedia. *Diatom* https://en.wikipedia.org/wiki/Diatom (accessed Sept 20, 2021).

Wikipedia. *Fur Formation* (https://en.wikipedia.org/wiki/Fur_Formation (accessed Oct 3, 2021).

CHAPTER 4

Communicating Algae Polycultures: Photobioreactors, the Phycosphere and Its Living Waters

Yogi Hale Hendlin, Johanna Weggelaar, Natalia Derossi and Sergio Mugnai

Abstract

Algae in the wild form consortia with other species which promote their own health and proliferate food sources. The recent increase in laboratory algae cultivation in commercial photobioreactors (PBR) so far has focused mainly on propagating single species of algae, rather than multi-species polycultures. Considering the current status of sterile PBR ecological habitat, the chapter investigates rearranging PBR set-up to take into account algal communication within and across species. These mutualistic species form the 'phycosphere': the microenvironment surrounding microalgal cells, potentiating the production of certain metabolites through interaction with cohabitating microorganisms. Better understanding the phycosphere prompts PBRs to become attentive to and incorporate algal-microbial consortia, to better arrange conducive habitats for algal flourishing as interspecies symbionts. From a multisolving approach, this may decrease the inputs needed for artificially maintaining growth-systems, moving from *status quo* sterility to multispecies co-maintenance. PBR polycultures also invite us to reconsider the role water plays within aquaculture, teaching us to appreciate water health and diversity, as has been done with soil.

Keywords

algae – bioreactors – consortia – communication – phycosphere – polyculture

1 Introduction

Microalgae form keystone species of aquatic ecosystems that power far-reaching food webs and biogeochemical cycling. Algae are highly adapted to these environments, flexibly maintaining homeostasis despite a constant

© Y. H. HENDLIN, J. WEGGELAAR, N. DEROSSI & S. MUGNAI, 2024 | DOI:10.1163/9789004683310_006
This is an open access chapter distributed under the terms of the CC BY-NC 4.0 license.

influx of foreign material and changing solution composition. Recent research has brought increasing attention to how algae interact with their external environment, as wild water environments are extremely variable in terms of physicochemical parameters as well as biological populations, given many species of (micro)organisms populate this habitat (Rossi et al., 2020; Nowruzi et al., 2022).

In terms of their industrial applications, microalgae have been and still are considered an important character of a biotechnology-fuelled bioeconomy, advertised as an irreplaceable asset to produce crucial chemicals such as biofuels and pharmaceuticals, or to clean polluted wastewater and sewage through their filtering ability. In order to produce biomass for the industry or to filter wastewaters, algae are monitored within cultivation systems designed to ensure their maximal growth. Various cultivation systems exist, ranging from low tech open ponds to highly sophisticated technologies. The current most utilized setup is to allow the growth of a specific and single microalgal strain as a monoculture in a protected environment called a photobioreactor (PBR). A PBR is a device made of transparent material – mostly plastic or glass – designed to expose algae to natural or artificial light so that they can photosynthesize and therefore proliferate. The PBR as a closed system allows constant monitoring of some physical and chemical parameters essential for commercial production, such as light, pH, nitrogen concentration and stirring rates (Yen et al., 2019 ; Sirohi et al., 2022). The artificially grown algal biomass is then regularly harvested in batches to avoid overpopulation and keep the algae in a continuous artificial state of growth.

However, recent studies highlight the importance (and complications) of mimicking a natural ecosystem in such a setup, suggesting the need to arrange and optimise the presence of multiple species together in biological consortia (Natrah et al., 2014; Yao et al., 2019; Mu et al., 2020; Wicker et al., 2023). Reasoning for this includes enhanced metabolism resulting from interactions between different species, growth patterns, and the reproductivity of microalgae. That is to say, algae, like most species inhabiting aquatic and terrestrial environments, humans included, are naturally better off within a community, itself ideally diverse. It is with and from a diverse biological community that microalgae best faces adversity.

To get more specific, many bacteria influence the development of algal blooms in nature (Tang et al., 2010), and the removal of biodegradable organic matter is promoted by microalgae-bacteria consortia in wastewater treatment ponds (Subashchandrabose et al., 2011). Since interactions between microalgae and bacteria exist in natural habitats and even are constitutive of some species, disrupting these multi-species signaling processes may end up undermining certain desired capabilities or characteristics of the microalgae, resulting in complications in healthy microalgae growth when attempted in confined environments such as bioreactors (Santos and Reis, 2014). But the investigation in this chapter is not merely to generate more efficient, less-input photobioreactors. Similar to Temple Grandin designing more humane slaughterhouses to give cows better lives, with the industrial benefit of efficiency (Grandin and Johnson, 2005), organizing multi-species photobioreactors too may present algae with better lives as well, and instill hopes of alternative ways of producing that don't have to compromise the needs of algae that are normally met in the wild. Investigating the role of interspecies communication in different microalgal species, and between microalgae and bacteria species, is one way to understand the crucial limits of closed single strain systems. This chapter addresses these roles, suggesting that microalgae do and should exist necessarily in multispecies consortia, that when broken, introduce novel problems for maintaining healthy microalgal growth requiring continual additional inputs to the detriment of algae themselves.

2 Communication and Interaction between Microalgae and Bacteria

Getting into the specifics of communication and interaction between microalgae and bacteria allows us to bring specific examples to support the general argument brought forward, namely that eliminating the possibility of interactions through the artificial constitution of axenic pure strains that characterize most photobioreactor growth systems lowers the resilience of microalgae and alters their ability to communicate.

Now, intraspecific communication between microalgae has been mostly investigated in relation to sexuality and mating (see for instance Frenkel et al., 2014). However, little is known about the way microalgae belonging to the same species (conspecifics) communicate in response to stress cues within their community, nor about microalgal capacities for establishing communities and colonies via effective chemical and electrical communication to survive hostile situations or the reduction of ecosystem diversity, known as dysbiosis. Indeed, the cultivation of microalgae in a confined environment such as a photobioreactor poses some questions related to the ecology of microalgae in terms of how individuals connect with other organisms. So far, most of the research conducted in this field documents issues arising from artificial microalgal extractive systems, but exploring the behavior of microalgae in their native environment is more challenging since little research actually examines communicative elements of microalgae in nature and the minimal biotic interactions necessary (and how these regulate microalgal colonies) for interspecifics (that is, algae not belonging to the same species) to thrive. Particularly, significant areas of inquiry deal with how colonial health and communication supervene upon individual microalgal cell growth and therefore adapt their growth rate and metabolism to the surrounding environment.

It is widely established that chemical communication among organisms, both intraspecific and interspecific, requires complex biological structures, including genes, RNA, proteins and other chemical messengers (Venuleo et al., 2017). Furthermore, algae, specifically the genus *Chara*, have been the first organisms not belonging to the animal kingdom where production of electrical action potentials in response to environmental modifications have been detected (Hope, 1961; Johnson et al., 2002). These non-animal electrical action potentials made possible the rise of the concept of "plant neurobiology," given the transmission of signals through action potentials similar to those of the animal brain (Baluška and Mancuso, 2009). From an evolutionary standpoint, algae possess ancient molecular pathways, which are widely used by land plants, showing that algae can illuminatingly be used as a model for investigating complex mechanisms found in the plant kingdom thanks to their simpler anatomy, structure and metabolic processes. Biochemical communication, then, seems a primitive ability developed by microalgae for establishing communities and colonies.

Microalgae are particularly effective in creating marine communities: they constantly interact with their surrounding environment, in response to physical stimuli such as temperature, chemical gradients, gravity, light and flow shear, which influence the motion of individual swimming microorganisms. Since these microorganisms rarely exist as individuals, environmental influences are reflected at the level of a population, causing cells to aggregate and interact to respond in a more effective way to environmental disturbances – forming what we call algae colonies, which have the common attribute of being two or more cells associated regularly such that this association is recognized as the morphological type of a genus and species (Starr, 1984). These patterned groupings can be therefore considered as a form of social behavior (Prakash and Croze, 2021). For instance, recent research highlights the capacity of microalgae to exhibit collective behavior such as swarming (Berge et al., 2012; Kage and Mogami, 2015), where complex networks are established among individuals, leading to a coordinated movement flow similar to the swarming of insects or birds' murmuration. However, the most studied and known phenomenon regarding collective behavior in the algal world is the development of the so-called 'algal bloom', some of them being harmful for human population and ecosystems in general (HAB, Harmful Algal Blooms, Hallegraeff, 2003).

As mentioned, microalgae almost never live alone, and many species of microalgae often co-colonize a specific ecosystem. Investigating how different species of microalgae communicate among each other is the basis for analyzing the relationships within an aquatic ecosystem. Such knowledge is necessary to ameliorate and improve the growth and health of microalgal communities inside confined environments such as PBRs. Due to the fact that the competition and resource scarcity paradigm still guides much of biological research, the most famous example of communication between different species of microalgae is known as allelopathy, which simply stands for inhibition. Allelopathy is commonly defined as the process involving chemical compounds released into the surrounding medium that have adverse effects, either directly or indirectly, on the growth of other microorganisms (Tan et al., 2019). The outcome of this type of communication can include death, paralysis (for motile cells) or inhibition of growth of the receiver (Legrand et al., 2003). It has thus been observed that algal cells can exert various forms of control over other competitive species by various mechanisms, such as photosynthesis impairment (Qian et al., 2018; Nikkanen et al., 2021) and damage of the cellular membrane, thus negatively affecting cell vitality and functioning (Chaïb et al., 2021).

However less studied, mutualistic interactions between microalgae and other microorganisms such as bacteria, conversely, yield equally interesting results. Scientific interest in cooperation between microalgae and bacteria has grown considerably in the last decade, mainly due to the practical applications of understanding microalgal-bacterial consortia beneficial for industrial activities such as wastewater treatment (Qi et al., 2021), but also for genuine scientific interest related to ecological studies (Cirri and Pohnert, 2019). For instance, Amin et al.'s (2009) ecological studies identify specific groups of heterotrophic bacteria capable of establishing close associations with some microalgae, influencing microalgal behaviours in various ways, from the stimulation of growth and morphogenesis to the germination of spores and colonization by forming biofilm communities (Lachnit et al., 2011). In addition to comingling with biofilm bacteria, microalgae can secrete exudates to influence heterotrophic bacteria (sometimes also called planktonic organisms, which feed on chemical nutrients rather than sun) in their surroundings. For this reason, the term '*phycosphere*' has been coined to describe the region of influence of microalgal exudates upon co-occurring organisms inhabiting the same space (Natrah et al., 2013; Kouzuma and Watanabe, 2015).

Other forms of interaction between bacteria and microalgae also exist. One of them is signal transduction, where the involved chemicals are more than mere nutrients, and rather act at a deeper level, being capable as such to activate or inhibit gene expression and physiological activities. Signal transduction between algae and bacteria goes beyond the typical boundaries set by systematic biology, often classified as 'interkingdom signaling'. In microalgae/bacteria mutualism, bacteria can secrete chemical signals that manage growth induction and morphogenesis of algae (Matsuo et al., 2005), while microalgae can repress the formation of excess biofilms (biofouling) on their surfaces.

Another form of microalgae and bacteria mutualism is horizontal gene transfer (HGT). Horizontal gene transfer is an evolutionary process in which genes are horizontally transferred between adjacent organisms, for example, microalgae and bacteria living in the phycosphere (Brembu et al., 2014; Moszczynski et al., 2012), rather than by the ("vertical") transmission of DNA from parent to offspring (reproduction). Such transfers blur genetic distinctions, allowing DNA material to move across species and contaminate the genetic material of another species (Burmeister, 2015). The ecological advantage of such a feature is that horizontally transferred genes can confer new key functions on microalgae to better survive under stress. For instance, microalgal-associated microbial communities are often enriched in gene functions involved in vitamin synthesis, the detection and attachment to host surfaces, biofilm formation, polysaccharide catabolism and various defense mechanisms (Song et al., 2021; Nagarajan et al., 2022).

On the other hand, microalgae can also be detrimental to other organisms by releasing compounds harmful to their grazers (Ianora and Miralto, 2010; Ratti et al., 2013) or to non-algal competitors for the same resources, such as bacteria and fungi (Burkholder et al., 1960; Kellam et al., 1988; Hagmann and Jüttner, 1996; Issa, 1999). The possibility that microalgae are involved in chemical interactions with multiple actors (i.e. 'multitrophic' interactions), for instance by producing compounds beneficial for individuals whose presence is disadvantageous for grazers as observed for other organisms (Vet and Dicke, 1992; De Vos et al., 2006; Stout et al., 2006; Nevitt, 2008), cannot be excluded.

Algal activity can also result in downregulating bacterial quorum sensing (QS, Zhou et al., 2016). QS here involves the coordination of the bacterial gene expression in a population density-dependent manner via the production and exchange of specific signal substances between individual cells. Algae are capable of interfering with this communication mechanism, thus repressing the growth and development of bacterial communities by mimicking bacterial QS signals (Teplitski et al., 2004).

To sum up, individual algae communicate with other conspecifics, with other algal species, and with other organisms in beneficial or detrimental ways which construct their individual, species, and group niches, increasing fitness and creating conducive ecologies.

3 The Role of the Phycosphere

As mentioned in the previous section, most of the interactions between microalgae and bacteria only affect the algal cell's external layer. To compare with terrestrial soil systems, it is well known how the contact zone surrounding a root is crucial for determining the health and the metabolism of the entire plant, and for the continued functioning of the soil ecosystem. This narrow zone surrounding plant roots, and massively influenced by them, is named the *rhizosphere* (literally 'the sphere of the ῥίζα', greek for 'root'). Considered one of the most complex ecosystems on Earth, the rhizosphere hosts microbial communities such as fungi and bacteria that play a fundamental role in the plant growth and metabolism (Mendes et al., 2013). In this perspective, soil is not seen as a mere container for the root system, simply providing support, water and nutrients, but a complex ecosystem in which an intricate network of living organisms enables emerging interspecies relationships.

Despite the highly divergent media characteristics between soil and water, the concept of the rhizosphere does translate from plant ecology to any aquatic environment where microalgae are present. Recently, a new term – *phycosphere*, 'the sphere of the φῦκος', greek for 'seaweed' this time – has been coined to define the microenvironment immediately surrounding microalgal cells. This area determines which metabolites are readily available and the parameters for interactions between microalgae and other microorganisms (mainly bacteria) (Seymour et al., 2017). This interaction allows chemical exchanges between bacteria and microalgae, with the latter continuously releasing organic compounds called exudates which feed the bacteria, while beneficial bacteria produce metabolites essential to algal growth and metabolism. Understanding this interface is essential to figure how global biogeochemical fluxes and oceanic nutrient fluxes work (Seymour et al., 2017; Fei et al., 2020). As we have already anticipated when discussing communication, the relationship between microalgae and bacteria is mediated and regulated by quorum sensing (QR), a process by which bacteria coordinate their gene expression and metabolism at the population level, and by quorum quenching (QQ), which suppresses the activity of antagonistic bacteria (Rolland et al., 2020). Therefore, in the phycosphere a delicate equilibrium between promoting and suppressing actions is established, regulating the ecosystem community in terms of mutualistic and parasitic behavior.

The moment we consider photobioreactors (PBRs) as an aquatic ecosystem, even though at a small scale, the interactions within the phycosphere must be taken into consideration, especially when a polycultural approach is adopted. Associated with microalgal cultivation and PBRs, by "polyculture" we wish to convey a community of diverse algal and bacterial species living in ecological homeostasis. Why adopt the term polyculture for microalgae? If, as mentioned in the introduction, microalgal cultivation has mainly focused on exploiting monocultures of highly productive microalgal strains (Andersen, 2005; Posten, 2009; Mostafa, 2012), such monocultures are difficult to maintain due to several constraints such as accidental contamination by wild microalgal strains, bacteria and pathogens (Singh and Patidar, 2018; Tan et al., 2019; Yin et al., 2020; Shaikh, et al., 2021). Recent research suggests that polycultures could promote ecosystem robustness (for example, see Newby et al., 2016). For this reason, a greater understanding of species interactions along with patterns of communities' change with time in a confined environment is needed to enhance the ecological health of PBR polycultures.

4 Algal Polycultures

A comparison can be made between PBRs and another important ecosystem, the agrarian one. Briefly overviewing monoculture versus polyculture systems in agriculture can provide possible implications of such different systems in algal cultivation. For terrestrial agriculture, monoculture systems involve growing a single (usually staple) crop in a field during the growing season. Monoculture production systems simplify crop management and allow for concentration of efforts and resources on maximizing economic return from a single crop. On the other hand, these systems do not reproduce the complex natural ecosystems because they tend to reduce soil biodiversity as well as, among many other social adverse effects, pauperize the soil, due to increased soil erosion, greater nutrient leaching and lower water-holding capacity.

On the contrary, polyculture is the growing of two or more crops together on the same piece of land during a growing season. Polyculture allows for spatial diversification of plant species, providing greater opportunity to beneficially use soil and environmental resources compared to monoculture. Polyculture systems in some cases can be challenging to manage because species growing together might thrive by feeding on similar species or have diverse resource needs. But mostly, they are more time intensive to harvest, resisting easily mechanized collection. However, polycultures have the potential to provide several advantages compared to monocultures, the most important ones being a greater tolerance to environmental and pest stress, providing insurance against total crop failure, and requiring and providing differential nutrients and rooting characteristics which have the potential for greater use of available light, water, and nutrient resources. Furthermore, polycultures promote and maintain biodiversity, as well as ensuring a mutualistic advantageous environment for all the living organisms. Finally, at soil level, polycultures take advantage of the complex relationships between different plant species and microorganisms, with distinct and diversified arbuscular mycorrhizal (AM) fungal community composition (Cloutier et al., 2020), symbionts that aid in plant nutrient acquisition, drought tolerance, pathogen protection, water uptake, and numerous other functions that affect plant health (Guzman et al., 2021). Reciprocally, where monoculture depletes the soil from its precious nutrients, polyculture allows to maintain or even enhance the soil quality. While monocultural practices manage to reach record yields in the short-term, often with the help of additional fertilizers and biostimulants, they prove to be deficient in the long-term when all available resources of the soil have been exploited. With its win-win principles between various organisms co-existing in the medium, polyculture ensures stable yields over the years and can also be regenerative by healing the soil and its inhabitants.

The question therefore is: can a polyculture approach be applied to PBRs, and how could it benefit microalgal cultivation? It is now well established that polyculture cultivation increases productivity of microalgae via both resource use efficiency and community stability (Newby et al., 2016; Thomas et al., 2019; Liu, 2021), since multiple microalgal species occupying different functional niches make use of the resources in a complementary way due to their different light absorption spectra, nutrient requirements, uptake rates and physiological complementarity (Novoveskà et al., 2016). When algal communities contain multiple species they become more resilient under varying conditions compared to monocultures' disturbance-reduced productivity (Newby et al., 2016). Along with these benefits, however, are the requirements for extensive knowledge of the whole community's functioning, planning and onset of optimal community co-habitant composition, and the proviso that polycultures may only be suitable for a set of well-studied conditions. But the polyculture concept can be also seen from a different perspective by associating microalgae and bacteria instead of different strains or species of microalgae alone. From this lens, the concept of phycosphere previously introduced can reveal a positive interaction between these two groups of organisms that can then enhance PBRs by mimicking – however reductively – a wild aquatic ecosystem.

This area of research is at the initial stages of long-term scientific investigation, but we can already find promising indications. For instance, Feng et al. (2021) found that the removal of pollutants from anaerobic digestion effluents is dramatically enhanced when algal-bacteria consortia rather than monocultures are introduced. Chaiwong et al. (2018) investigated the treatment performance of an algal-bacterial PBR (AB-PBR) treating a septic tank effluent compared to a solely algal one. Here, the efficiency and effectiveness of the AB-PBR was higher than the traditional one with a single algal strain. Several studies have proven the possibility to improve H_2 production when using co-cultures of algae and bacteria, with some of them focusing on the use of the alga *Chlamydomonas* (reviewed by Fakhimi et al., 2020). Bélanger-Lépine et al. (2020) found that a native microalgae-bacteria consortium isolated locally grew better in wastewater when compared to pure algal strains, and that different fatty acid profiles were produced. The same results obtained for Fito and Alemu's (2019) evaluation of the potential of an algal-bacteria consortium to manage municipal wastewater. Wastewater treatment and removal of pollutants have been the major fields of research until now regarding the potential of algae-bacteria consortia. In the past years, investigation of microalgal-bacterial consortia for biotechnological applications has been promoted. Even though the physiological mechanisms behind the interaction between microalgae and bacteria in practical applications is not yet well understood, promising results suggest increased capacities for water cleaning. Zhang et al. (2020) attentively review these aspects, such as an improved capacity for carbon and nutrient removal by enhanced flocculation, heavy metal removal via biosorption and adsorption, degradation of organic hazardous compounds, and even production of biofuels.

5 Reconsidering Water

The comparison between agrarian and aquatic farming practices suggests that the benefits of polyculture have not properly reached the biotechnological sector of algae production in particular, nor of aquaculture in general. In fact, the way we approach water and land as cultivation mediums differs. If soil health is primordial for land farming, as shown for instance by the "Three Sisters" system, this is not the case for water quality. The "Three Sisters" is a

Native American indigenous practice consisting of intercropping three different plants (beans, squash, and corn) that will symbiotically benefit each other. Such an arrangement is not only oriented towards the increase of crop production efficiency, it more importantly aims at preserving soil health and ensuring an ecologically-tuned management of nutrients (Kapayou et al., 2023). In other words, the plants like it. This form of traditional ecological knowledge is now part of a broader movement that aims to support the transition of conventional agriculture towards regenerative agroecology.

We shall begin to wonder if a similar consideration of water health can be found in aquaculture. Amongst others, research done in Australia around indigenous principles for water quality shows that local communities hold water in deep esteem, having valuable knowledge of water health. For instance, such communities use the term "living water" to describe permanent water in a dry land showing distinct physical properties and some specific cultural significance (Australian Government Initiative, 2023). A water environment is thus far from being characterized by its chemical formula; it is rather an ecosystem in itself. Contrasting with the traditional wisdom around soil, which is revived in contemporary approaches to agriculture, traditional knowledge around water has long been overlooked – if 'living soil' is by now common sense, the same cannot *yet* be said about 'living water'.

Similarly to soil throughout the modernization of agriculture, water has been and still is approached as a mere medium for transporting certain quantities of nutrients for assimilation by the crop of interest. Water is thus taken as a substance *through which* certain qualities are conveyed, and not as holding qualities in itself. On top of this, the general discourse over polluted or clean water, often simplifying the transition from the first to the second by advertising various water-cleansing strategies, encourages a binary conception of water, reducing a whole range of possible qualities to two: clean or polluted. Biological, physical and chemical properties of water, its various degrees of fertility and adaptation to wide spectra of living organisms are obliterated by two mainstream categories.

PBR technologies are based on the assumption that a closed system can be perfectly monitored. In this sense, PBRs nourish the dream that with the right amount of light and nutrients and the right agitation rate, an optimal state could be reached allowing maximum algae growth without any external disturbance. For that purpose, distilled or deionized water is often used as a safe medium and all nutrients, including CO_2, are added afterwards to the water. In her 'History of Surrounds', sociologist and historian of science Hannah Landecker (2016) recounts the systematic replacement of the mysteries of biological life enabling reproduction of biological entities *in vitro* (take for instance biological fluids, or sea water), by known synthesized chemicals in order to increase control of the culture. The limits of monoculture brought up in this chapter begin to hint at the failure of biotechnological will to control, by highlighting its inherent contradictions, namely the increased maintenance and the high quantity of external inputs needed to keep microalgal cultivations alive in such controlled settings.

In a way, these limitations speak of the inadequacy of considering water as a *tabula rasa* to which valuable nutrients and chemicals can be added to feed the culture in question. A more ecological approach to algal cultivation based on polyculture could invite new considerations over the role of water quality and its ecosystem diversity in intra- and inter-species algal communication. Beginning to consider water too as a complex system in which networks of living entities allow currently unappreciated sets of relationships could help us slowly move away from the sterility paradigm of current microalgae culture.

6 Conclusions

Photoautotroph-bacteria co-culture investigations so far indicate higher success rates in societal applications than monocultures, opening a window of hope into believing that an alternative to extractivism and to efficiency logics of growth does exist. Algal management in PBRs thus far in commercial units have sought laboratory axenic conditions with single species, rather than polycultures with multiple strands of algal organisms, let alone algal-bacterial consortia. Yet, even the extant research on these interspecies consortia for commercial applications likely underestimates the true necessary symbionts which allow algae to perform their water-cleaning ecosystem services during metabolism. Other organisms besides bacteria may also be indispensable for optimal algal health and productivity – part of the phycosphere that deserves additional analysis. Reproducing *in situ* environmental conditions in closed laboratory systems is always difficult with organisms but additionally challenging with species like algae that are constitutively multi-species. Nonetheless, recent experiments show the effective mobilization of interspecies algal-bacteria PBR to better tackle real-world applications. Through better understanding the phycosphere, more sophisticated forms of PBRs incorporating algal-microbial consortia may require fewer inputs, and teach us how to retrieve our ability to appreciate water health and diversity.

References

Amin, S.A., D.H. Green, M.C. Hart, F.C. Küpper, W.G. Sunda, C.J. Carrano. 2009. Photolysis of iron-siderophore chelates promotes bacterial-algal mutualism. Proc Natl Acad Sci USA 106(40): 17071–6.

Andersen, R.A. 2005. Algal Culturing Techniques, 1st Edition. Elsevier.

Australian Government Initiative, Indigenous principles for water quality, https://www.waterquality.gov.au/anz-guidelines/guideline-values/derive/cultural-values/principles. Retrieved on 14th April 2023.

Baluška, F., S. Mancuso. 2009. Plant neurobiology: from sensory biology, via plant communication, to social plant behavior. Cogn Process 10 (Suppl 1), 3–7.

Bélanger-Lépine F., M. Lemire-Lamothe, A. Tremblay, S. Rondeau, P. Marchand, Y. Huot, S. Barnabé. 2020. Cultivation of an algae-bacteria consortium in a mixture of industrial wastewater to obtain valuable products for local use. Industr Biotech 16: 33–42.

Berge, T., L. Poulsen, M. Moldrup, N. Daugbjerg, P.J. Hansen. 2012. Marine microalgae attack and feed on metazoans. ISME J 6: 1926–1936.

Brembu, T., P. Winge, A. Tooming-Klunderud, A.J. Nederbragt, K.S. Jakobsen, A.M. Bones 2014. The chloroplast genome of the diatom *Seminavis robusta*: new features introduced through multiple mechanisms of horizontal gene transfer. Mar Genomics 16: 17–27.

Burkholder, P.R., L.M. Burkholder, L.R. Almodovar. 1960. Antibiotic activity of some marine algae of Puerto Rico. Botanica Marina 2: 149–156.

Burmeister, A.R. 2015. Horizontal gene transfer. Evol Med Public Health 2015: 193–4.

Chaïb, S., J.C.A. Pistevos, C. Bertrand, I. Bonnard. 2021. Allelopathy and allelochemicals from microalgae: an innovative source for bio-herbicidal compounds and biocontrol research. Algal Res 54: 102213.

Chaiwong, C., T. Koottatep, N. Surinkul, C. Polprasert. 2018. Performance and kinetics of algal-bacterial photobioreactor (AB-PBR) treating septic tank effluent. Water Sci Technol 78: 2355–2363.

Cirri, E., G. Pohnert. 2019. Algae–bacteria interactions that balance the planktonic microbiome. New Phytol, 223: 100–106.

Cloutier, M.L., E. Murrell, M. Barbercheck, J. Kaye, D. Finney, I. García-González, M.A. Bruns. 2020. Fungal community shifts in soils with varied cover crop treatments and edaphic properties. Sci Rep 10, 6198.

De Vos, M., W. Van Zaanen, A. Koornneef, J.P. Korzelius, M. Dicke, L.C. Van Loon, C.M. Pieterse. 2006. Herbivore-induced resistance against microbial pathogens in Arabidopsis. Plant Physiol 142: 352–363.

Fakhimi, N., D. Gonzalez-Ballester, E. Fernández, A. Galván, A. Dubini. 2020. Algae-bacteria consortia as a strategy to enhance H_2 production. Cells 9(6), 1353.

Fei, C., M.A. Ochsenkühn, A.A. Shibl, A. Isaac, C. Wang, S.A. Amin. 2020. Quorum sensing regulates 'swim-or-stick' lifestyle in the phycosphere. Environ Microbiol 22: 4761–4778.

Feng L., J. Song, H. Gu, X. Zhen. 2021. Mechanism of contaminant removal by algae-bacteria symbiosis in a PBR system during the treatment of anaerobic digestion effluents. Agrl Water Manag 247: 106556.

Fito, J., K. Alemu. 2019. Microalgae – bacteria consortium treatment technology for municipal wastewater management. Nanotechnol Environ Eng 4: 4.

Frenkel, J., W. Vyverman, G. Pohnert. 2014. Pheromone signaling during sexual reproduction in algae. Plant J, 79: 632–644.

Grandin, T., C. Johnson. 2005. Animals in translation: Using the mysteries of autism to decode animal behavior. 1st Edition, Harcourt.

Guzman A., M. Montes, L. Hutchins, G. DeLaCerda, P. Yang, A. Kakouridis, R.M. Dahlquist-Willard, M.K. Firestone, T. Bowles, C. Kremen. 2021. Crop diversity enriches arbuscular mycorrhizal fungal communities in an intensive agricultural landscape. New Phytol, 231: 447–459.

Hagmann L., F. Jüttner. 1996. Fischerellin A, a novel photosystem-II-inhibiting allelochemical of the cyanobacterium *Fischerella muscicola* with antifungal and herbicidal activity. Tetrahedron Letters 37: 6529–6542.

Hallegraeff, G. 2003. Harmful algal blooms: a global overview. In: Manual on Harmful Marine Microalgae, 25–50. Edited by G.M. Hallegraeff, D.M. Anderson and A.D. Cembella. UNESCO publishing.

Hope, A.B. 1961. The action potentials of cells of *Chara*. Nature, 191: 811–812.

Ianora, A., A. Miralto. 2010. Toxigenic effects of diatoms on grazers, phytoplankton and other microbes: a review. Ecotoxicology, 19(3): 493–511.

Issa, A.A. 1999. Antibiotic production by the cyanobacteria *Oscillatoria angustissima* and *Calothrix parietina*. Env Toxic Pharmac, 8: 33–37.

Johnson, B.R., R.A. Wyttenbach, R. Wayne, R.R. Hoy. 2002. Action potentials in a giant algal cell: a comparative approach to mechanisms and evolution of excitability. J Undergrad Neurosci Educ, 1(1): A23–7.

Kage, A., Y. Mogami. 2015. Individual flagellar waveform affects collective behavior of *Chlamydomonas reinhardtii*. Zoolog Sci., 32(4): 396–404.

Kapayou, D.G., Herrighty, E.M., Hill, C.G. et al. (2023). Reuniting the Three Sisters: collaborative science with Native growers to improve soil and community health. Agric Hum Values 40, 65–82.

Kellam, S.J., R.J.P. Cannell, A.M. Owsianka, J.M. Walker. 1988. Results of a large-scale screening programme to detect antifungal activity from marine and freshwater microalgae in laboratory culture. British Phycological Journal 23: 45–47.

Kouzuma, A., K. Watanabe. 2015. Exploring the potential of algae/bacteria interactions. Curr Opin Biotechnol, 33: 125–9.

Lachnit, T., D. Meske, M. Wahl, T. Harder, R. Schmitz. 2011. Epibacterial community patterns on marine macroalgae are host-specific but temporally variable. Environ Microbiol, 13: 655–665.

Landecker, H. 2016. It is what it eats: Chemically defined media and the history of surrounds. Studies in History and Philosophy of Science, 57: 148–160.

Legrand, C., K. Rengefors, G.O. Fistarol, E. Graneli. 2003. Allelopathy in phytoplankton – biochemical, ecological and evolutionary aspects. Phycologia 42: 406–419.

Liu, W., D. Fu, T. Pan, R.P. Singh. 2021. Characterization and polyculture analysis of microalgae strains based on biomass production and nutrient consumption, and bacterial community in municipal wastewater. Water 13, 3190.

Matsuo, Y., H. Imagawa, M. Nishizawa, Y. Shizuri. 2005. Isolation of an algal morphogenesis inducer from a marine bacterium. Science 307: 1598.

Mendes, R., P. Garbeva, J.M. Raaijmakers. 2013. The rhizosphere microbiome: significance of plant beneficial, plant pathogenic, and human pathogenic microorganisms. FEMS Microbiology Reviews, 37: 634–663.

Mostafa, S.S. 2012. Microalgal biotechnology: prospects and applications. Plant Science, 12: 276–314.

Moszczynski, K., P. Mackiewicz, A. Bodyl. 2012. Evidence for horizontal gene transfer from bacteroidetes bacteria to dinoflagellate minicircles. Mol Biol Evol. 29(3): 887–92.

Mu R., Y. Jia, G. Ma, L. Liu, K. Hao, F. Qi, Y. Shao. 2020. Advances in the use of microalgal – bacterial consortia for wastewater treatment: Community structures, interactions, economic resource reclamation, and study techniques. Water Environment Research, 93: 1217–1230.

Nagarajan, D., D-J. Lee, S. Varjani, S.S. Lam, S.I. Allakhverdiev, J.S. Chang. 2022. Microalgae-based wastewater treatment – Microalgae-bacteria consortia, multi-omics approaches and algal stress response. Sci Total Env, 845: 157110.

Natrah, F.M.I., P. Bossier, P. Sorgeloos, F.M. Yusoff, T. Defoirdt. 2014. Significance of microalgal – bacterial interactions for aquaculture. Reviews in aquaculture, 6: 48–61.

Nevitt, G.A. 2008. Sensory ecology on the high seas: the odor world of the procellariiform seabirds. J Exp Biol 211: 1706–1713.

Newby, D.T., T.J. Mathews, R.C. Pate, M.H. Huesemann, T.W. Lane, B.D. Wahlen, S. Mandal, R.K. Engler, K.P. Feris, J.B. Shurin. 2016. Assessing the potential of polyculture to accelerate algal biofuel production. Algal Research, 19: 264–277.

Nikkanen, L., D. Solymosi, M. Jokel, Y. Allahverdiyeva. 2021. Regulatory electron transport pathways of photosynthesis in cyanobacteria and microalgae: Recent advances and biotechnological prospects. Physiol Plant, 173: 514–525.

Novoveská, L., A.K.M. Zapata, J.B. Zabolotney, M.C. Atwood, E.R. Sundstrom. 2016. Optimizing microalgae cultivation and wastewater treatment in large-scale offshore photobioreactors. Algal Research, 18: 86–94.

Nowruzi, B., M.A. Shishir, S.J. Porzani, U.T. Ferdous. 2022. Exploring the interactions between algae and bacteria. Mini Rev Med Chem. 22(20): 2596–2607.

Posten, C. 2009. Design principles of photo-bioreactors for cultivation of microalgae. Engineering in Life Sciences, 9(3), 165–177.

Prakash, P., O. Croze. 2021. Photogyrotactic concentration of a population of swimming microalgae across a porous layer. arXiv: 2107.09346.

Qi, F., Y. Jia, R. Mu, Q. Guo, Q. Meng, G. Yu, J. Xie. 2021. Convergent community structure of algal – bacterial consortia and its effects on advanced wastewater treatment and biomass production. Sci Rep 11, 21118.

Qian, H., J. Xu, T. Lu, Q. Zhang, Q. Qu, Z. Yang, X. Pan. 2018. Responses of unicellular alga *Chlorella pyrenoidosa* to allelochemical linoleic acid. Sci Total Env., 625: 1415–1422.

Ratti, S., A.H. Knoll, M. Giordano. 2013. Grazers and phytoplankton growth in the oceans: an experimental and evolutionary perspective. PLoS One 8: e77349.

Rolland, J.L., D. Stien, S. Sanchez-Ferandin, R. Lami. 2016. Quorum sensing and quorum quenching in the phycosphere of phytoplankton: a case of chemical interactions in ecology. J Chem Ecol., 42(12): 1201–1211.

Rossi, S., F. Casagli, M. Mantovani, V. Mezzanotte, E. Ficara. 2020. Selection of photosynthesis and respiration models to assess the effect of environmental conditions on mixed microalgae consortia grown on wastewater. Bioresource Technology, 305: 122995.

Santos, C.A., A. Reis. 2014. Microalgal symbiosis in biotechnology. Appl Microbiol Biotech, 98: 5839–5846.

Seymour, J., S. Amin, J.B. Raina, R. Stocker. 2017. Zooming in on the phycosphere: the ecological interface for phytoplankton – bacteria relationships. Nat Microbiol 2, 17065.

Shaikh, S.M., M.K. Hassan, M.S. Nasser, S. Sayadi, A.I. Ayesh, V. Vasagar. 2021. Comprehensive review on harvesting of microalgae using polyacrylamide-based flocculants: potentials and challenges. Separation Purification Tech, 119508.

Singh, G., S. Patidar. 2018. Microalgae harvesting techniques: a review. J. Env. Manag., 217: 499–508.

Sirohi, R., A.K. Pandey, P. Ranganathan, S. Singh, A. Udayan, M.K. Awasthi, A.T. Hoang, C.R. Chilakamarry, S.H. Kim, S.J. Sim. 2022. Design and applications of photobioreactors – a review. Bioresource Technology, 349: 126858.

Song, W., B. Wemheuer, P.D. Steinberg, E.M. Marzinelli, T. Thomas. 2021. Contribution of horizontal gene transfer to the functionality of microbial biofilm on a macroalgae. ISME J 15, 807–817.

Starr R.C. 1984. Colony formation in algae. In: Linskens H.F., Heslop-Harrison J. (eds) Cellular Interactions. Encyclopedia of Plant Physiology (New Series), vol 17. Springer, Berlin, Heidelberg.

Stout, M.J., J.S. Thaler, B.P. Thomma. 2006. Plant-mediated interactions between pathogenic microorganisms and herbivorous arthropods. Ann Rev Entomol, 51: 663–689.

Subashchandrabose, S.R., B. Ramarkrishnan, M. Megharaj, K. Venkateswarlu, R. Naidu. 2011. Consortia of cyanobacteria/microalgae and bacteria: biotechnological potential. Biotech Adv 29, 896–907.

Tan, K., Z. Huang, R. Ji, Y. Qiu, Z. Wang, J. Liu. 2019. A review of allelopathy on microalgae. Microbiology, 165(6): 587–592.

Tang, Y.Z., F. Koch, C.J. Gobler. 2010. Most harmful algal bloom species are vitamin B_1 and B_{12} auxotrophs. PNAS 107, 20756–20761.

Teplitski, M., H. Chen, S. Rajamani, M. Gao, M. Merighi, R.T. Sayre. 2004. *Chlamydomonas reinhardtii* secretes compounds that mimic bacterial signals and interfere with quorum sensing regulation in bacteria. Plant Physiol, 134: 137–146.

Thomas, P.K., G.P. Dunn, A.R. Good, M.P. Callahan, E.R. Coats, D.T. Newby, K.P. Feris. 2019. A natural algal polyculture outperforms an assembled polyculture in wastewater-based open pond biofuel production. Algal Research 40: 101488.

Venuleo, M., J.A. Raven, M. Giordano. 2017. Intraspecific chemical communication in microalgae. New Phytol, 215: 516–530.

Vet, L.E., M. Dicke. 1992. Ecology of infochemical use by natural enemies in a tritrophic context. Ann Rev Entomol 37: 141–172.

Wicker, R.J., E. Kwon, E. Khan, V. Kumar, A. Bhatnagar. 2023. The potential of mixed-species biofilms to address remaining challenges for economically-feasible microalgal biorefineries: a review. Chem Eng J, 451: 138481.

Yao, S., S. Lyu, Y. An, J. Lu, C. Gjermansen, A. Schramm. 2019. Microalgae – bacteria symbiosis in microalgal growth and biofuel production: a review. J Appl Microbiol, 126: 359–368.

Yen, H.W., I.C. Hu, C.Y. Chen, D. Nagarajan, J.S. 2019. Design of photobioreactors for algal cultivation. In: Biofuels from algae, 2nd Edition. Edited by Pandey et al., Elsevier.

Yin, Z., L. Zhu, S. Li, T. Hu, R. Chu, F. Mo, D. Hu, C. Liu, B. Li. 2020. A comprehensive review on cultivation and harvesting of microalgae for biodiesel production: environmental pollution control and future directions. Bioresource Technology, 301: 122804.

Zhang, B., W. Li, Y. Guo, Z. Zhang, W. Shi, F. Cui, P.N.L. Lens, J.H. Tay. 2020. Microalgal-bacterial consortia: From interspecies interactions to biotechnological applications. Ren Sust Energy Rev, 118: 109563.

Zhou, J., Y. Lyu, M. Richlen, D.M. Anderson, Z. Cai. 2016. Quorum sensing is a language of chemical signals and plays an ecological role in algal-bacterial interactions. CRC Crit Rev Plant Sci., 35(2): 81–105.

Note: This paper is a revised version of our open-access article: Mugnai, S., Derossi, N., & Hendlin, Y. (2023). Algae communication, conspecific and interspecific: the concepts of phycosphere and algal-bacteria consortia in a photobioreactor (PBR). *Plant Signaling & Behavior*, 18(1), 2148371. https://doi.org/10.1080/15592324.2022.2148371.

CHAPTER 5

Algae in the Human World: Beauty and Taste Come First

Ole G. Mouritsen and J. Lucas Pérez-Lloréns

Abstract

Humans have interacted with algae for millennia. This paper describes the human journey in the company with algae from the earliest days of our species until today where the need for a green transition and sustainable eating behaviour has put renewed focus on algae as a material of many uses not least as foodstuff. The evolution of *Homo* has been shaped by our ancestors being seashore dwellers with plenty of access to marine foodstuff that contains critical nutritional elements for the evolution of the large human brain. Being at the bottom of the food web, algae are the source of nutrients, e.g., the precious super-unsaturated fatty acids, that pile up through the food chains, but it starts with the marine algae. Algae have during times been a rich source for human activities as a material of unique composition and multiple uses. Algae are currently in focus as a green, sustainable food source because algae are at the base of the trophic web, feeding directly off the sun. Macroalgae (seaweeds) in particular have influenced human life conditions both on evolutionary timescales as well as in recent centuries and all the way into the Anthropocene. No wonder that these organisms have entered human mythology, folklore, poetry, art, and gastronomy. This paper will focus on two often overlooked facets of algae, which have been of key importance for their interwoven relationship with humans: their beauty and their taste.

Keywords

seaweeds – food – taste – flavour – gastronomy – aesthetics – sustainability – evolution

© OLE G. MOURITSEN AND J. LUCAS PÉREZ-LLORÉNS, 2024 | DOI:10.1163/9789004683310_007
This is an open access chapter distributed under the terms of the CC BY-NC 4.0 license.

1 Algae and Humans in Evolutionary Perspective

On the time scale of the age of our planet, which was formed about 4.5 billion years ago, algae appear relatively early on the scene, already about 3 billion years ago, in the form of the blue-green algae (cyanobacteria). We call these organisms microalgae because they are mostly unicellular or form small colonies. Microalgae were teeming in the period from 2 to 0.5 billion years ago. They changed the Earth and its ecosphere completely by their capacity of performing photosynthesis which had the effect that molecular oxygen started to accumulate in the atmosphere. During the period from 3 to 1 billion years ago, because of photosynthetic activity, the partial pressure of oxygen in the atmosphere raised by a factor of 10^{10}. It is likely to have been the most dramatic change ever on our Blue Planet and was a complete game changer for the evolution of life. Molecular oxygen facilitated the biochemical processes that led to those special molecules, e.g., higher sterols like cholesterol, sitosterol, and fucosterol, that are essential for forming higher eukaryotic life in the case of animals, plants, and seaweeds, respectively. These sterols are absolutely essential for providing the necessary mechanical stabilization and lateral organization of the plasma cell membranes. Eukaryotic life was proliferating from about 1 billion years ago.

Seaweeds are multicellular algae, and they are a highly heterogeneous and diverse group of organisms (i.e., a hotchpotch) that thrives in the marine realm. Some are very small whereas others are large as trees. Together with ferns and lichens they were described by the German botanist von Zalusian in 1592 as *ruda et confusa*. The phylum of red algae (Rhodophyta) is the oldest, first appearing around 1.5 billion years ago and then entering a phase of rapid proliferation about 700 million years ago. The green algae (the phylum Chlorophyta) appeared somewhat later, around 900 million years ago. Finally, the brown algae (the class Phaeophyceae) arose about 500 million years ago. Whereas the higher plants (the kingdom Plantae) evolved from green and red algae, the brown algae are on a completely different branch of the tree of life, departing at least 2 billion years ago from the branch that let to red and green algae and ultimately higher plants (Fig. 5.1).

On this scale of algae and seaweed evolution, the evolutionary history of our species is incredibly short. The human lineage is believed to have emerged in its early primitive forms only 4–6 million years ago. One way of tracking the evolutionary path of the lineage in relation to seaweeds as food may be via brain development which requires special nutrients that are abundant in seafood, such as zinc, magnesium, iodine, taurine, vitamin B_{12}, and super-unsaturated fatty acids (Cornish et al. 2017). A major growth of the frontal region of the

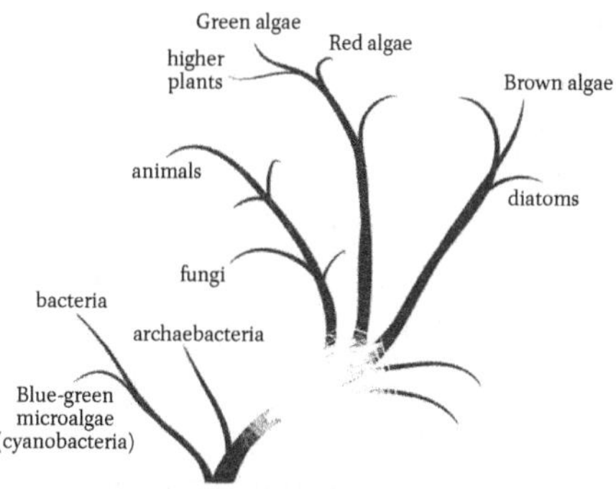

FIGURE 5.1
The 'tree of life' (phylogenetic tree). The tree in this illustration is incomplete, but shows the relative placement of macroalgae and a single group of microalgae. The undefined area in the middle marks the transition between prokaryotes (Bacteria, Archaea) and eukaryotes
COURTESY OF JONAS DROTNER MOURITSEN

neocortex is thought to have occurred sometime during the Pleistocene Period, i.e., 2.5–2 million years ago (Aiello and Wheeler 1995; Hawks et al. 2000; Glazko and Nei 2003; Schoenemann 2006; Preuss 2011). The neocortex is believed to be the site of cognition and intelligence. The actual structure and the size of the frontal cortex are considered to constitute the most significant differences between us and our closest primate relatives (Teffer and Semendeferi 2012; Hofman 2014; Rilling 2014).

Evolutionary biologists see the period 1.8–1.9 million years ago as a time of transition from ape species (Australopithecines) to the hominins (*Homo erectus*) with a smaller brain than modern humans. It has been argued that an important driving force for this evolution was that, during this transition period, the ancestors of *Homo erectus* (the habilines) became hunter-gatherers and changed from herbivores into omnivores. It has been proposed that the hominins learned to use fire and heat to prepare food 1.9 million years ago, which provided for a diet much denser in energy and proteins (Wrangham and Conklin-Brittain 2003; Wrangham 2009; Comody et al. 2011). This in turn fuelled the development of a larger brain, possibly supported by access to large quantities of super-unsaturated fatty acids from marine sources, which are instrumental for building the large neural circuitry of a complex brain (Crawford 2007; Cunnane et al. 2014; Mouritsen 2016). *Homo sapiens* who are thought to date back about 200,000 years (Wrangham 2009) is unique in eating a diet that is rich in cooked and non-thermally processed food (Organ et al. 2011), and this is a universal trait for humans across all cultures and continents.

2 Entangled Life Paths: Seaweeds and Humans

It is impossible to say when the evolving human lineage first became close to seaweeds as a source of food, the reason being that seaweeds, having no durable and hard parts, quickly disintegrate and leave very little archaeological traces, not to speak about evidence regarding how they have impacted human life. It is however compelling to suggest, as alluded to above, that access to seaweeds and marine organisms feeding on seaweeds and microalgae has been a determining factor for human brain development and hence human evolution (Cornish et al. 2017), leading to a brain/body mass ratio that is about ten times larger than for all other animals of the size of a human, except the dolphin (Crawford 2007; Mouritsen 2016).

Seaweeds collectively contain all the chemical components required in a diet for human growth and nutrition, including proteins, minerals, vitamins, dietary fibres, as well as the super-unsaturated fatty acids, the latter although in relatively low concentrations compared to fatty fish. The caloric content of seaweeds is, however, rather low, and seaweeds would in any case only be part of a full diet, although potentially a very important part. Most likely, our ancestral coastal dwellers would have supplemented their diet by other marine organism like fish, crustaceans, molluscs, and other shore-based organisms as well as sugar-rich fruits and starchy plants from the dry land.

The first firm record of seaweeds being used by humans for food and/or medicine was found at a site in Monte Verde (Chile) dating 14.000 years back (Dillehay et al. 2008). The archaeological excavations documented use of twenty different marine seaweed species in a settlement located several kilometres inland from the coast of Chile, suggested that the inhabitants were familiar with the use of seaweeds as food and medicine. Traces of seaweeds found on stone tools show that the seaweeds were worked in some fashion, possibly by chopping them into pieces or by grinding them for medicinal purposes (Dillehay et al. 2008; Mouritsen 2013). These findings have been taken as supporting evidence for a theory of the peopling of the Americas according to which the first inhabitants came to North America from Asia by land over the Beringia land bridge at least 14,000 years ago and then migrated further along the eastern coastline to South America. This migration path has been called 'the kelp highway' (Erlandson et al. 2007). The Monte Verde evidence of an advanced exploitation of marine algae far inland implies that the population must, in an earlier period, have acquired significant practical knowledge about finding food along the seashore at different times of the year and in a variety of coastal settings.

3 Human Utilization of Seaweeds

The many ways humans in prehistoric and historic times all around the world have used seaweeds have been extensively reviewed in books and recent review articles (see e.g., Mouritsen 2013; Pereira 2016; O'Connor 2017; Pérez-Lloréns et al. 2018; 2020; 2023; Mouritsen et al. 2021). Examples of uses range from foodstuff, medicinal aids, building material, insulation, fuel, containers, currency, source for minerals and fine chemicals, pollution clean-up, and combating climate change – to ceremonial and religious purposes as well as jewellery, toys, and firecrackers. In a recent review, the history of human interaction with seaweeds was presented as testimony to the way humankind has, repeatedly, been 'saved by seaweeds' (Mouritsen et al. 2021; Pérez-Lloréns et al. 2023). Under extreme conditions, such as famine, diseases, environmental disasters, and wars, humans have turned to seaweeds in times of 'needs most' and created new opportunities for their uses to mitigate disasters. Another recent review has highlighted how prominently seaweeds have featured in mythology, folklore, poetry, and human daily life (Pérez-Lloréns et al. 2020).

In the following we will focus attention on human uses of seaweeds as foods and highlight some often-overlooked facets of seaweeds, which have been of key importance for their interwoven relationship with humans in relation to aesthetics and sensory perception: beauty and taste. Let's try to reconnect to the seaweeds by the way we sense them by our eyes and our palate.

4 A Whole Universe of Colours, Shapes, and Morphologies

Considering that there are about 12,000 different species of seaweeds and that they are genetically very different, it may not be surprising that they display a wide range of visual appearances. However, their beauty must be seen and admired in their natural aqueous environment. Most people may only know seaweeds from the dead and possibly decomposing and smelly stuff that is washed ashore. This is clearly reflected in a statement by an anonymous Roman writer who wrote *refunditur algâ* (the sea detests seaweeds and casts them ashore) and supplemented by the Roman poet Horace's (65–27 BCE) statement, *et genus et virtus, nisi cum re, vilior algâ est* (high birth and meritorious deeds, if not linked to wealth, are as useless as seaweed), the stage is set. A proper balance is provided by a statement by the Chinese writer Sze Teu (600 BCE): "Seaweeds are a delicacy fit for the most honoured guests, even for the king himself."

The Mexican journalist Vincente García Torres (1811–1894) wrote the following about seaweed colours and shapes with the view of enlightening the general public:

> There are undoubtedly no plants more beautiful than seaweeds. An adornment to our collections, they retain with their shapes and bright

> colours all the appearance of life itself, they are extremely varied in their forms, dimensions and colours, etc. As for their shape, they present the simplest, the most elegant, and the most fanciful. As for their size, compare the extreme smallness of a protococus to those immense macrocystis from the Pacific Ocean that are said to reach up to 500 metres in length; in terms of their colour, all the shades of green and red through to olive green. Wherever there is moisture there are seaweeds.
>
> VINCENTE GARCÍA TORRES, *Instrucción para el pueblo* [Education for the People] (1852)

As here highlighted, we content that seaweeds should be admired in their natural element, the water, and not from those specimens washed ashore. However, many seaweeds live far away from the shore or thrive in inaccessible and turbulent waters. It is mostly at coastal regions subject to substantial tidal variations that seaweeds can be inspected by common people without access to diving equipment.

Still, some of their beautiful colours, shapes and morphologies can be enjoyed from specimens that have been properly harvested, dried, pressed, and mounted on special paper as used for collections in an herbarium (*exsicata*). Obviously, the three-dimensional character is lost, but the morphology can still be seen or imagined. Kept in the dark, the colours are surprisingly well preserved, possibly protected by the natural antioxidants present in large amounts in seaweeds. Such seaweeds preparations continue to inspire artists and designers in their work.

> Our beaches, after a stormy day, are strewed with specimens of this plant [*Plocamiun cartilagineum* or sea-comb, a red seaweed] sometimes large enough to cover the palm of the hand, and now and then of sufficient size to spread over the surface of a half sheet of post paper. Some years since, the poor people who lived on various parts of our coasts, were accustomed to make for sale very pretty landscapes of sea-weed; and this was a very suitable plant for their purpose, as it retained its colour, and well represented a tree. This practice is less general now, but little groups of shells and sea-weeds are still occasionally offered for sale, and this elegant plant is generally selected to ornament the picture. Dr. Johnston [a British phycologist] says, that fancy work with this sea-weed was once a favourite amusement with the princesses, the daughters of George III.
>
> ANNE PRATT, *Chapters on the Common Things of the Sea-Side* (1850)

5 In the Wild and in Aquaculture

Seaweeds can form enormous beds of vegetation at appropriate substrates in the sea, and in particular brown seaweeds grow to large sizes constituting virtual underwater forests. Some species grow from miniscule sizes to tens of metres in a season, and some giant kelps can muster a growth rate of two centimetres an hour (Mouritsen 2013). In the novel *Twenty Thousand Leagues Under the Sea* (Jules Verne 1868), an enthralled professor Aronnax masterfully describes the structure and biodiversity he observes during a walk through a submerged forest:

> The forest was made up of big treelike plants, and when we entered beneath their huge arches, my eyes were instantly struck by the unique arrangement of their branches – an arrangement that I had never before encountered.
>
> None of the weeds carpeting the seafloor, none of the branches bristling from the shrubbery, crept, or leaned, or stretched on a horizontal plane. They all rose right up toward the surface of the ocean. Every filament or ribbon, no matter how thin, stood ramrod straight. Fucus plants and creepers were growing in stiff perpendicular lines, governed by the density of the element that generated them. After I parted them with my hands, these otherwise motionless plants would shoot right back to their original positions. It was the regime of verticality [...].
>
> Here the range of underwater flora seemed pretty comprehensive to me, as well as more abundant than it might have been in the arctic or tropical zones, where such exhibits are less common. But for a few minutes I kept accidentally confusing the two kingdoms, mistaking zoophytes for water plants, animals for vegetables. And who hasn't made the same blunder? Flora and fauna are so closely associated in the underwater world!
>
> JULES VERNE, *Twenty Thousand Leagues Under the Sea* (1868)

As alluded to in the above quote, the life forms under water are in so many ways different from the terrestrial life we are familiar with.

In Fig. 5.2 is shown part of a forest of oarweed (*Laminaria digitata*) stretching their fronds towards the sunlight. The massive sizes of algal mats can be gauged from aerial views like the one in Fig. 5.3 showing green algal bloom at the tidal channels in a saltmarsh in Cádiz Bay.

FIGURE 5.2
Underwater forest of oarweed (*Laminaria digitata*)
PHOTO COURTESY OF GEORGE STOYLE, SCOTTISH NATURAL HERITAGE

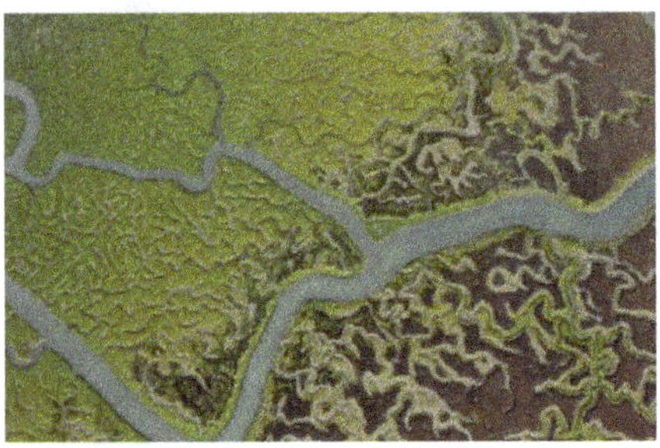

FIGURE 5.3
Aerial view of the salt marshes at low tide in Cádiz Bay (Spain) covered by green seaweeds
PHOTO COURTESY OF EDEA RESEARCH GROUP.

FIGURE 5.4
Louis Druehl, famous phycologist and founder of the small company Canadian Kelp in Bamfield, Vancouver Island (Canada) with a freshly harvested piece of macrokelp, *Macrocystis pyrifera* (Linnaeus)
PHOTO COURTESY OF JONAS DROTNER MOURITSEN

Some seaweeds for human consumption are still harvested from wild populations, including some of the giant kelps. An example is illustrated in Fig. 5.4 showing small-scale, cottage-industry (Canadian Kelp Resources, Bamfield) harvesting of macrokelp (*Macrocystis pyrifera*) off the coast of Vancouver Island. This species is supposed to be the largest seaweed on the planet, growing to magnificent sizes up to 60 metres.

However, most of the seaweed (more than 95%) used in the world for human consumption is produced in aquaculture, mostly in Southeast Asia. The most prominent example is purple laver (*Porphyra/Pyropia* spp.) grown for *nori* production. *Nori* is the paper-thin sheets used for example for sushi. *Nori* is the single most valuable marine crop amounting to a yearly commercial value of about 7 billion US dollars. *Nori* enjoys a long history particularly in Japan (Mouritsen 2013), as beautifully mirrored in Japanese *haiku* (Bashō et al. 2013):

> *kami atsuru*
> *mi no otoroi ya*
> *nori no suna*
>
> failing health
> chewing dried seaweed
> my teeth grate on sand
>
> MATSUO BASHŌ (1644–1694)
> (translated by JANE REICHHOLD)

In Fig. 5.5 are shown illustrations of ancient harvesting along with modern aquaculture of *Porphyra/Pyropia* in Japan. Before the middle of the 20th century, harvesting of wild resources was very unpredictable due to not only varying weather and water conditions but also ignorance regarding the biology of *Porphyra/Pyropia*. This is the reason why *Porphyra* was called 'the grass of gamblers or card players' since its life cycle was unknown and there was no guarantee of good harvests; it was like a game of roulette (Pérez-Lloréns et al. 2018). Only after the British phycologist Dr. Kathleen Mary Drew-Baker (1901–57) in 1949 discovered the peculiarities of the life cycle of this species did it become possible to farm it in a rational manner. Drew-Baker discovered that this seaweed species has an alternating generation shift where every second generation is microscopically small and can only live on the surface of shell fish. Since the discovery of the life cycle, Dr. Drew-Baker earned the eternal gratitude of Japanese aquaculture producers. In 1963 they erected a statue to her memory at the Sumiyoshi shrine, overlooking the Ariake Sea where 50% of the *Porphyra/Pyropia* cultures are located, and on 14 April (the anniversary of her birth) each year people involved in the *nori* industry meet there to celebrate the Drew Festival, during which they refer to Dr. Drew-Baker as the 'Mother of the Sea.' The ceremony involves placing Drew-Baker's academic cap and gown on her monument while a British flag is raised, and then laying a tribute of *nori* from the current crop before the shrine, a unique example of ritualized culinary visibility. Dr. Drew-Baker died at age 51 in 1957, apparently

FIGURE 5.5
Harvesting of *Porphyra/Pyropia* spp. for *nori* production two centuries apart in Japan. Top: Classical Japanese woodblock print. Bottom: Small farm in Kisarazu, Chiba Prefecture (Japan)
PHOTO COURTESY: OLE G. MOURITSEN

completely unaware of her vital contribution to the development of the most valuable aquaculture industry in the world: the cultivation of *nori*.

We shall later come back to the taste and texture of *nori*.

Another important seaweed that is farmed in great amounts for human consumption is the large brown seaweed (kelp) *Saccharina japonica* (and other *Saccharina* species, like *S. longissima*) which is known as *konbu* in Japanese. As we shall see later this species plays a key role when using seaweeds to impart umami taste. *Konbu* is farmed around the coasts of Hokkaido, and it comes in several subspecies of different qualities and flavours, often reflected in major price differences. After harvesting, the fronds are dried in the sun as illustrated in Fig. 5.6 after which they are packaged and stored at least two years and often up to ten years. During this time the *konbu* develops more mellow marine flavours (Mouritsen 2013; Pérez-Lloréns et al. 2018).

Some seaweeds are harvested in the wild or grown in aquaculture not only for their flavour, nutritional value, or technological applications but also for their looks and beauty. An example is carrageen (*Chondrus crispus*) that has commercial uses as source of hydrogels (carrageenans) but is also grown for the use in seaweed salads. In Fig. 5.7 is shown a selection of a product called *hana tsunomata* that are commercial varieties of carrageen (*Chondrus crispus* Stackhouse) in four natural colours. They are grown in land-based aquaculture installations at Acadian Seaplants (Nova Scotia, Canada) and are sold almost

FIGURE 5.6
Drying *konbu* (*Saccharina longissima* (Miyabe) C.E. Lane, C. Mayes, Druehl & W. Saunders) at Hokkaido, Japan
PHOTO COURTESY OF ROBERT THOMSON

FIGURE 5.7
Hana tsunomata, commercial varieties of carrageen (*Chondrus crispus* Stackhouse) in four natural colours grown in aquaculture at Acadian Seaplants, Nova Scotia (Canada)
PHOTO COURTESY OF JONAS DROTNER MOURITSEN

exclusively to the demanding Japanese market for uses in seaweed salad mixtures where they add an obvious aesthetic element.

6 In Museums and Collections

In the first half of the 19th century and later in the Victorian era, English laymen developed an interest in, and fascination with, natural history, leading them to devote their leisure time to the independent study of Nature, often at the seashore (Boalch 2006). Typically chaperoned by the local schoolteacher or medical doctor, English ladies went along the shore at low tide collecting strange objects on the foreshore including bits of seaweed which they dried, pressed carefully, and mounted on paper as shown in Fig. 5.8 (Pratt 1850). These activities were often ridiculed in public, as illustrated by the Punch caricature in Fig. 5.9. In most cases, the Linnaean classification system for naming the species was not well known and the collected species carried vernacular names varying from location to location, and often carrying fanciful names such as mermaid's shaving brush, sea-girdles, peacock's tail, sea lace, sea thong, sea grape, Midshipman's pickle, sea hangers, sea wand, red ware, sea furbelows, sea-rumpet, ever-lasting bladder-thread, hen-ware, honey-ware, whiplash, sea comb, etc.

FIGURE 5.8
Folded book with pressed seaweed samples, probably collected by a young girl on the English Channel Islands in early nineteenth century. On the back of the book is inscribed the small poem quoted verbatim in the text. From the collections at the Natural History Museum, London
PHOTO COURTESY OF JACOB THUE

FIGURE 5.9
Caricature of the collection of seaweeds and other strange objects on an English beach in Victorian times
DRAWING BY JOHN LEECH FROM THE SATIRICAL MAGAZINE PUNCH, 1856

The interest in seaweeds seems also to have inspired poets to point out the injustice inherent in referring to them as weeds. In the little poem below, the seaweeds themselves ask to be called 'Ocean's gay flowers:'

Algae, bright order!
By Cryptogamists defended –
Translate marine plants
as Linnaeus intended.
You collect and admire us,
we amuse leisure hours;
"Then call us not weeds,
we are Ocean's gay flowers."

This poem, attributed to an unknown poet Curtis, was often written on folding books like the one displayed in Fig. 5.8 (Mouritsen 2013) and typically quoted by girls on the English Channel Islands at the beginning of the 19th century. The term 'cryptogamists' in the poem derives from Cryptogamia that refers to an archaic botanical classification system.

Hunt (2005) has analysed popular writings from the mid-Victorian literature in England about seaweeds and how they reveal contemporary ideas about property, philanthropy, and natural theology, including such issues as social history, gender, and natural history. One of his findings is that this kind

FIGURE 5.10
Old Californian postcard showing man with washed ashore bullkelp (*Nereocystis luetkeana*)
PUBLIC DOMAIN (HTTPS://BIT.LY/3PQ3DKD)

of 'seaweeding' was predominantly a women's activity and mainly based on aesthetic engagement. It was looked upon by professional (male) natural scientists as a somewhat questionable endeavour.

Both in England as well as on the European continent in those countries where seaweeds were commonly known among people, e.g., Brittany, seaweeds were around 1900 often prominently featuring on postcards, e.g., at Christmas time. The postcards were either early black and white photographs or imaginative colour drawings, often including people in contact with seaweeds. An amusing example is shown in Fig. 5.10, showing a man carrying the stipe and bulb of a giant kelp (*Nereocystis luetkeana*) with some remains of fronds worn by being washed up at the foreshore.

One of the most beautiful books on seaweeds is *The Seaweed Collector: A Handy Guide to the Marine Botanist* by Shirley Hibberd (1872). It contains plenty black and white illustrations as well as eight tissue-guarded colour woodcut plates. It helps identify different species, as well as helping gatherers classify and sort their finds. It used to be more common for sea visitors to pick up and press seaweeds as scientific specimens or even souvenirs. Most commonly, the dried and pressed specimens were kept as objects of beauty safely under glass or even framed and hung as art, as was quite usual in the Victorian era.

Natural museums around the world host collections of dried seaweeds mounted on special paper (*exsicata*) but most often these collections are not open to the public because they are very fragile, but presumably also because common museum visitors are more interested in spectacular exhibits like dinosaurs and exotic animals. It is thought-provoking that the algal world is almost invisible to the public at natural history museums, considering that the algae together represent by far the largest biological 'kingdom' on the planet. As an example, the Natural History Museum in London, is host to 600,000 seaweed specimens, the oldest dating from the 1700's and some collected during

FIGURE 5.11
Selection of images from the algal herbarium collections at the Natural History Museum, London. From top left to bottom right: *Alaria esculenta*, *Claudea elegans*, *Cladodonta lyallii*, and *Scinaia furcelleta*
PHOTO COURTESY OF JACOB THUE

famous explorations, but none of them are normally among the exhibits. Some of the specimens from this museum not accessible to the public are shown in Fig. 5.11 that shows a selection of species that are particularly aesthetically pleasing to look at (Mouritsen 2013). These collections in London also include historical material, e.g., the original, annotated specimens of European purple laver (*Porphyra umbilicalis*) from Dr. Kathleen Mary Drew-Baker's pioneering studies of *Porphyra* in the 1940ies. Some of the museum's specimens have for obvious reasons served as inspiration for textile designers.

7 In the Kitchen

Sight, as well as smell, are the first senses to go into action when they encounter food. Through our sight we evaluate the dish aesthetically. Everyone knows the expression 'you eat with your eyes first,' since the first impression a diner receives is the sight of the ingredients that make up the dish (colour, form, size, brightness, surface texture, arrangement, etc.). Appearance and smell allow us to anticipate, through experiences stored in our brain, the flavour of what we are about to savour. Therefore, appearance will produce expectations that will affect our capacity for decision: will we accept or reject a dish. Thus, we can understand the emphasis that chefs place on the presentation of dishes (plating) as a prelude to the savouring of a fine meal. Chefs play with the contrast of colours, visual textures, shapes, proportions, arrangement, and harmony of the various ingredients that make up the dish. It is especially true in the Japanese cuisine since, at least from the Edo Period (1603–1867) where etiquette in imperial banquets demanded that guests should show a sense of gratitude and satisfaction at the mere sight of food on their plates. This sense of appreciation was called *katachi no aji* or 'the flavour of the shape' (Pérez-Lloréns et al. 2018).

Seaweeds with their wide variety of shapes, colours, flavours and textures (see below) are a very attractive ingredient that many Western high-end restaurants (e.g., Noma, El Bulli, Central) have incorporated quite recently in their menus (as their Japanese counterparts). The art of preparing and cooking seaweeds in the best way, to enhance its organoleptic properties as well as taking advantage of its nutritional properties has been recently coined as phycogastronomy (Pérez-Lloréns et al. 2018; Mouritsen et al. 2018; 2019a), a term that also applies to the culinary uses of microalgae (Perez-Lloréns 2020). Such avant-garde culinary tendency has permeated little by little to other casual or midrange restaurants and also to household cuisine. However, to achieve a wider seaweed popular consumption it is simply not enough to just put seaweeds on the table. Even if seaweeds have been declared as future food, health food, and brain food (Dhargalkar and Pereira 2005), and even if seaweeds are promoted as edible, nutritious, and healthy (Cornish et al. 2017), and can even be prepared as tasty food, it is difficult to change people's food preferences. This is where the collaboration between chefs and scientists comes into play (Mouritsen et al. 2018). Seaweed dishes have to be tasty for people to start eating them and continue to do so on a repeated and regular basis (Mouritsen 2017). The boom of phycogastronomy, reflected by the increasing amounts of edible seaweed imported by Western countries during recent years (FAO 2018), is mostly the result of the rise in popularity of seaweed cookbooks (see Mouritsen et al. 2018 for a complete list), TV cooking shows, or blogs and specialized magazines (Berger 2017).

Therefore, seaweeds are frequently used to garnish seafood dishes, salads, soup, or desserts. Some species have strains with different colours that are very eye-catching such as the Rhodophytes *Meristotheca papulosa* (*tosaka-nori* in Japanese) that comes in three different colours, white (*shiro-tosaka*), green (*ao-tosaka*), and red (*aka-tosaka*), or *Chondrus crispus* (*hana-tsunomata* meaning *Chondrus* flower) in three colours (pink, yellow and green) as shown in Fig. 5.7. The latter one was produced and marketed by the Canadian company Acadian Seaplants to satisfy the important Japanese market, which is very demanding regarding the quality of seaweeds, since Japanese customers appreciate not only the flavour or the nutritional qualities, but also the colour, shape, freshness, hygiene, mouthfeel, etc. Its crispy texture, mild taste, and

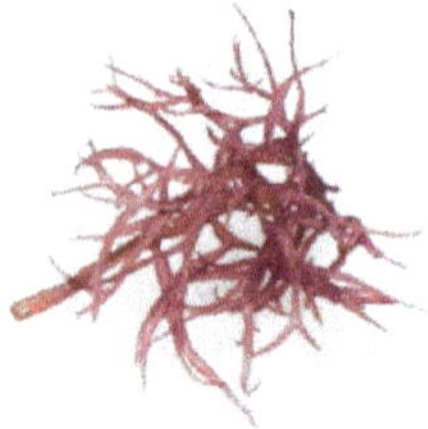

FIGURE 5.12
The delicate red, Japanese seaweed *funori* (*Gloiopeltis* spp.) has a beautiful stringy and fractal appearance and a delicious crunchy texture
PHOTO COURTESY OF JONAS DROTNER MOURITSEN

tri-chromatic appearance give a touch of elegance to dishes (Mouritsen 2013; Pérez-Lloréns et al. 2018). Another aesthetically attractive seaweed that is furthermore rather crispy is the Japanese seaweed *funori* (*Gloiopeltis* spp.) as shown in Fig. 5.12. *Funori* enters salads and is being used as a topping or garnish with seafood dishes. The species *ogonori* (*Gracilaria* spp.) serves similar purposes. It is interesting that *Graciliaria* is now an invasive species in European waters and considered a nuisance whereas Japanese consider this species to be very attractive and delectable.

The eye-catching appeal of fresh and colourful seaweeds was chosen to create the poster announcing the gastronomic fair Madrid Fusion 2017, one of the most important events and exhibitions on the Spanish and international gastronomic scene (Mouritsen et al. 2019a). In fact, Ferran Adrià (El Bulli) who is considered as one of the most influential chefs at the end of the 20th century is so happy with seaweeds that he includes them in what he considers to be 'food with soul,' that is, products which have a particular relevance in El Bulli's cuisine (Pérez 2006). Also, the renowned food critic Richard Cornish said that "seaweed is the culinary trend taking diner's palates to another dimension".

An example of combining the lavish shapes, colours, as well as aromas of seaweed with humble food like potatoes and cabbage is illustrated in Fig. 5.13. Usually, the aroma of seaweeds is not highlighted in food because of a sometimes-overpowering marine aroma. In this case it is opposite since the so-called sea truffle ('truffle seaweed') (*Vertebrata lanosa*) and pepper dulse (*Osmundea pinnatifida*) here elicit respectively truffle aroma and a peppery spiciness.

Whereas the delicate species described above all belong to the red seaweeds (Rhodophyta phylum) and are mostly used in the kitchen from rehydrated dried seaweeds, there is a special one from the green phylum (Chlorophyta), *Caulerpa lentillifera*, that is eaten fresh. It is shown in Fig. 5.14, and it is seen that this species appropriately is called sea grapes. The small grape-like spheres are extremely crispy, and they literally explode between the teeth, imparting a very juicy and fresh marine flavour.

FIGURE 5.13
Savoy cabbage and potatoes garnished with so-called sea truffle ('truffle seaweed') (*Vertebrata lanosa*) and pepper dulse (*Osmundea pinnatifida*). Dish created by chef Klavs Styrbæk
PHOTO COURTESY BY JONAS DROTNER MOURITSEN

FIGURE 5.14
Simple presentation of fresh sea grapes (*Caulerpa lentillifera*)
PHOTO COURTESY OF SUSHISHO MASA
(CC BY 2.0; HTTPS://BIT.LY/3UUOUYH)

8 The Motherlode of Deliciousness: Umami and Texture from Seaweeds

Whereas the flavour and texture of seaweeds are highly appreciated in Southeast Asia and Polynesia and gradually becoming accepted as food in the Western world, a quite common attitude towards seaweeds in the West is accurately described by them being 'weeds' or 'wracks,' that is something useless and possibly unpleasant and smelling stuff cast ashore. Most people do not realize or know that seaweeds encompass a very large group of organisms that is very diverse regarding flavour and texture. This attitude toward seaweeds would be like judging the edibility of the entire collection of plant-based foods based on the impression of decaying apples and plums on the ground in an orchard. The 'unfairness' of describing seaweeds as 'weeds' transpires clearly from the poem quoted above.

It comes a surprise to many people when they learn that one of the five basic tastes, umami (meaning delicious taste), which all humans appreciate and crave for evolutionary reasons (Mouritsen 2012; Mouritsen and Styrbæk 2014), was in fact first proposed as a basic taste and then coined umami after it was discovered in 1909 (Ikeda 2002 [1909]) that a particular brown Japanese seaweed, *konbu* (*Saccharina japonica*), contains very large amounts of a compound (glutamate) that elicits this unique and delicious taste.

Similarly, the common attitude towards the mouthfeel (texture) of seaweeds among many westerners is often described by such derogatory terms as slimy, chewy, hard etc., i.e., associated with an unattractive mouthfeel. Again, the many thousands of different species have very different textures, and of course their texture, in the same way as for plants, will depend on how they are prepared or cooked. Those who were unfamiliar with potatoes would probably also find it unpleasant to eat raw potatoes.

8.1 *Umami*

The Japanese soup stock, *dashi*, is made from an aqueous extract of *konbu* (*Saccharina japonica*) and a special fish product called *katsuobushi* (cooked, salted, dried, smoked, and fermented bonito) or *niboshi* (dried small fish). Alternatively, the *katsuobushi* can be substituted by dried shiitake to produce a vegan *dashi* (Mouritsen and Styrbæk 2014). *Dashi* is much more than a conventional soup stock since it is the centre around which the Japanese cuisine revolves (Tsuji 1980). Japanese describe the taste of *dashi* by the term *umai*, meaning delicious. When the Japanese chemist Kikunae Ikeda in 1908 analysed the composition of *konbu* he discovered that a very large fraction (up to 3wt% dry weight) consisted of a salt of one of the most common amino acids, glutamic acid. He proposed that this salt, mono-sodium glutamate, is the source of the deliciousness and he coined the term 'umami' to describe this unique taste quality (Ikeda 1909). Moreover, he suggested that this new taste is a basic taste, adding a fifth taste modality to the four classical ones, sweet, sour, salty, and bitter (Mouritsen and Styrbæk 2014). It took almost a hundred years before it became commonly accepted that umami is a basic taste and only after the discovery of an umami receptor in our tastebuds. Ever since Ikeda's discovery, *konbu* has inextricably been associated with umami. In Fig. 5.15 is shown a piece of *konbu* aesthetically rolled up to be conveniently used for preparing a *dashi* in the home kitchen.

The umami potential of different subspecies of *konbu* in terms of their content of free glutamate is very different and it often correlates with the price. Many western chefs have begun to use *konbu* and *dashi* in their creations and an obvious quest, as part of the locavore movement, has been to search out local seaweeds as potential umami providers (Figueroa et al. 2021; Milinovic et al. 2021). Recent studies have however shown that *Saccharina japonica* is somewhat unique with respect to a high glutamate content, and most other studied brown, red, and green species have been shown to contain very little free glutamate (Mouritsen et al. 2019b). A notable exception is the red species dulse (*Palmaria palmata*) that can have a glutamate content similar to the poorer qualities of *konbu* (Mouritsen et al. 2013). Even those brown species of the Laminariaceae family, such as oar weed (*Laminaria digitata*), sea tangle (*Laminaria hyperborea*), and sugar kelp (*Saccharina latissima*), that often, but

FIGURE 5.15
Rolled piece of the Japanese brown seaweed, *konbu* (*Saccharina japonica*) to be used for soup stock (*dashi*)
PHOTO COURTESY BY JONAS DROTNER MOURITSEN

FIGURE 5.16
Rolled sushi (*maki-sushi*) with *nori* seaweed paper wrapped around cooked rice and shiitake
PHOTO COURTESY OF JONAS DROTNER MOURITSEN

somewhat misleadingly, are called Atlantic *konbu*, also contain very little free glutamate (Mouritsen et al. 2019b).

However, a seemingly low umami potential judged from low contents of glutamate can either be enhanced by multisensorial interaction with certain aroma compounds (Frøst et al. 2021) or more powerfully by other substances (free nucleotides) that are very prominent in fish, shellfish, as well as fungi. This principle of so-called umami synergy (Mouritsen and Khandelia 2012) is the one that is active in *dashi* preparations (*konbu+katsuobushi/niboshi* or *konbu*+shiitake) and possibly better known in Western cuisine by good food pairings like eggs-bacon, tomatoes-meat, and champagne-oysters (Schmidt et al. 2020).

It is noteworthy that the seaweed species *Porphyra/Pyropia* spp. used to produce *nori* as described above is unique in the sense that the same species can in provide both free glutamate and free nucleotides (inosinate) and hence in itself, without pairing with other foodstuff, can elicit umami synergy. This is one of the reasons why *nori* is used both as a condiment for rice dishes as well as a wrapping for sushi, so-called *maki-sushi* (Mouritsen 2009), as shown in Fig. 5.16.

The principle of umami synergy, which is based on a now understood molecular mechanism on the receptor level (Zhang et al. 2008; Mouritsen and Khandelia 2012) is probably the most remarkable and powerful science underlying the use of seaweeds in international cuisine.

8.2 *Texture*

Although umami can be a dominant sensory experience with eating various seaweeds, it is generally the case that fresh and mildly treated seaweeds (in particular when dried and stored cold and dark) have rather mild and subtle tastes and aromas. Food cultures with preference for seaweeds in their diet therefore often savour seaweeds for their textures, that is the way the feel in the oral cavity – the so-called mouthfeel (Mouritsen and Styrbæk 2017). This is reflected in the language and the number of terms a language contains to describe the sense of mouthfeel. As an example, the Japanese language have about four times as many terms to describe aspects of texture as western languages (Drake 1989). Specifically in Japanese, several different terms are used to describe mouthfeel, such as *hagotae* (crunchiness, resistance to chewing), *kuchi atari* (palatability, mouthfeel), and *shitazawari* (tongue-feel). In case of the sea grapes in Fig. 5.14 the sensation would onomatopoetically be called *puchi-puchi*.

Although many western chefs somewhat dogmatically ask for fresh and wet seaweeds, possibly influenced by the same tradition when it comes to vegetables, it is striking that fresh and wet seaweeds are seldom used in Asian cuisines. Most seaweeds are traded salted or dried, and if well cared for during transport and storage, the dry seaweeds easily rehydrate and swell to their original fresh state and can be used as such. Moreover, during storage some of the overpowering and less desirable marine aromas tend to evaporate.

The texture of fresh and rehydrated seaweeds varies from species to species and is strongly dependent on time of harvest and age of the seaweed and can be described as fine, cartilaginous, meaty, crunchy, spongy, velvety, juicy etc. (Bruhn et al. 2019). When roasted, fried, and sometimes salted, seaweeds that are tough and chewy in the fresh or dry state can become crispy and crunchy. A popular product in many western countries is the green and very crunchy *hiyashi wakame* salad that is made of blanched and salted wakame (*Undaria pinnatifida*) with different dressings. A similar salad can easily be made from so-called Atlantic *wakame* (*Alaria esculenta*). It appears that customers favour such seaweed salads not only for their green and fresh look but for their crunchy mouthfeel.

Possibly the best-known seaweed product in the western world is *nori*, well known from sushi such as the rolled sushi, *maki-sushi*, an example of which is shown in Fig. 5.16. *Nori* is produced from the red seaweed *Porphyra/Pyropia* spp. in a manner similar to paper production. The major difference is that the pulp from the alga is not added a cohesive glue as paper pulp made from plant fibres but is 'glued' together by the alga's own adhesive, being polysaccharides (hydrogels) like carrageenan. After drying and sometimes roasting, the paper-thin sheets of *nori* are very crisp. However, they easily adsorb moist and become chewy and soggy. Therefore, *maki-sushi* should be enjoyed right after has been prepared and before the cooked rice has wetted the *nori*. Only in this way does one experience the correct texture of the seaweed.

Dried *Porphyra/Pyropia* spp. has also found its way into high-end gastronomy as illustrated in Fig. 5.17. In this dish a poached fresh oyster is presented in a broken gel made from seawater and topped with dry seaweed imparting a crunchy

FIGURE 5.17
Dried wild *nori* (*Porphyra/Pyropia* spp.) over poached oysters. Dish created by chef Koji Shimumura
PHOTO COURTESY OF EDITION KOJI SHIMUMURA

FIGURE 5.18
Cavi-art, a commercial seaweed product invented by Jens Møller in Denmark. The product is made by spherification using different brown seaweed materials exploiting their capacity to form hydrogels via their alginate content
PHOTO COURTESY BY JENS MØLLER PRODUCTS APS

texture and an umami-rich taste to the oyster, in perfect umami-synergy with the free nucleotides and free glutamate in the mollusc.

Use of extracts from brown seaweeds in the form of the polysaccharide alginate has in recent decades been one of the pillars of molecular gastronomy (Barham et al. 2010). Upon binding to calcium ions from an appropriate aqueous solution, sodium alginate forms gels and can be used to for objects and capsules of various sizes and shapes. The technique is called spherification and was spearheaded within gastronomy by the chefs at Restaurant El Bulli. The spheres can be arranged to encapsulate various liquid with different and often surprising flavours when the spheres are brought to explode on the palate. In this way the special chemical contents of brown seaweeds are used to impart interesting textures to otherwise fluid foodstuff.

It is interesting to note that long time before alginate and other seaweed-based hydrogels made it into molecular gastronomy, a product called Cavi-art was invented and launched on the market by the Danish industrialist Jens Møller almost 30 years ago (Mouritsen 2017). The first series of Cavi-art products were meant to be a vegan substitute for fish roe, with a good flavour profile, interesting texture, and long shelf life. The product was based on spherification (Fig. 5.18) using different brown seaweed materials and their capacity to form hydrogels due to their alginate content. Later, Cavi-Art spheres (now called pearls) were produced, incapsulated with a range of flavourful and spicy liquids. When Jens Møller first launched his product line, the fact that it was based on seaweeds only appeared in small print on the label on the backside of the package. In recent years, the design of the label has changed: it now carries the word seaweed in large letters on the very front. This example from the market demonstrates that seaweeds are becoming a positive word in the context of food.

9 Seaweeds for the Future: Beauty and Taste

Lessons from the history of human uses of seaweeds can be used as reminders and inspiration, as well as a guide to show us how to refresh our entangled life paths with the ancient seaweeds. As they have done so many times before when we have faced crises, they may once again help us to face current and future challenges, not least regarding how to deal with changing climates and the pressing issues of sustainable and healthy eating (Doumeizel and Aass 2020; Mouritsen et al. 2021). Seaweeds should be considered as part of the green transition and be counted in, along with plants and fungi, as some of the most sustainable food sources on the planet (Mouritsen and Schmidt 2020).

Despite of humans' close relationships with seaweeds, most of us are not aware of their many services to us, both during evolution and not least in historical times as well as present days. As pointed out in this paper we believe that one powerful way of reconnecting to this truly amazing and ancient group of organisms is by paying closer attention to way we sense them by our eyes and our palate.

Acknowledgments

The work by OGM was supported by the Carlsberg Foundation (Grant no. CF21-0711).

References

Aiello, L.C., P. Wheeler. 1995. "The expensive tissue hypothesis: the brain and the digestive system in human and primate evolution." Current Anthropology 36: 199–221.

Barham, P., L.H. Skibsted, W.L.P. Bredie, M.B. Frøst, P. Møller, J. Risbo, P. Snitkjaer, L.M. Mortensen. 2010. "Molecular gastronomy: A new emerging scientific discipline." Chemical Reviews 110: 2313–2365.

Bashō, M., J. Reichhold, S. Tsujimura. 2013. Bashō: the complete haiku. New York: Kodansha America, Inc.

Berger, M. 2017. "In the Weeds." Accessed May 11, 2019. www.cooksillustrated.com/science/785-articles/feature/inthe-weeds.

Boalch, G.T. 2006. "Seaweed collecting in Britain in Victorian times." In World seaweed resources. An authoritative reference system. Ver. 1.0, DVD ROM edited by A.T. Critchley, M. Ohno and D.B. Largo. UK: ETI Inf. Services Ltd.

Bruhn, A., G. Brynning, A. Johansen, M.S. Lindegaard, H.H. Sveigaard, B. Aarup, L. Fonager, L.L. Andersen, M.B. Rasmussen, M.M. Larsen, D. Elsser-Gravesen, and M.E. Borsting. 2019. "Fermentation of sugar kelp (*Saccharina latissima*)- effects on the sensory properties, and content of minerals and metals." Journal of Applied Phycology 31: 3175–3187.

Comody, R.N., G.S. Weintraub, and R.W. Wrangham. 2011. "Energetic consequences of thermal and nonthermal food processing." Proceedings of the National Academy of Sciences USA 108: 19199–19203.

Cornish, M.L, A.T. Critchley, and O.G. Mouritsen. 2017. "Consumption of seaweeds and the human brain." Journal of Applied Phycology 29: 2377–2398.

Crawford, M.A. 2007. "A role for lipids as determinants of evolution and hominid brain development." Biologiske Skrifter Danske Videnskabernes Selskab 56: 7–24.

Cunnane, S., K. Stewart, and I. Tattersall. 2014. "The role of freshwater and marine resources in the evolution of the human diet, brain and behaviour." Journal of Human Evolution 77: 1–216.

Dhargalkar, V.K., N. Pereira. 2005. "Seaweed: promising plant of the millennium." Science and Culture 71: 60–66.

Dillehay, T.D., C. Ramírez, M. Pino, M.B. Collins, J. Rossen, J.D. Pino-Navarro. 2008. "Monte Verde: seaweed, food, medicine, and the peopling of South America." Science 320: 784–786.

Drake, B. 1989. "Sensory textural/rheological properties: A polyglot list." Journal of Texture Studies 20: 1–27.

Erlandson, J., M.H. Graham, B.J. Bourque, D. Corbett. 2007. "The kelp highway hypothesis: Marine ecology, the coastal migration theory, and the peopling of the Americas." The Journal of Island and Coastal Archaeology 2: 161–174.

FAO, 2016. "The State of World Fisheries and Aquaculture." FAO Fisheries and Aquaculture Department, Rome.

Figueroa, V., M. Farfán, J.M. Aguilera. 2021. "Seaweeds as novel foods and source of culinary flavours." Food Reviews International. https://doi.org/10.1080/87559129.2021.1892749.

Frøst, M.B., L. Duelund, M.A. Petersen, A.L. Hartmann, O.G. Mouritsen. 2021. "Odour-induced umami – olfactory contribution to umami taste in seaweed extracts (dashi) by sensory interactions." International Journal of Gastronomy and Food Sciences 25: 100363.

Glazko, G.V., M. Nei. 2003. "Estimation of divergence times for major lineages of primate species." Molecular Biology and Evolution 20: 424–434.

Hawks, J., K. Hunley, S.-H. Lee, M. Wolpoff. 2000. "Population bottlenecks and Pleistocene human evolution." Molecular Biology and Evolution 17: 2–22.

Hibberd, S. 1872. The Seaweed Collector: A Handy Guide to the Marine Botanist. London: Groombridge.
Hofman, M.A. 2014. "Evolution of the human brain: when bigger is better." Frontiers in Neuroanatomy 8: 1–12.
Hunt, S.E. 2005. "'Free, bold, joyous': the love of seaweed in Margaret Gatty and other mid-Victorian writers." Environment and History 11: 5–24.
Ikeda, I. 2002. "New seasonings." Chemical Senses 27: 847–849. [Translation from original article in Journal of Chemical Society of Japan 1909; 30: 820–836].
Milinovic, J., P. Mata, M. Dinez, J.P. Noronha. 2021. "Umami taste in edible seaweeds: The current comprehension and perception." International Journal of Gastronomy and Food Science 23: 100301.
Mouritsen, O.G. 2009. Sushi. Food for the Eye, the Body & the Soul. New York: Springer.
Mouritsen, O.G. 2012. "Umami flavour as a means to regulate food intake and to improve nutrition and health." Nutrition and Health 21: 56–75.
Mouritsen, O.G. 2013. Seaweeds. Edible, Available & Sustainable. Chicago: Chicago University.
Mouritsen, O.G. 2016. "Deliciousness of food and a proper balance in fatty-acid composition as means to improve human health and regulate food intake." Flavour 5: 1.
Mouritsen, O.G. 2017. "Those taste weeds." Journal of Applied Phycology 29: 2159–2164.
Mouritsen, O.G., H. Khandelia. 2012. "Molecular mechanism of the allosteric enhancement of the umami taste sensation." FEBS Journal 279: 3112–3120.
Mouritsen, O.G., K. Styrbæk. 2014. Umami: unlocking the secrets of the fifth taste. New York: Columbia University Press.
Mouritsen, O.G., K. Styrbæk. 2017. Mouthfeel: How Texture Makes Taste. New York: Columbia University Press.
Mouritsen, O.G., P. Rhatigan, J.L. Pérez-Lloréns. 2018. "World cuisine of seaweeds: science meets gastronomy." International Journal of Gastronomy and Food Science 14: 55–65.
Mouritsen, O.G., P. Rhatigan, J.L. Pérez-Lloréns. 2019a. "The rise of seaweed gastronomy: phycogastronomy." Botanica Marina 62: 195–209.
Mouritsen, O.G., L. Duelund, M.A. Petersen, A.L. Hartmann, M.B. Frøst. 2019b. "Umami taste, free amino acid composition, and volatile compounds of brown seaweeds." Journal of Applied Phycology 31: 1213–1232.
Mouritsen O.G., C.V. Schmidt. 2020. "A role for macroalgae and cephalopods in sustainable eating." Frontiers in Psychology 11: 1402.
Mouritsen, O.G., P. Rhatigan, M.L. Cornish, A.T. Critchley, J.L. Pérez-Lloréns. 2021. "Saved by seaweeds: phyconomic contributions in times of crisis." Journal of Applied Phycology 33: 443–458.
O'Connor, K. 2017. Seaweed: a global history. London: Reaction Books.

Organ, C., C.L. Nunn, Z. Machanda, R.W. Wrangham. 2011. "Phylogenetic rate shifts in feeding time during the evolution of Homo." Proceedings of the National Academy of Sciences USA 108: 14555–14559.

Pereira, L. 2016. Edible seaweeds of the world. Boca Raton: CRC Press.

Pérez, J. 2006. "Algas a la carta". Accessed June 11, 2021. http://www.diariosur.es/prensa/20060827/sociedad/algas-carta_20060827.html.

Pérez-Lloréns, J.L. 2020. "Microalgae: From staple foodstuff to avant-garde cuisine". International Journal of Gastronomy and Food Science 21: 100221.

Pérez-Lloréns, J.L., I. Hernández, J.J. Vergara, F.G. Brun, A. León. 2018. Those curious and delicious seaweeds. A fascinating voyage from Biology to Gastronomy. Cádiz: Editorial UCA.

Pérez-Lloréns, J.L., O.G. Mouritsen, P. Rhatigan, M.L. Cornish, A.T. Critchley. 2020. "Seaweeds in mythology, folklore, poetry, and life." Journal of Applied Phycology 32: 3157–3182.

Pérez-Lloréns, J.L., A.T. Critchley, M.L. Cornish, O.G. Mouritsen. 2023. "Saved by seaweeds (II): Traditional knowledge, home remedies, medicine, surgery, and pharmacopoeia." Journal of Applied Phycology. https://link.springer.com/article/10.1007/s10811-023-02965-6.

Pratt, A. 1850. Chapters on the Common Things of the Sea-Side. London: Printed for the Society for promoting Christian knowledge.

Preuss, T.M. 2011. "The human brain: rewired and running hot." Annals of the New York Academy of Sciences 1225(S1): E182–E191.

Rilling, J.K. 2014. "Comparative primate neuroimaging: insights into human brain evolution." Trends in Cognitive Sciences 18: 46–55.

Schmidt, C.V., K. Olsen, O.G. Mouritsen. 2020. "Umami synergy as the scientific principle behind taste-pairing champagne and oysters." Nature Scientific Reports 10, 20077 (2020).

Schoenemann, P.T. 2006. "Evolution of the size and functional areas of the human brain." Annual Reviews of Anthropology 35: 379–406.

Teffer, K., K. Semendeferi. 2012. "Human prefrontal cortex: evolution, development, and pathology." Progress in Brain Research 195: 191–218.

Tsuji, S. 1980. Japanese Cooking: A Simple Art. Tokyo: Kodansha International.

Verne, J. 1867. Vingt Mille Lieues Sous les Mers. Paris: Pierre-Jules Hetzel.

Wrangham, R.W. 2009. Catching fire: how cooking made us human. New York: Basic Books.

Wrangham, R.W., N. Conklin-Brittain. 2003. "Cooking as a biological trait." Comparative Biochemistry and Physiology A 136: 35–46.

Zhang, F., B. Klebansky, R.M. Fine, H. Xu, A. Pronin, H.T. Liu, C. Tachdjian, X.D. Li. 2008. "Molecular mechanism for the umami taste synergism." Proceedings of the National Academy of Sciences USA 105, 20930–20934.

CHAPTER 6

An Investigation of Algae's Applications, Inspired by Indigenous and Vernacular Craft Traditions

Kathryn Larsen

Abstract

Algae was once a valuable resource for crafts around the world. In Japan, funori (*Gloipeltis tenax, Gloiopeltis complanate, Gloiopeltis furcate*) seaweed glue historically aided in the creation of kimono fabric, shoji sliding doors, and ceramic finishes. The strong fronds of bull kelp (*Durvillaea potatorum*) are fashioned into leather-like water-carriers by the aboriginal people of lutruwita (Tasmania), in a tradition that continues to present day. In Denmark, Irish moss (*Chondrus crispus*) was used as a base to create a matte, even ceiling paint. This historical use also reveals a cultural connection to algae, in countries like Denmark and in Japan, in the way certain species were chosen for craft as opposed to being used as a potential food source.

However, with the rise of industrialized production, many of these vernacular applications of seaweed have become obscure, used only in restoration and local heritage projects. When the average person looks at seaweed today, they typically only see food, or worse, a rotting mass on a beach, instead of a potential craft material.

By using algae material-based design as a resource and a tool for education, we can begin to build a cultural connection with algae again in Western Europe.

Keywords

algae – material driven design – seaweed – heritage – seagrass – natural building – seagrass – eelgrass

1 Introduction

For the last three years, I have been experimenting with both seagrass and algae in both design and the built environment. My material experiments, often displayed for the public and for students, have led me to make two conclusions based on a phenomenological approach.

© KATHRYN LARSEN, 2024 | DOI:10.1163/9789004683310_008
This is an open access chapter distributed under the terms of the CC BY-NC 4.0 license.

The first conclusion is that average person in Western Europe typically does not view algae as a craft-based material. My interactions with visitors at a pop-up installation with my work, usually proceeds in the following manner. A visitor will usually first ask what the material is, if they have not read the informational label. When they are informed it comes from seaweed, they then ask if they can eat what I have created. The second reaction, usually after asking permission, is to sniff and inspect the material, touching it and exploring the texture. If it smells like old seaweed or is slimy, they react strongly in revulsion. If it is scentless, or mild in scent, they react strongly in shock and surprise that it does not smell bad.

From these experiences, the second conclusion I have deduced, is that in Western Europe, we have become culturally removed from algae as a material. In many of the countries where my work has been exhibited, either seagrass or seaweed has, at one point in time, been utilized as a material for architecture or a craft-based application. However, visitors are usually unaware of this fact. Today, the role of seaweed is not typically one we experience in a tactile manner, apart from perhaps eating a roll of sushi. I am currently based in the Netherlands, and there is no local dried seaweed sold on the shelves of grocery stores, apart from sushi nori.

Algae, however, still plays an important role in our daily lives. It is broken down into carrageenan, agar-agar, beta-carotene and alginate in an extraction process.[1] These powders are then added to thicken products like toothpaste, color foods, or are marketed as a vegan form of gelatin. However, without a google search, the average person would never realize these extracts listed on the back of a packaging label originate from algae. We are consuming seaweed without even realizing it, devoid of their original context.

As the effects of climate change intensify, sustainable seaweed farming has become an increasingly popular proposed solution. However, one of the issues that inhibits seaweed farming in the west at the present, is the way seaweed is perceived by the general public. For example, in Denmark, most people interact with it while it is rotting on the beach in clumps of seagrass. Because of this, algae are generally seen as a smelly nuisance. In fact, Ole Mouritsen, from the University of Southern Denmark, believes that the aversion towards accepting seaweed into the local diet stems from the fact that "people don't like the idea

---

1 "Ocean Planet: Lesson Plan 1." *NASA*, NASA, https://seawifs.gsfc.nasa.gov/OCEAN_PLANET/HTML/education_lesson1.html.

of eating something washed up"[2] He goes further on to say, "[y]ou wouldn't go to an orchard and eat the rotting fruit on the ground."[3]

However, I believe we could potentially change our relationship with algae by integrating local seaweed into design and craft again. This can renew our cultural connection to algae and begin to change perceptions towards using it, especially if the materials have minimal scent and a pleasing texture. In particular, I have looked at how algae-based glues from Japanese and Danish history can be used in clay plaster and paint, and how algae can be used to create a leather-like material, inspired by the ongoing tradition of aboriginal Australian bull kelp water carriers.

1.1 *Japanese Algae Glue*

The way algae was historically utilized as a craft material in Japan reveals a strong cultural motivation behind how and why certain species were utilized. Funori seaweed glue historically aided in the creation of kimono fabric, shoji sliding doors, and ceramic finishes.[4] But how and why were the species for funori, a combination of *Gloipeltis tenax, Gloiopeltis complanate*, and *Gloiopeltis furcate*[5] chosen?

One of the reasons might be because few Japanese dishes commonly utilize funori as food. For example, hegisoba from Niigata prefecture utilizes the seaweed glue as the binder in soba noodles.[6] In a country so reliant on seaweed as a food, algae for glues were specifically chosen from species considered less tasty. To utilize the wrong species of seaweed on a craft application could be seen as a waste.

In 1968, Constance I. MacFarlane wrote, "In some areas, harvesting of this [*Gloiopeltis*] seaweed is decreasing because of competition with new chemical sizings."[7] In many ways this predicted the future of funori, which became

2 Taylor, Michelle Warwicker & Anna-Louise. "Seaweed: Should People Eat More of It?" *BBC News*, BBC, 4 May 2012, https://www.bbc.com/news/magazine-17870743.

3 Taylor, Michelle Warwicker & Anna-Louise. "Seaweed: Should People Eat More of It?" *BBC News*, BBC, 4 May 2012, https://www.bbc.com/news/magazine-17870743.

4 Swider, Joseph R., and Martha Smith. "Funori: Overview of a 300-Year-Old CONSOLIDANT." *Journal of the American Institute for Conservation*, vol. 44, no. 2, 2005, pp. 117–126, https://doi.org/10.1179/019713605806082329.

5 Swider, Joseph R., and Martha Smith. "Funori: Overview of a 300-Year-Old CONSOLIDANT." *Journal of the American Institute for Conservation*, vol. 44, no. 2, 2005, pp. 117–126, https://doi.org/10.1179/019713605806082329.

6 "Hegisoba." *Hegisoba* - 【郷土料理ものがたり】, http://kyoudo-ryouri.com/en/food/603.html.

7 Macfarlane, Constance I. Nova Scotia Research Center, Halifax, Canada, 1968, pp. 5–6, *THE CULTIVATION OF SEAWEEDS IN JAPAN AND ITS POSSIBLE APPLICATION IN THE ATLANTIC PROVINCES OF CANADA*.

increasingly expensive over time and was eventually replaced by synthetic glues. Funori is still used in some traditional craft applications, especially for kimono fabric. It is also used sparingly in conservation work, due to the high cost.[8]

MacFarlane also wrote about the usage of *Chondrus* and *Iridaea cornucopia* to produce funori glue in 1968, however noted that these seaweeds are inferior to produce the glue and are thus, cheaper. Today, Japanese tsuchi-nori, or "glue walls", as well as Shikkui lime plaster utilizes powdered Tsunomata glue. Tsunomata glue is derived from Chondrus species.[9] In these contemporary plaster blends, the weak seaweed glue helps with the moisture retention of the plaster, as well as the workability of the mixture. It also helps aid in trowel-polishing the finish.[10]

1.2 *Experimenting with Seaweed Clay Plaster*

To prepare for this experiment, I took a clay plastering course at Tierrafino Clay in Amsterdam. In this course, I prepared several control tests, and learned the proper method of clay plastering, water-to-mix ratio, and finishing techniques. However, the Tierrafino clay plaster products are already prepared with sand and fibers in a way to prevent cracking.

Starting over with a clay mixture from scratch comes with the risk of the mix cracking if the proper ratio is not achieved. The plaster cracks as it dries and shrinks. One of the ways to prevent this is to include straw fibers in the mix. Seaweed glue plaster, however, does not always include straw fiber in the mix. This could be because the seaweed glue improves the workability of the mix according to Kyle Holzhueter, an expert in Japanese plastering.[11]

1.3 *Process*

In order to test algae glue as a binder for clay plaster in The Netherlands, I first tried to find a species that could act as a control similar to funori. As the Tsunomata glue is composed of Chondrus species, as well as some versions of funori, I settled on *Chondrus crispus*, or Irish moss, as the seaweed base. However, as previously mentioned, the *Chondrus* species for funori were considered inferior to *Gloiopeltis*. To prepare the seaweed glue base, I used

8 "Traditional Uses of Funori – Tri-Funori." *Tri-Funori*, https://tri-funori.com/traditional-uses-of-funori/.

9 "Chondrus Yendoi." *Algaebase*, https://www.algaebase.org/search/species/detail/?species_id=3486.

10 "Nori-Tsuchi: Clay, Sand and SEAWEED GLUE." 左官 Japanese Plastering, japaneseplastering.blogspot.com/2014/06/nori-tsuchi-clay-sand-and-seaweed-glue.html.

11 "Nori-Tsuchi: Clay, Sand and SEAWEED GLUE." 左官 Japanese Plastering, japaneseplastering.blogspot.com/2014/06/nori-tsuchi-clay-sand-and-seaweed-glue.html.

powdered dried Irish moss, blended 1:1 with water, and brought to a boil on a stovetop.

To create the starting ratio of the plaster, I used author Kyle Holzhueter's advice, from Japanese Plastering's article on Nori-Tsuchi.[12] He states to mix fine, sifted clay and sand in a 1:1 ration, and then add seaweed glue at a rate of 1kg per bucket of dry clay-sand. It was not clear what the quantity of 1 bucket was.

The ratio I tested was: 600g dry clay, 600g sand, and seaweed glue mix created from 85g of dried, powdered Irish moss and 85ml water. I added the plaster mix with glue to 100ml water, and then added more water until the plaster was workable. The seaweed glue increased the viscosity of the clay mix, which felt thicker and spongier than regular clay plaster mix. This could be due to the gelling effect of carrageenan in Irish moss.

In the first test, I experimented with both one, and two layers of plaster, applied in ¾ centimeter thickness approximately. Plaster drying time is dependent on the warmth and humidity in the room, and samples can be sped up on a warm sunny day. However, as the tests were done in autumn on a cloudier day, this slowed the drying time. After three hours, the plaster was set enough that I could begin to set the finish with a trowel without disturbing the strength of the layers. I continued this process of trowel finishing periodically until the sample was fully dry. The entire drying process took 12 hours.

1.4 *Results*

The first plaster sample cracked extensively. There appeared to be a difference in whether or not two layers were applied, as the single layer cracked more than the two-layer sample. I theorized also that the application of plaster could also be a factor, as when the fiberless clay plasters at Tierrafino were applied in samples, the thickness of plaster was very important. The less fiber and sand in the mix, the thinner the application should be.

I made a second test of the seaweed glue plaster mix after 24 hours, reusing the same mix as was before. According to the Tierrafino course, if the clay plaster is not fully dried, it can be continued to be used with adding a small amount of water if needed. If the clay plaster fully dries, it might need to sit with the water and begin to rehydrate over time. The plaster mix was still very wet, and so I added only 1 tbsp of water to the mix before plastering again.

With the second sample, I applied the mix in quarter centimeter thickness. As before, I set the finish with trowel every three hours until fully dry. This mix

---

12 "Nori-Tsuchi: Clay, Sand and SEAWEED GLUE." 左官 Japanese Plastering, japaneseplastering.blogspot.com/2014/06/nori-tsuchi-clay-sand-and-seaweed-glue.html.

cracked along the side slightly after 8 hours due to a plastering error, however the rest of the sample was smooth.

1.5 *Plastering Conclusion*

The way seaweed clay plaster is applied and finished with trowel has a significant impact on the quality of the dried sample. Although Chondrus crispus was used as the base for the seaweed glue, it is a fairly weak version of seaweed glue. It is difficult to know if this was a factor in the plaster strength. For my thesis at TU Delft, I will experiment further using invasive *Sargassum natans* and *Sargassum fluitans*, which contain alginate, as both a potential glue, as well as a potential fiber. I hope that by including the seaweed as a potential fiber in the mix can improve the ease of plastering.

1.6 *Danish Algae Glue*

Another reason I chose Chondrus crispus as a glue for my initial algae plaster tests, was because it was used commonly as a glue in Denmark for paint. Before the 1960s, Danish wall paint was typically composed of a glue base, and a pigment for the color. In the case of mosfarve, literally moss color, the pigment was chalk. Mosfarve was traditionally only used for whitening ceilings, as Irish moss produces a very weak glue. However, this white glue provides a strong advantage in ceiling paint. The effect it produces is an even, matte white on flat surfaces, and it can easily be removed by washing the surface with cold water.[13] The seaweed used in Denmark for paint is native around the coastline, however for some reason, become commonly referred to in Danish as "Icelandic moss" instead of Irish moss.[14] It is also referred to as "carrageenan seaweed", due to its high carrageenan content.

There is very little literature on why Irish moss over other seaweeds was chosen to make paint in Denmark, especially as it has such a weak binding application. It could be due to the high carrageenan content in the Irish moss, as opposed to agar-agar. The carrageenan made the paint into a gel, which could be applied to ceilings without any risks of drips, according to the traditional painters at the Danish Open Air Museum. The carrageenan in Irish moss was historically used as the base for paper marbling across Europe in the 19th century, and just as funori was used for sizing kimono textiles, Irish moss

13 "Limfarve Til Vægge Og Lofter – SLKS." *Information Om Bygningsbevaring*, Kulturstyrelsen, https://slks.dk/fileadmin/user_upload/SLKS/Omraader/Kulturarv/Bygningsfredning/Gode_raad_om_vedligeholdelse/9.9_Limfarve_til_vaegge_og_lofter_.pdf.

14 "Limfarve Til Vægge Og Lofter – SLKS." Information Om Bygningsbevaring, Kulturstyrelsen, https://slks.dk/fileadmin/user_upload/SLKS/Omraader/Kulturarv/Bygningsfredning/Gode_raad_om_vedligeholdelse/9.9_Limfarve_til_vaegge_og_lofter_.pdf.

was also used to size yarns for weavers.[15] Perhaps from this already-existing artisanal use of the seaweed, it evolved to be used as a paint base in Denmark.

1.7 *Experimenting with Mosfarve*

As a control, I attempted the traditional recipe for mosfarve. However, by mistake I purchased kappa carrageenan, an extract of Irish moss, instead of powdered Irish moss. As the traditional recipe calls for 1 part boiled Irish moss glue to 5 parts wettened chalk,[16] I boiled the correct ratio of kappa carrageenan and added it to the chalk. The result was a boiled chalk jelly completely unsuitable as a paint application. Not to be deterred, I decided to try again to avoid wasting the carrageenan.

This time I rehydrated the ratio of kappa carrageenan, simply stirring it in an equal amount of water, to dissolve it into a glue. The chalk must be wettened before application. This is done by filling a container with water and carefully adding the chalk to it, and letting it settle on the bottom. The container must not be stirred. After an hour or two, the clear water on top of the chalk can gently be poured off, and the seaweed glue stirred in.

The second attempt with the hydrated kappa carrageenan was a success, however the smell and texture of the paint would likely be different had Irish moss been used instead of the extract. Applying the paint at first was difficult-while hydrated, the chalk appears clear when the paint mixture is applied to a paper surface. It is first after drying that the chalk begins to appear as a matte white. Clean brush strokes are necessary for an even application, so the first trial appeared streaky after application, as opposed to subsequent trials.

After application, it became very clear why the paint was initially only used as a ceiling paint. Touching the paint causes the chalk to rub off of the binder, onto other surfaces. This would not be suitable for a wall application or an application with lots of handling.

It is possible to combine this paint with pigment. In order to test it, I combined the mosfarve paint base with varying amounts of blue and green strains of dried spirulina microalgae. The result was a pastel green and blue, in varying shades. However, mixing the two separate spirulina paints did not result in an even blue-green application. The pigment clumped separately in their own mixes, resulting in a very textured and separated color.

---

15 Aschoff, Christian. "Die Retro-Bibliothek: Nachschlagewerke Zum Ende Des 19. Jahrhunderts." *Retro*, https://www.retrobibliothek.de/retrobib/seite.html?id=103142.

16 Vadstrup, Søren. *Center for Bygningsbevaring I Raadvad Anvisninger Til* … 2010, https://www.bygningsbevaring.dk/uploads/files/anvisninger/14-ANVISN_Mosfarve.pdf.

1.8 *Dulse Seaweed Pigment*

After using the carrageenan extract, I decided to test if other species of red algae were also suitable as paint. I used dulse, *Palmaria palmata,* as it is native to both Denmark and the Netherlands, as a base for the new paint. As a species, it is easier to purchase in the Netherlands than Irish moss, as it has been popularized for having a bacon-like flavor when fried.

At first, I tested dulse as a pigment by itself. I rehydrated powdered dulse and tested the color after different points of boiling it. When the dulse-water mix was used without heat, a reddish-pink color was produced. With exposure to heat, and the longer heat was added, the mix became browner in color. Eventually over time as the mixture boiled, the particles of dulse began to clump and thicken together. This was not optimal for a smooth application on paper.

1.9 *Dulse Seaweed Watercolor*

After testing the pigment ability of dulse, I decided to experiment with using it as a base paint. While researching Danish paint applications, one of the recommendations was to not use mosfarve on any sort of crown molding or textured surfaces, as the paint will gather and clump.[17] Instead, using an animal glue-based paint would be more appropriate.

1.10 *Process*

I decided to combine a small amount of gelatin with the dulse powder and add a plasticizer to reduce the clumping in the paint base. When making watercolor paints, honey is commonly used as a plasticizer, however I decided to test with glycerol. I used 85 grams of powdered dulse, 7 grams of gelatin, 240ml water, and 5ml glycerol as a base. This is a recipe that also works as a bioplastic when strained, poured and dried into a frame.

To combine the recipe together, I mixed all the dry components in a pot first. Then I added glycerol and mixed it into the dry ingredients. Finally, I added the water and brought it to a boil for 5 minutes.

After the base was done boiling, the seaweed and gelatin liquid was strained using a coffee filter into a container. At this point, I added natural pigment in the form of blue and green spirulina. I did not measure the amount of pigment added but added an amount until it was evenly dispersed in the mix.

---

17 "Limfarve Til Vægge Og Lofter – SLKS." Information Om Bygningsbevaring, Kulturstyrelsen, https://slks.dk/fileadmin/user_upload/SLKS/Omraader/Kulturarv/Bygningsfredning/Gode_raad_om_vedligeholdelse/9.9_Limfarve_til_vaegge_og_lofter_.pdf.

The mix applied very evenly and smoothly as a watercolor. There was a small level of pigment separation and streakiness, however this is likely due to using spirulina as a pigment over a different kind of pigment. The blue spirulina was less streaky when applied as opposed to the green spirulina.

The finish of the dried samples had a high level of gloss and shine. This is likely due to the inclusion of gelatin in the mix. Unlike the test with mosfarve, when the two paints were blended together, the pigments combined to form different shades of blue-green. The paint also built nicely in layers, as opposed to the mosfarve.

The paint was useable for 3 hours, however after the 3-hour mark, it began to gel. By adding some warm water, the mix became usable again, however, was diluted in color. Because of this, it is best to use the paint when it is freshly mixed.

1.11 *Paint Conclusion*

Although mosfarve has a very nice matte finish, it is a weak paint, and slightly difficult to apply evenly to an inexperienced user. Using dulse seaweed alone as a pigment and a paint is dependent on the temperature applied to it, and it is very watery in texture. However, by combining the dulse seaweed with gelatin and a plasticizer like glycerin into the base, the mixture is transformed into a very functional watercolor-esque paint. When wet, it has a strong smell of seaweed, however when dried, there is only a very faint smell of dulse and salt when sniffed directly. It would be interesting to test other plasticizers like honey in the mix in the future, as well as other pigments, to see how well it stands as a base.

1.12 *Indigenous Applications of Algae*

Several indigenous communities have special relationships with algae. *Nereocystis leutkeana* (Bull kelp), *Macrocystis* (Giant kelp) and *Porphyra umbilicalis* (Purple laver) are all important species for the First Peoples of the Northwest Coast of North America. The purple laver and giant kelp were both used for eating. The fronds of the bull kelp were also cooked down into soups, however the stipe (stem) of the bull kelp served as one of the most important plant-based materials for First Peoples. In contrast to other cultures with seaweed, indigenous communities tend to use different parts of kelp as a base material, as opposed to making extracts or glues from it. They found creative ways to manipulate and process the seaweed, to work it into different applications.[18]

---

18 Turner, Nancy J. "COASTAL PEOPLES AND MARINE PLANTS ON THE NORTHWEST COAST."

For example, the freshly cut stipes were cured and used as fishing line. The hollow stipes were also used as hoses, for adding water into steam pits. Stipes and hollow bulbs were even used to carry and store food items, like molasses. Several of these seaweed-based traditions continue to this day.[19]

Perhaps most interestingly, are the different ways bull kelp was processed. Kelp was steamed, bent and dried into shapes. According to ethnobotanist Nancy J. Turner (2019), to make fishing line, "curing methods varied: some Coast Salish peoples alternately soaked the kelp in fresh water and dried over a smoking fire; the Nuu-chah-nulth dried the kelp and soaked in dogfish or whale oil. Kelp lines were dried for storage, but had to be soaked in water before use, or they were too brittle. After soaking, they were strong and flexible once again."[20] These different methods give many insights into how kelp can be processed to be used in material applications.

Another community with a special connection to bull kelp as a material is the aboriginal peoples of lutruwita (Tasmania). This species of bull kelp (*Durvillaea potatorum*) is not to be confused with the bull kelp (*Nereocystis leutkeana*) of the First Peoples of the Northwest Coast of North America, as it has differing properties. *Durvillaea potatorum* is incredibly strong and woodsy when dried. Because of this, the aboriginal community uses the fronds while damp to craft water carriers, using the seaweed as a leather-like textile.[21] This practice has existed for hundreds of years, and is continued today by aboriginal artisans.[22]

1.13 *Working with Algae as Leather*

After coming into contact with the contemporary work of Julia Lohmann, I became intrigued with the idea of using kelp as a material in itself. Lohmann herself makes an analogy: that kelp can be used as leather.[23] By studying the indigenous processing techniques of different macroalgae, I learned that kelp

19 Turner, Nancy J. "COASTAL PEOPLES AND MARINE PLANTS ON THE NORTHWEST COAST."

20 Turner, Nancy J. *Plant Technology of First Peoples in British Columbia*. Royal British Columbia Museum, 2019.

21 Thurstan, Ruth H., et al. "Aboriginal Uses of Seaweeds in Temperate Australia: An Archival Assessment." *Journal of Applied Phycology*, vol. 30, no. 3, 2018, pp. 1821–1832, https://doi.org/10.1007/s10811-017-1384-z.

22 Maiden, Vicki Matson-Green and Siobhan, and Bernice Condie. "Bernice Condie Keeps Her Ancestors' Traditions Alive." *Bernice Condie Keeps Her Ancestors' Traditions Alive – ABC (None) – Australian Broadcasting Corporation*, 8 July 2008, https://www.abc.net.au/local/stories/2008/07/09/2298488.htm.

23 Aalto University. *Julia Lohmann: 'We Know Too Much and Do Too Little.'* 2020, Accessed 14 Oct. 2021.

fronds are often worked while wet and dried into a hard material. It appeared to be similar to how rawhide was processed, so I also researched this process.

From my own experiences, I found that soaking the kelp is an important step, and so is oiling it to create a smooth, even-textured finish. After these steps, the kelp can be processed into leather-like materials, by sewing and molding. However, when the algae dry, it is hard and brittle again. I wanted to create a true textile out of the material and preserve the strength and flexibility of the wettened version of the kelp, even after being dried. After several series of experiments in my kitchen, I came up with a repeatable vegan process to turn almost any type of kelp into a flexible, leather-like material. The strength of the final kelp leather is dependent on the growing conditions and species of kelp.

I initially called my resulting material work "kombu leather", after using various Japanese kelp species as the source of many of my experiments. However, at this point, I was ignorant of the Japanese perception of kombu as a valuable food item. After a conversation with a Japanese seaweed farmer, I realized that in a Japanese context, my work appeared as if I was playing with my food, and not respecting it. Because of this, I decided to change how I sourced my seaweed for my experiments.

My chance to integrate this came in the form of a residency for Non-Extractive Architecture, in Venice, Italy. The lagoon of Venice is an ecosystem in itself, with many different native species. However, invasive species like *Sargassum muticum* and *Undaria pinnatifida* (wakame) have invaded the canals.[24] Although these species are edible, they also absorb arsenic[25] while growing, making their levels potentially unsuitable for consumption. This is why I chose to use fronds of the lagoon wakame in my experiments for my residency.

The wakame leather smelled strongly while wet, perhaps even worse than other species I have worked with, potentially due to lagoon water pollution. It smelled so strongly that others could not bear to be in the same room as it while it was drying. However, once dry, the smell was imperceptible. Only by sniffing the material closely, could one get a hint of the lagoon water smell. Although I was able to sew the material into a patchwork textile strong enough to upholster a stool, the wakame leather was weaker than other seaweed leathers I have produced. Thus, in the future, I would only use it as a lightshade material. The color of the algae changes first from brown to bright green by

24 Curiel, D., et al. "The Introduced Alga Undaria Pinnatifida (Laminariales, Alariaceae) in the Lagoon of Venice." *Hydrobiologia*, vol. 477, no. 1, 2002, pp. 209–219.

25 Caliceti, M., et al. "Heavy Metal Contamination in the Seaweeds of the Venice Lagoon." *Chemosphere*, vol. 47, no. 4, 2002, pp. 443–454, https://doi.org/10.1016/s0045-6535(01)00292-2.

being heated. Then, with repeat exposure to sunlight, the algae eventually become colorless and clear. If the color should be preserved, then keeping it out of direct sunlight should be the goal.

2 Conclusion on Algae as a Material

Aboriginal and vernacular traditions around the world held algae in high regard as a material for use in craft and design. They each had their own traditions and processing techniques in order to utilize it and had specific cultural reasons for utilizing certain species over others. Specifically, algae were used as glue, utilized in both paint and plaster, and were a valuable raw material for many indigenous communities.

Today, much of western Europe has forgotten its own cultural connection to seaweed. However, by continuing to use algae in material ways in design and craft, we have a way to encourage new educational and emotional connections to algae.

For example, after I made my stool of algae and seagrass in Venice, I was inundated with many questions from the public, who found it difficult to believe that seaweed could become such a material. However, most asked if it was possible to eat it. Sea lettuce was used in several Italian dishes like Zeppolini di Alghe,[26] so the Italian people automatically associated my design with food first, and secondly as material.

For my thesis at TU Delft, I will continue prototyping with seaweed-based plaster, mussel shell cement, and seaweed glues and paints. At the end of my material experimentation, I will put my prototypes in an exhibition in the faculty, and once again examine the response to the new material tests I have made. However, my research does not stop with material experimentation. I will attempt to tie my experiments together in a design for a building, that is intrinsically connected to the Dutch landscape and mussel farming traditions of the Netherlands. By rooting algae in regional traditions like this, I hope that we in Western Europe can begin to accept it visibly in our day-to-day lives again.

References

Aalto University, director. *Julia Lohmann: 'We Know Too Much and Do Too Little.'* 2020, Accessed 14 Oct. 2021.

26 Pascarella, Marianna. "Frittelle Di Alghe." *Ricette Semplici e Veloci Di Marianna Pascarella*, https://www.ricettedalmondo.it/frittelle-di-alghe.html.

Aschoff, C. "Die Retro-Bibliothek: Nachschlagewerke Zum Ende Des 19. Jahrhunderts." *Retro*, https://www.retrobibliothek.de/retrobib/seite.html?id=103142.

Caliceti, M., E. Argese, A. Sfriso, B. Pavoni. 2002. "Heavy Metal Contamination in the Seaweeds of the Venice Lagoon." Chemosphere, 47: 443–454.

"Chondrus Yendoi." Algaebase, https://www.algaebase.org/search/species/detail/?species_id=3486.

Curiel, D., P. Guidetti, G. Bellemo, M. Scattolin, M. Marzocchi. 2002. "The Introduced Alga *Undaria Pinnatifida* (Laminariales, Alariaceae) in the Lagoon of Venice." Hydrobiologia, 477: 209–219.

"Hegisoba." Hegisoba – 【郷土料理ものがたり】, http://kyoudo-ryouri.com/en/food/603.html.

"Limfarve Til Vægge Og Lofter – SLKS." Information Om Bygningsbevaring, Kulturstyrelsen, https://slks.dk/fileadmin/user_upload/SLKS/Omraader/Kulturarv/Bygningsfredning/Gode_raad_om_vedligeholdelse/9.9_Limfarve_til_vaegge_og_lofter_.pdf.

Macfarlane, C.I. 1968. "The cultivation of seaweeds in Japan and its possible application in the Atlantic provinces of Canada". Nova Scotia Research Center, Halifax, Canada.

Matson-Green, V., S. Maiden. 2008. "Bernice Condie Keeps Her Ancestors' Traditions Alive." ABC (None) – Australian Broadcasting Corporation, 8 July 2008, https://www.abc.net.au/local/stories/2008/07/09/2298488.htm.

"Nori-Tsuchi: Clay, Sand and SEAWEED GLUE." 左官 Japanese Plastering, japanese-plastering.blogspot.com/2014/06/nori-tsuchi-clay-sand-and-seaweed-glue.html.

"Ocean Planet: Lesson Plan 1." *NASA*, NASA, https://seawifs.gsfc.nasa.gov/OCEAN_PLANET/HTML/education_lesson1.html.

Pascarella, M. "Frittelle Di Alghe." Ricette Semplici e Veloci Di Marianna Pascarella, https://www.ricettedalmondo.it/frittelle-di-alghe.html.

Thurstan, R.H., Z. Brittain, D.S. Jones, E. Cameron, J. Dearnaley, A. Bellgrove. 2018. "Aboriginal Uses of Seaweeds in Temperate Australia: An Archival Assessment." Journal of Applied Phycology, 30: 1821–1832.

"Traditional Uses of Funori – Tri-Funori." Tri-Funori, https://tri-funori.com/traditional-uses-of-funori/.

Turner, N.J. 2001. "Coastal peoples and marine plants on the Northwest coast". Int. Assoc. Marine Science Libraries and Information Center, 69–76. https://core.ac.uk/download/pdf/4167045.pdf.

Turner, N.J. 2019. Plant Technology of First Peoples in British Columbia. Royal British Columbia Museum.

Vadstrup, S. 2010. Center for Bygningsbevaring I Raadvad Anvisninger Til, https://www.bygningsbevaring.dk/uploads/files/anvisninger/14-ANVISN_Mosfarve.pdf.

Warwicker, M., A.L. Taylor. 2012. "Seaweed: Should People Eat More of It?" *BBC News*, BBC, 4 May 2012, https://www.bbc.com/news/magazine-17870743.

CHAPTER 7

Uses of and Considerations on Algae in Medieval Islamic Geography

Mustafa Yavuz

Abstract

Recent studies in the History of Botany put forth that the books translated to and authored in Arabic have circulated from the East of the Caspian Sea, to the centre of Iberian Peninsula, strengthening the 'traditional uses' of plants and alike. An ancient genre of writing called the 'book on the Materia medica' was especially the most favourite in Medieval Islamic Geography. In these books, algae have been mentioned among the kinds of medicinal plants. In this study, I investigate several Materia medica books among which I shall first focus on Avicenna's *Canon of Medicine*, then have a look at Aliboron's *Book of Pharmacy* since these two were contemporary sources from the 11th century. I shall also investigate two illustrated sources, an Arabic copy of Dioscorides' *Materia Medica* and al-Ghafiqi's *Book on Simple Drugs*, both from the 13th century. In doing so, on the one hand, I will be able to compare the drawings of algae, and on the other I will shed light on the transfer of knowledge on algae. These two methods will result in the *textual apparatus* and the *illustrative apparatus* in my study.

Keywords

algae – mosses – medicinal manuscripts – history of medicine – history of botany – history of pharmacy

1 Introduction

Until the discovery of 'inferior' organisms, in the classical meaning there were two kingdoms of taxonomy: *the plants* and *the animals*. In this division, plants were always primitive compared to animals. According to the Pseudo-Aristotelian *De Plantis*, the difference between animals and plants was the manifestation of vitality (being alive). Therefore, plants were regarded to

© MUSTAFA YAVUZ, 2024 | DOI:10.1163/9789004683310_009
This is an open access chapter distributed under the terms of the CC BY-NC 4.0 license.

be occult or not evident when it comes to the vitality compared to animals.[1] De plantis was regarded as the bible of plant-science and therefore translated to Arabic in the ninth century (Yavuz and Herraiz Oliva, 2020). Understandably, for many centuries during the Medieval, plants had been considered as the key organisms referring to the difference in the vitality between living and non-living things. Therefore, the Great Chain of Being was sequenced as minerals, plants, and animals etc. Similar to many other creatures that are neither animals, nor minerals, *algae* were considered as a part of the plant phenomena. What was the position of algae in the diversity of minerals, plants, animals, humans, angels etc. as viewed by medieval scholars? In other words, what kind of role were algae playing for a medieval natural philosopher, for a physician, or for a polymath?

The aim of this study is an investigation on Arabic medicinal sources authored by Medieval Scholars, in order to describe and understand how they received information on algae from the ancient lore and how much they transmitted to us. Regarding this aim, the structure of this study will be as follows. Section 2 introduces the types of books that deal with the *Plant Phenomena in Medieval Islamic Geography*. Considerably, algae are mentioned among the Simple Drugs since they were very similar to plants rather than animals or minerals. In fact, as we understand from Materia medica books from the antiquity and the Medieval, any living being, or any part of living beings may be regarded as a source of medicine. In Section 3, I offer a brief comparison of non-illustrated manuscripts namely *Works of Ibn Sina and al-Biruni*. First focusing on Ibn Sina's (Avicenna) *Canon of Medicine*, it being the prestigious book of medieval medicine in the East and in the West. Then I will look at al-Biruni's (Aliboron) *Book of Pharmacy*. These two books were contemporary sources from the 11th century. In Section 4, I will examine the two illustrated sources, an Arabic copy of Dioscorides' Materia medica and al-Ghafiqi's *Book on Simple Drugs*, both from the 13th century. In doing so, on the one hand, I will be able to compare the drawings of *Algae in the Illustrated Manuscripts*, and on the other hand, I will shed light on the transfer of knowledge on the algal phenomena. Finally, I will try to identify any algal genera or species if this is possible. Sections 3 and 4 are aligned in a chronological order to avoid anachronism. These sections provide the informative basis for discussion in Section 5 where I evaluate what is original or transmitted regarding our findings on the

1 In the famous piece of Ps.Aristotelian corpus De Plantis, Nicholas of Damascus says: "Vita in animalibus et plantis inventa est, in animalibus manifesta apparens, in plantis vero occulta, non evidens" (Meyer, 1841, 5).

uses of algae in eleventh and thirteenth century pharmaco-medicine in the Islamic civilization.

I assume that all these sections which provide a great deal information from Medieval Islamic Pharmaco-Medicine often with references to the antiquity, will provide epistemological and ontological basis on which we may question our knowledge of and in relation to the algae, in the contemporary world.

2 Plant Phenomena in Medieval Islamic Geography

Recent studies in the history of botany put forth that the ancient Greek studies on plants have been revived in Arabic, through a transfer or a transform of knowledge. Consequently, from agricultural studies to the philosophical ones, there were discussions on the plant phenomena here and there, which yielded the formation of (علم النبات / Ilm al-Nabat) *the Plant Science*.[2] The books translated to and authored in Arabic have circulated from the East of the Caspian Sea to the centre of Iberian Peninsula, strengthening the 'traditional uses of' and 'knowledge of' plants and alike.

As discussed by Yavuz & Herraiz Oliva (2020), we can group medieval plant studies under four (or more) categories, all of which have their roots in the antiquity. Among these, the first genre *Kitab al-Filaha / Book of Agriculture* focused on the plantation, growth and harvesting of plant products, and in doing so; preserved, provided, and distributed the knowledge for the nutritional needs of humankind. The second genre named *Kitab al-Nabat / Book of Plants* did the same for the general investigation of plant phenomena and their dissimilarity from the animals by discussing different plant parts and habitats of plants, thus satisfying the philosophical needs of intellectuals. The third genre consisted in books on occult sciences, which were never studied in traditional Islamic Medieval education system. The fourth genre, *Kitab al-Adwiyyat al-Mufradah / Book of Materia medica* provided knowledge on medicinal plants which corresponded to a guidebook for pharmacists and herbalists. This ancient genre of books was especially the most favourite one in the Medieval

2 It should be noted for the general audience that, here the correct word is knowledge instead of science. However, for practical purposes, refraining the anachronism, I prefer to use "science" through this chapter.

Islamic Geography.[3] In these books, we see that algae have been mentioned among the kinds of medicinal plants.

When we examine Arabic lexicons and dictionaries for an equivalent of algae, we find the term طحلب tuhlub (طحالب tahaleb in plural). This word was used by authors of Medieval Islamic Geography in order to denote *water-moss, seaweed*, and even some kinds of aquatic plants. It is possible in some cases to identify a genus or species of algal phenomena with an investigation of Arabic sources of Islamic Medieval based on two different apparatuses. The first is the *textual apparatus* which is mostly an investigation and a comparison of Arabic texts. As mentioned below, in this study the text from four different books will be examined. Probably, these texts will have parallel approaches especially when they refer to the ancient Greek or Roman authors. The *illustrative apparatus* is the analysis and comparison of botanical illustrations if any. Since it has been harder to draw a botanical element exactly as it is, rather than to describe it by words, the precise copy of such a manuscript cost a lot moreover, most books from medieval ages lack illustrations.

3 Works of Ibn Sina and al-Biruni

3.1 *Ibn Sina*

He is known shortly as *Avicenna* by Christians and Jews, and *Ibn Sina* by Muslims. To mention his full name, Abu Ali al-Husayn ibn Abd Allah ibn al-Hasan ibn Ali ibn Sina (980–1037), we should also add that he had a reputation as 'Prince of Physicians', since he was one of the most famous physicians and philosophers in Medieval Islamdom. The most important work that Ibn Sina authored in the field of medicine is undoubtedly his encyclopaedic book called *al-Qanun fi al-Tibb* (*Canon of Medicine*).

Written in Arabic, the lingua franca of the Islamic geography, this work was the standard medical textbook used and commented on, for many centuries in Islamdom. It was translated into Latin, the common language of the Christian geography in the early periods. Thus, it was accepted as a major source among physicians until the eighteenth century (McGinnis 2010, p. 251). The *Canon of Medicine* consists of five books with the following content: *Kulliyat* (*Principles*) is the general discussion of the scientific and philosophical foundations of

3 I prefer to use Islamic Geography; however, it is possible to use Islamdom as well. This is to mention that in a certain territory, the rulers or the governing characters were Muslim, there were many religious groups like Muslims, Christians, Jews, Sabaens, Nabatans, Zoroastrians etc. involved in the intellectual and medicinal studies.

medicine and anatomy. *Mufradat* (*Simple drugs*) focuses on approximately 800 herbal, animal and mineral substances used as independent medicine, namely materia medica. *Mualajat* (*Pathology*) which mentions the specific or localized ailments for various diseases. *Hummiyat* (*Fevers*) deals with more general diseases, such as fever, that affect the whole body. Finally, *Murakkabat* (*Compound drugs*) is about pharmacology, drugs, and consists of medical prescriptions. In his *Canon of Medicine*, Ibn Sina created a system of medicine based on the three pillars: *Natural Philosophy* from Aristotelian, knowledge of practical and theoretical medicine from Galenic, and finally pharmaco-botanical knowledge from Dioscoridean traditions. In his synthesis of medicine, nevertheless, Ibn Sina consulted opera of many other physicians and cited by name including (but far from being limited to) *Hippocrates*, *Abu Bakr al-Razi* (Rhazes in Medieval Latin), and *Ali ibn Abbas al-Majusi* (Haly Abbas in Medieval Latin).

al-Qanun fi al-Tibb already had a great number of circulating handwritten copies in Islamic geography, until it was printed for the first time in Rome, in the sixteenth century. From the time of Pope Gregory XIII, press permission for the printing of any book was given by the highest political authority in the country (Witcombe 2004, pp. 73–74). As can be seen from the phrase *cum licentia superiorum* on the front cover, the necessary permission and support for the printing of this important work was provided by the Papal Authority (Figure 7.1). Even in 1772, there were still more than eight hundred copies of *al-Qanun* in the warehouses of the Medici Printing House (Toomer 1996, pp. 20–25). This number may be an indication of how many copies of the work that had been produced, if it was not a commercial mistake. In the sixteenth century, the preparation for Arabic-print-edition of *al-Qanun*, required both a text review and some technical skills accompanied by certain equipment. Among the names who were involved in this process, the *al-Qanun Printing Project*, we should first mention *Giovanni Battista Raimondi* (1536–1614) as the manager of the printing house and especially the head of the project. *Robert Granjon was* the person who prepared the Arabic fonts.

The typography preparation of a work to be printed in the Arabic alphabet is more difficult compared to that in Latin, since most letters of the Arabic script require separate types for their initial, middle, trailing and independent positions. This *al-Qanun Printing Project* was initiated in 1584 by *Cardinal Ferdinand de Medici*, under the auspices of *Pope Gregory XIII*. After nine years of preparation and five successive popes, the publication of the work was completed during the reign of *Pope Clemens VII* (Siraisi 1987, pp. 148–150). The project was led by Raimondi; however, *Giovan Battista Lucchese* and *Paolo Orsini of Constantinople* were in the team as well. In addition, Patriarch

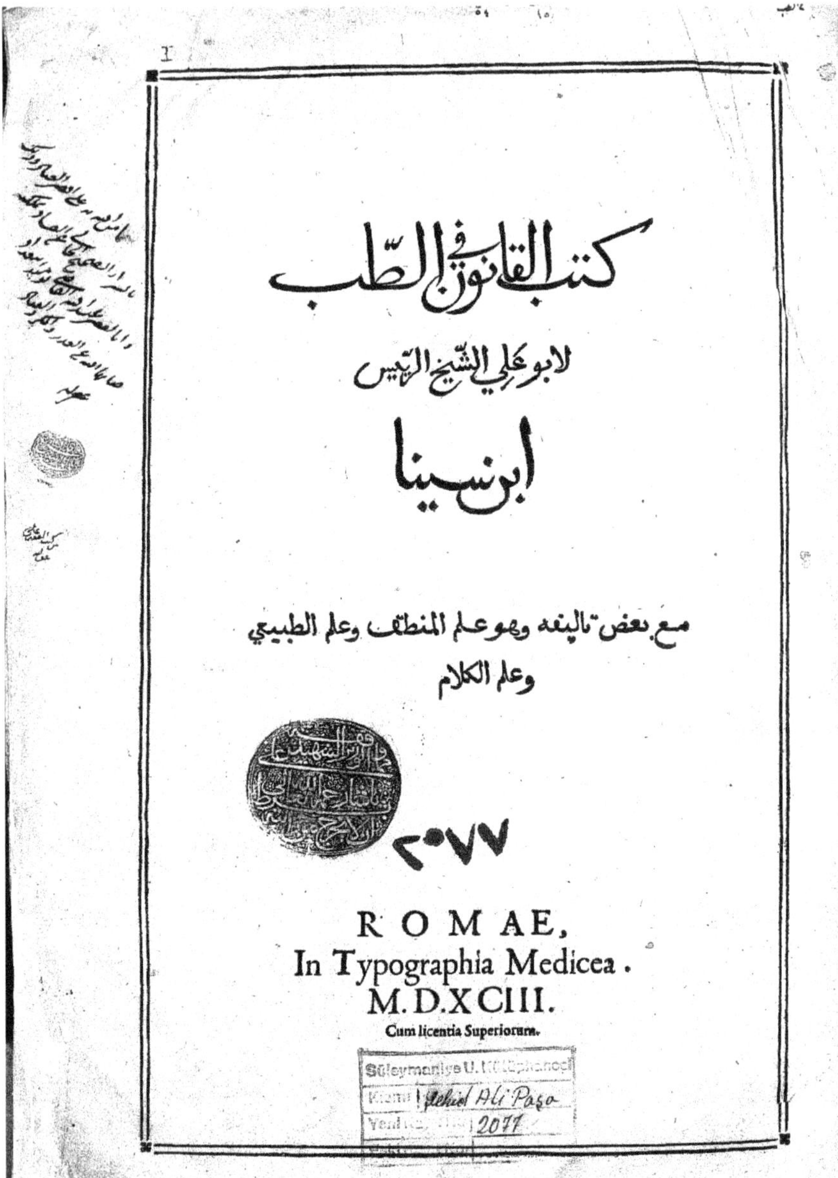

FIGURE 7.1 Front cover of 1593 Rome Edition (Süleymaniye Library, Şehid Ali Paşa 2077)

Ignatius Ni'matallah Asfar Mardini[4] provided the core manuscripts and was present in the proofreading and translation (Casari 2016, Vol. 86, pp. 221–224).

Tuhlub, the seaweed or alga is mentioned in الكتاب الثاني *the second book*, الجملة الثانية، في بيان الأدوية المفردة *second part, on the materia medica*, حرف الطاء under the drugs starting with the letter *Ta*. In this study, the Arabic texts I have included below are derived from the Rome edition (1593, p. 183). I have also compared the text from the relevant chapter of al-Dinnawi (1999, p. 501). Then explanations and corrections in the footnotes had been added wherever it was necessary. The differences from 1593 edition are given in footnotes with an indication of letter D. The English translation is mine otherwise noted, I have included some words in brackets (), to support the meaning.

English Translation:	Arabic Text:
Tuhlub	طحلب 11.300
Quiddity: The riverine (variety is of) water and earth (element type). The marine (variety) is very astringent. And the rock-moss which is *Hazaz al-Sakhr* (tetter of rocks), we have mentioned (it above).	الماهية:[5] النهري مائي أرضي، والبحري أشد قبضة. وأما طحلب الصخر وهو حرار الصخر[6] وقد ذكرناه.
Nature: (It is) cold.	الطبع: بارد.
Properties: It is styptic for blood (or bleeding) in any place it is painted. Marine (variety) is most severe (in this respect).	الخواص: حابس للدم في كل موضع طلاء، والبحري أشد.

4 Mar Ignatius Nemet Aloho I (Mardin, ca. 1515 – Bracciano, 1587), was the Syro-Orthodox Patriarch of Antioch (modern Antakya, Turkey) between 1557–1576. Settled in Rome in 1577, he personally delivered the manuscripts of *al-Qanun fi al-Tibb* to Cardinal Ferdinand (Ferdinando I de Medici, Grand Duke of Toscany). *Ni'matallah* played a key role in the connection between Rome and the Levant in the sixteenth century onwards. He was a physician, astronomer, and theologian; thus, he engaged in the calendar reform promoted by Pope Gregory XIII, the establishment of a printing press in Arabic and Chaldean, and he worked to unite the Syrian Orthodox and Roman Churches. More information on his life and works are available at Toomer (2016, pp. 20–25); Borbone & Farina (2014) and Casari (2017) studies; moreover, the involvement of *Typographia Medicea* in the Renaissance – Levant connection, there is an outstanding work authored by Farina & Fani. (2016).

5 D: الماهية: معروف، و Quiddity: (It is) known and.

6 In 1593 Rome edition, this drug is given (in plural form) as حرار الصخور. Probably due to a typo error, the name is misgiven. It must be corrected as حزاز الصخر (singular) or حزاز الصخور (plural). A literal translation of these words means tetter of rock(s), which denotes saxicolous lichens and/or mosses.

Tumours and Blisters: It is applied (in cases of) hot tumours, erysipelas tumours, and eczemas. The lenticular weed, (if used) with fine flour, is similar (in this effect).	الأورام والبثور: يجعل على الأورام الحارة والحمرة والنملة، وكذلك العدسي من الطحلب مع السويق.
Organs of Articulation: (It is also applied) on vehement gout and severe arthralgia. If / when decocted with old olive oil, it softens the nerves.	آلات المفاصل: على[7] النقرس الحار وأوجاع المفاصل الحارة، وإذا أغلي بالزيت العتيق لين العصب.
Excretory Organs: (When) Bandaged with it, the haematocele (or hydrocele) of the intestines is atrophied.	أعضاء النفض: يضمد به قيلة الأمعاء فيضمرها.

3.2 *Al-Biruni*

He is known as *Aliboron* by Christians and Jews, and *al-Biruni* by Muslims. His full name was Abu al-Rayhan Muhammad ibn Ahmad al-Biruni. He was born in Khwarazm (in Uzbekistan) in 973 and he died in Ghazna (in Afghanistan) ca. 1050 (Kennedy, 1970, pp. 147–158). In fact, al-Biruni was famous for his works on astronomy, mineralogy, and ethnography. However, his final opera was about materia medica. He died soon after completing his *Kitab al-Saydanah fi al-Tibb*, which is "The Book of Pharmacy in Medicine."

Kitab al-Saydanah is organized in a form which consists of an introduction and five fasl (فصل). In the first *fasl*, al-Biruni discusses possible etymological origins of the word *Saydanah* and he explains the reasons to use this word in the title of the book. In the second *fasl* he compares foods, drugs, and poisons, while in the third he discusses the substitution between different kinds of materia medica. The fourth *fasl* is devoted to mention Arabic as a proper language for philosophy and medicine. In the last *fasl*, al-Biruni mentions his sources, namely Dioscorides, Galen, Paul of Aegina, and Oribasius, whose books were available in Arabic. Al-Biruni mentions 890 items of Materia medica in Arabic, Greek, Latin, Syriac, Persian, and Sanskrit. *Kitab al-Saydanah fi al-Tibb* has been studied less than Avicenna's *al-Qanun fi al-Tibb*. Two codices are known to us, the first Bursa[8] codex consists of 134 folios and the second Baghdad codex[9] that of 209. An edition and an English translation were produced by

7 D: و على and on/over.
8 Bursa Inebey Manuscript Library, Kurşunlu Collection No: 149.
9 Baghdad Iraqi Archaeology Department No: 191.

Hakim Mohammed Said (1973), with the assistance of Rana Ehsan Elahie, Kamal M. Habib and L.A. D'Silva (vol. 1) together with a commentary by Sami Khalaf Hamarneh (vol. 2). This work is well-known as Hamdard edition. The differences of Bursa manuscript from Hamdard edition are given in footnotes with an indication of letter **B.**

English Translation:

Tuhlub

Dioscorides: "The greenery (vegetation) similar to lentil and lying over the water, is duckweed.[10] *Tuhlub al-Bahri* (seaweed) is the thing which is found on stones and reef that are close to the sea." (It is) thin, similar to hair in thinness, it does not have a stalk. Moisture and dew occur on it. It is said that *Tuhlub al-Barri* is named as *Jamah ghawk*.[11]

Sanawbari[13] said about it in *"Nehr Quyaq"*:
"Whence the frogs cry: Quyaq! Quyaq! So that we reply,
They settle in the residue of summer tuhlub, vestment tunic."

Arabic Text:

[85^{r}]طحلب 662

ديسقوريدس: اما الخضرة الشبيهة بالعدس القائم فوق الماء فهو عدس الماء. واما الطحلب البحري فهو شيء يكون على الحجارة والجرف[12] الذي يقرب من البحر، دقيق شبيه بالشعر في الدقة لا ساق لها، يقع عليه من الندي والطل. ويقال الطحلب البري يسمى جامه غوك.

وقال الصنوبري في نهر قويق.
إذا ما الضفادع ناديته قويق قويق[14] الي أن يجيبا،
فيأوين منه بقايا كسين من طحلب الصيف ثوبا قشيبا.

10 In Arabic Materia medica books, عدس الماء (*Adas al-Ma, Water-lentille*) is probably a member of Lemna L. genus, namely common duckweed. Genus Lemna has species that are known as free-floating aquatic plants, but not algae.

11 The phrase: جامه غوك means "raiment of the frog".

12 B: الجوف hollow, cave, cavity.

13 His name is Abu Bakr Muhammad ibn Ahmad ibn Husayn al-Dabbi al-Sanawbari. He was born in Antioch about 895 he died in 945. He is often considered as the first Muslim poet of nature, landscapes, trees, gardens, etc. Probably al-Biruni was interested in his opera (Said, 1973).

14 B: writes فويق فويق however, it must be قويق قويق meaning the sound and call of frogs.

Galenos: Bryon thalassion, Tuhlub al-Barri.[15] In another topic, mentions (as) water-lentil and says that it is *tuhlub*. The one that Dioscorides proceeded was the lichen of the rocks.[16] As for *tuhlub* according to the Arabs, it is this (thing) that floats on water.

Dhu al-Rummah[18] said:

"A fountain, environs (of which) covered with tuhlub, overflowing,
Where the frogs and the fish clamour."

Ibn Durayd[19] said: When tuhlub gets old, it dries then it becomes peat. *Darast al-Sahrat* (is the rock that) becomes green by the accumulation of Tuhlub.

جالينوس: برواون ثلاسيون الطحلب البري.[17] وفي موضع آخر ذكر عدس الماء وقال هو الطحلب. كان الذي يسير اليه ديسقوريدس هو حزاز الصخور. فأما الطحلب عند العرب هو هذا الذي يطفو فوق الماء.

وقال ذوالرمة:

عينا مطحلبة الأرجاء طامية،
فيها الضفادع والحيتان تصطخب.

ابن دريد: إذا قدم عهد الطحلب فيبس فهو الخث. ودرست الصخرة اخضرت من طحلب ركبها.

---

15 It should be Tuhlub al-Bahri, see the footnote 17.

16 In the Arabic texts, حزاز الصخور (hazaz al-sukhur, tetter of rocks) or حزاز الصخر (*hazaz al-sakhr, tetter of rock*) denote saxicolous lichens, in allusion to the skin disease called lichen planus today. For further discussions see Yavuz (2018) and Yavuz (2020) studies.

17 Arabic text mentions الطحلب البري (Tuhlub al-Barri, Terrestrial tuhlub) here, however, it should be الطحلب البحري (Tuhlub al-Bahri, Marine tuhlub), since Galen mentions θάλασσα, the sea.

18 Dhu al-Rummah, Abu al-Harith Ghaylan ibn Uqbah ibn Rabiah (d. 736), was a famous Bedouin poet. According to Sami Khalaf Hamarneh (Said 1973), he often visited spice shops that this explains his knowledge on spices he mentioned in his poems. More information on his poetry and especially his romantic verses has been given by Abbott (1972, pp. 164–202).

19 Ibn Durayd, Abu Bakr Muhammad ibn al-Hasan al-Azdi (d. 933), was one of the most celebrated hexicographers, and grammarians (Said 1973).

Al-Jahiz:[20] Dried tuhlub does not ignite by fire, and neither does some wood in Kerman,[21] and (of this wood) it was a cross on the neck of a monk; the people were fascinated with it and claimed that it was from the wood of the (original) Cross. It (Kerman) was said to be near Khabis.[22]

الجاحظ: الطحلب المجفف لا يحترق بالنار وكذلك عود بكرمان وقد كان منه صليب في عنق راهب يفتتن به الناس ويزعم أنه من خشبة الصليب. قيل هو بقرب خبيص.

Abu Tammam:[23]
"You showed me the skin of water, which
I thought to be lots of tuhlub."
Tuhlub around the water is green. What I saw on top of the water, have layered like a spider's web, is *armad* (عرمض). It is said that *awafiq* (عوافق) is the greenery (vegetation) that rises on water, and it is *armad* and *tuhlub*.

ابو تمام:
ابديت لي عن جلدة الماء الذي
قد كنت اعهده كثير الطحلب
الطحلب ما حول الماء اخضر. والعرمض[24] ما رأيته على رأس الماء مثل نسج العنكبوت قد طبقه. قيل العوافق[25] الخضرة التي تعلو الماء وهو العرمض والطحلب.

Paulos (Aeginata): Tuhlub al-Bahri (seaweed) is found on trees (like) oak, walnut, and male pine from which the tar is derived.[26]

بولس: الطحلب البحري يوجد في شجر البلوط والجوز والصنوبر الذكر الذي منه القطران.

20 Al-Jahiz, Abu Uthman Amr ibn Bahr ibn Mahbub (d. 869), the famous prolific author who studied Aristotelian philosophy and wrote especially on zoological, rhetorical, and sociological topics (Said 1973).

21 Kerman or Kirman (Persian: کرمان) is the capital city of Kerman Province, Iran.

22 Khabis (Persian: خبیص) is a village in of Firuzabad County, Fars Province, Iran.

23 Abu Tammam, Ḥabib ibn Aws al-Tai (d. ca. 850), was an Arabian poet and Muslim convert. For more information on his biography, Thatcher (1911, pp. 1–81) should be visited.

24 Persian, a green film, or moss that floats on the surface of stagnant water.

25 This word عوافق may be a broken version of غلفق it may be caused by a copy-error.

26 Paul of Aegina (d. c. 690) a Byzantine Greek physician. His opera is called Ἐπιτομῆς Ἰατρικῆς βιβλία ἑπτά, Medical Compendium in Seven Books. In his compendium VII.3.128, we read Bryon thalassion, the algae and Bryon sphlanchnon, the mosses and lichens:

"Bryon thalassion, refrigerat et adstringit inde im positum feruentes phlegmonas iuvat. Bryon sphlanchnon, quod in quercibus et populis albis et piceis inucnitur, distrahendisimul et mollien di habet potestatem mediocriter. Praecipue quod cedris adheret" (Brunfels 1531, p. 14). Regarding this text we can conclude that the Arabic authors have mentioned mosses/lichens instead algae or in other words, Bryon sphlanchnon was referred instead Bryon thalassion as a mistake.

Poem by al-Buhturi:[27]	شعر للبحتري:
"Until the dawn appears from its sides,	حتى تبدى الصبح من جنباته،
Like water shining from behind the tuhlub."	كالماء يلمع من وراء الطحلب.
Abu Hanifah:[28] *Armad* and *ghalfaq* are the vegetation that spread over water, if present in the periphery, it is tuhlub.	ابو حنيفة: [85v] العرمض[29] والغلفق[30] هو الاخضر الذي يتغشى الماء وإذا كان في جوانبه فهو الطحلب.
Al-Mashahir:[31] *Armad* is thicker than tuhlub.	المشاهير: العرمض ما غلظ من الطحلب.

4 Algae in the Illustrated Manuscripts

This section is divided into two titles. The first is about Dioscoridean tradition which focuses on the Arabic translation of Materia medica for the entries of algae, while the second is about the illustrated work of an Andalusian physician: al-Ghafiqi.

4.1 *Dioscoridean Tradition*

Pedanios Dioscorides was born in Anazarba (Adana, Turkey) in the first century. He authored a book called Περί ύλης ιάτρικης (Peri Hyles Iatrikes) which is well known in the Latin language as *De Materia Medica*.[32] Based on the previous

27 Al-Buhturi, Abu Ubadah al-Walid ibn Ubayd ibn Qahtan al-Tai (d. 897), a picturesque poet and follower of Abu Tammam (Said 1973).

28 Dinawari, Abu Hanifah Ahmad ibn Dawud ibn Wanand (d. 895), Persian polymath who authored Kitab al-Nabat, Book of Plants.

29 B: الغومض.

30 Persian, water-moss.

31 Al-Mashahir is a book, al-Biruni sometimes mentions "author of al Mashahir", or "compiler of al-Mashahir", however, at the moment, it is not clear which book is this and who is its author.

32 It was first translated to English by Goodyer (1655), printed by Gunther (1934) and there are many prints and reprints until Osbaldeston's (2000) illustrated translation. However, the recent translation of Beck (2005) also gives the Greek titles which are useful to follow the nomenclature of materia medica.

writings, local traditions, and personal experience. Dioscorides compiled his treatise consisting of more than 800 chapters mostly on plants, minerals, and animals (Riddle, 1980; Scarborough, 2012). De Materia medica is the result of his extensive journey as a military physician of the Roman army (Riddle 1985, pp. 2–4; Scarborough and Nutton, 1982: pp. 213–217), we can trace his geographical references especially in Asia Minor. It is not necessary to mention his prestigious impact on Medieval Arabic and Renaissance Latin authors who wrote on medicinal plants. Thus, his De Materia medica has been the leading figure and the studies on it constructing a tradition in the history of botany and that of pharmacy. The content has been continuously copied, translated, and transmitted from one language to another. Circulation of knowledge on medical substances in and around the Mediterranean basin has been accelerated after Dioscorides. Riddle (1985) has a masterpiece on Dioscorides' biography. However, Touwaide (1999, 2009) has excellent analyses on Dioscorides' De Materia Medica, and the Dioscoridean Tradition. Dubler (1991, Vol. II, pp. 349–350) discusses a brief but informative account of Dioscorides in Islamic Medieval.

De Materia medica consists of five books, divided into different subjects: The first book covers aromatics, oils, ointments, trees, liquors, gums, and fruits. The second, animals, cereals, milk and milk products, herbs, spices, grains, resins, oils, ointments, trees, and fruits. The third, roots, weeds, herbs, juices, seeds. The fourth roots, weeds, and herbs. Finally, the fifth book mentions drinks, vines, wines, and inorganic materials.

In this section of the study, I have examined four manuscripts which are Arabic copies of Dioscorides' De Materia Medica, that should be cited as follows: Süleymaniye Library, Ayasofya 3702 and 3703 (will be shortened as A 3702 and A 3703), Bibliotheque National de France, Arabe 2849 (BnF 2849) and finally the British Library, Oriental Manuscripts, Or 3366 (Or 3366). The differences among the codices are given in the footnotes. In the right column, I have added the Arabic text which is derived from the above-mentioned manuscripts. In the middle column, I supply an Arabic-English translation. In the left column, I have added the Greek-English translation of the relevant text by Beck (2005). In doing so, I expect that the reader would find ease in comparing what the Arabic translation says and whether or not the Arabic text follows the original one. Dioscorides' De Materia medica has three different entries: φακός ὁ ἐπί των τελμάτων, βρύον θαλάσσιον, and φῦκος θαλάσσιον.

Beck's Translation:
φακός ὁ ἐπί των τελμάτων

The duckweed, which grows on stagnant waters, is a marsh-plant like the lentil, and since it has a cooling property, it is a suitable plaster by itself as well as with barley groats for every kind of inflammation, for erysipelas, and for gout. It also heals children's hernias.

Arabic-English Translation:
Faqus

It is Tuhlub. It is found on stagnant waters. It is greenery (vegetation) similar to lentil in its form. It is found in thickets on stagnant water. It is cold. Thus, if / when it is bandaged alone or with fine flour, it is suitable for hot tumours and for gout. If / when it is bandaged for the haematocele (or hydrocele) of the infected intestines of infants, (it is suitable) either adhered or bandaged.

Arabic Text:
ذكر فاقس

هو الطحلب. الموجود على الما القائم. وهو الخضرة الشبهة بالعدس في شكلها. الموجودة في الاجام و[33]المياه القائمة. وهو بارد. ولذلك إذا يضمد به وحدة او مع السويق وافق الحمرة والأورام الحار والنقرس. إذا ضمد به قيلة الامعاء العارضة للصبيان ألصقها[34] اي[35] أضمرها.

Beck's Translation:
βρύον θαλάσσιον

The sea lettuce, it grows on rocks and on shells by the sea. It is lettuce-like, slender, stemless, quite astringent, and effective for

Arabic-English Translation:
Bruun balasiyus

It is Tuhlub al-Bahri (seaweed). It is the thin that becomes (groves) on rocks and reefs that are near the sea. It is thin, similar to the thinness of the hair. It does not

Arabic Text:
ذكر بروون بالاسيوس[36]

وهو طحلب البحري. هو شي يكون على الحجرة[37] والجرف التي يكون[38] بالقرب من البحر. وهو دقيق شبيه في دقة بالشعر. وليس له ساق. وهو قائم[39]

33 BnF 2849: على.
34 BnF 2849: - A 3703: الرقها ascended.
35 BnF 2849: -.
36 A 3703 starts with an epi-title ذكر الطحلب البحري – On Tuhlub al-Bahri. Then the secondary title reads as تروون بالاسيوس -truun balasiyus instead بروون بالاسيوس -bruun balasiyus. This title is a broken version of βρυον θαλασσιον.
BnF 2849: The title is بروون البحري.
37 BnF 2849: الحجارة.
38 A 3703, BnF 2849: -.
39 BnF 2849: قباض.

inflammations and for gouts that need cooling.

have stalk. It is actually very cooler. It heals hot tumours and those that need cooling because of gout.

مبرد[40] جدا. ويصلح للأورام الحارة وللمحتاج الي التبريد من النقرس.

Beck's Translation:

φῦκος θαλάσσιον

Seaweed, there is one kind that is broad, another that is longish and somewhat purple, and another that is curly growing in **Crete**, near the shore; it is very colourful and not prone to decay. All of them have an astringent property and are good in plasters for gout and for other inflammations. But

Arabic-English Translation:

Fuqus

Fuqus al-Bahri: It is counted in types among which (there is) the one that is broad, the one that is long, and its colour is reddish / purplish, the one that is curly, growing in two places in the island called **Qriti**.[41] It is very beautiful in colour, and it does not decay. These types all have astringent potency and (they) suit in case if (they) are bandaged for gout and for similar hot tumours. It

Arabic Text:

ذكر فوقس

فوقس البحري: وهو عده أصناف منه[42] ما هو بالعرض[43]، ومنه ما هو بالطول[44] ولونه الي الحمرة، ومنه جعد. وينبت عند الأرضين[45] في الجزرة التي يقال لها[46] قريطي حسنا الزهر[47] جدا وليس يعفن.[48] وقوة هذه الأصناف كلها قابضة ويصلح ان[49] تضمد[50] بها للنقرس[51] وسائر الأورام

40 BnF 2849: - .

41 Crete, we find that before Dioscorides, Theophrastos mentions Cretan algae in his Historia Plantarum 4.6.5. We learn that in Crete, φῦκος grows on the rocks near the land and people use the dye obtained from this seaweed to dye wool and clothes. Theophrastos devotes a chole chapter (H.P. 6) to the kinds of aquatic plant phenomena including most algal species and perhaps some lichens of the Mediterranean, which should be investigated in a further study.

42 A 3703, BnF 2849: فمنه.

43 BnF 2849: الى العرض.

44 BnF 2849: الى الطول.

45 BnF 2849: الأرض.

46 BnF 2849: - .

47 A 3702, A 3703: لزهر.

48 Both codices A 3702 and A 3703 reads something like "بعض – some", however, this word is not meaningful according to the context. Probably, due to a copyist error, it was broken.

49 BnF 2849: -.

50 A 3703: ليتضمد, BnF 2849: ليضمد.

51 BnF 2849: النقرس.

they must be used while moist and before they have dried up. **Nicander** says that the purple is also an antidote for poisonous animals. Some thought that this is the one the women use, that being a tiny root, and which is called similarly seaweed.

is necessary to use these types (while) moist before they dry. **Niqidres**[52] may claim that the one whose colour is red / purple is suitable for the most harmful of poisonous creatures, and some people think that this class / type is the one used by women, rather, it is a small root / origin that associates these types by name only.

الحارة. وينبغي ان يستعمل
هذه الأصناف وهي رطبة
قبل ان يجف.
وقد يزعم نيقيدرس ان الذي
لونه الي الحمرة يصلح اضرر
ذوات السموم، و من الناس
من ظن ان هذا الصنف هو
الذي يستعمله النسا و الذي
يستعمله النسا انما هو اصل
صغر يشرك هذه الأصناف
بالاسم فقط.[53]

After giving the textual material, I will look at the images in the manuscripts if any. Information on the illustrative account can be summarized as follows: Dioscorides' φακός (IV, 87) is mentioned in folio 109^{a} of BnF Gr. 2179 (Figure 7.10), the Arabic translation فاقس is in 96^{b} of A 3702, 92^{a} of BnF Ar. 2849, 141^{b} of Or 3366 (Figure 7.4), and in 126^{b} of Bodleian d 138 (Figure 7.7). βρύον θαλάσσιον (IV, 98) is given in 112^{b} of BnF Gr. 2179 (Figure 7.11), the Arabic translation بروون بالا سيوس is in 98^{b} of A 3702 (Figure 7.2), 92^{b} of BnF 2849, 143^{b} of Or 3366 (Figure 7.5), and in 129^{a} of Bodleian d 138 (Figure 7.8). Dioscorides' φῦκος θαλάσσιον (IV, 99) is mentioned in folio 112^{b} of BnF Gr. 2179 as well as

52 Here Dioscorides mentions Νίκανδρος ὁ Κολοφώνιος / Nicander of Colophon who was a famous poet in the 2nd century BC. His book Theriaca is about venoms and venomous animals. In Theriaca 845, he mentions: "*ἢ ἔτι καὶ φοινίσσον ἁλὸς καταβάλλεο φῦκος*" where he enumerates various plants as remedies against the strokes of some animals like moray, sea-snake, and stingray. φῦκος – according to Nicander – is a seaweed known as its purple colour or dye. A recent study by Overduin (2014) lays the Greek text of Theriaca with commentary.

53 The sentences in red appear neither in A 3702 nor A3703, but only in BnF 2849. This shows us that there were several lines of transmission in the Dioscoridean tradition in the Islamdom. Further studies would allocate a stemma of the known codices. In this text, the bold text shows repeating words, and this repetition may be a copyist mistake that we sometimes come across in handwritten books.

the Arabic translation فوقس is in 99ª of A 3702 (Figure 7.3), 92ᵇ of BnF 2849, 137ª of Or 3366 (Figure 7.6), and 129ᵇ of Bodleian d 138 (Figure 7.9). Below are the images from some Dioscoridean manuscripts, both in Arabic and Greek. Please note that some codices are not illustrated at all, and some do not have figures of every item.

4.2 *Al-Ghafiqi*

Abu Jafar Muhammad ibn Ahmad al-Ghafiqi (d 1165) was a famous physician and pharmacist of al-Andalus. We do not have much information on his biography. Moreover, his encyclopaedic compilation does not have many copies compared to those of Ibn Sina or Ibn al-Baytar. al-Ghafiqi's *Kitab al-Adwiyyat al-Mufrada* (Book of Simple Drugs) was studied by Di Vincenzo (2009) in her PhD thesis and the first part was edited. Al-Ghafiqi's opera was facsimiled and introduced by a number of specialists as well (Ragep et al 2015). According to the codicological data derived from the Osler codex, this manuscript was transcribed in 654 AH / 1256 AD. There are more than 450 illustrations in the manuscript. Al-Ghafiqi uses the abjad order to mention the Materia medica (Gacek 2015). In the entries, he uses abbreviations instead of famous sources known to him. Among these, د means Materia medica of Dioscorides, ج that of Galen and ح al-Hawi (Continens of Rhazes). If there is a second letter, this letter depicts the number of the book (actually sections of the book). For instance, د د depicts Dioscorides' 4th book, while ج و Galen's 6th. In the table below, the right column gives the Arabic text and the left an English translation.

In al-Ghafiqi's Kitab al-Adwiyyat al-Mufrada, there is a chapter on algae which is as follows under the title طحلب – Alga. The main text is copied from Osler codex (O: McGill University, Osler Library 7508), whereas the differences in other codices are given in the footnotes. Here are the abbreviations of codices. **M** denotes the codex in Library of Majlis-i Malek, 5958 (Iran), and **T** of that in Bibliotheque Nationale 18177 (Tunis).

English Translation:

Tuhlub

Dioscorides IV: Faqus, Tuhlub al-Nahri (river-moss) is the greenery (vegetation) similar to lentil in its form, found in thickets on standing (stagnant) water.

Al-Hawi (Continens of Rhazes) VIII: Its temperament is a humid (and) cold temperament, and it bears both characteristics as if it is in the third degree.

Dioscorides: Likewise, if / when it is bandaged alone or with fine flour, it is compatible with erysipelas, hot tumours, and gout. (It is) Bandaged for the haematocele (or hydrocele) of the infected intestines of infants. As for Tuhlub al-Bahri (seaweed), it is something found on stones and cliffs that are near the sea. And it (seaweed) is thin, similar to hair in thinness, it doesn't have a stalk.

Arabic Text:

طحلب

د د فاقوس[54] الطحلب النهري هو[55] الخضرة الشبهة بالعدس في شكلها الموجودة في الاجام على الماه[56] القائمة.

ح ح مزاج هذا مزاج رطب بارد وبير[57] من الخصلتين كانه في الدرجة الثالثة.[58]

د وكذلك[59] إذا ضمد[60] به وحدة او مع السويق وافق الحمرة والأورام الحارة و النقرس[61] و اذا ضمدت[62] به قيلة[63] الامعاء العارضة للصبيان اضمرها و اما[64] الطحلب البحري فهو شيء يكون[65] علي الحجارة و الجرف[66] التي بقرب[67] البحر و هو[68] دقيق شبيه في دقته الشعر و ليس له ساق.

---

54 T: هو فاقوس و Faqus and it is.
55 T: و هو and it.
56 M, T: المياه.
57 T: و هو يبر.
58 T: ثانيا.
59 M: لذلك. هو بارد و T: و لذلك
60 T: تضمد.
61 T: و ينفع النقرس it relieves gut.
62 T: ضمد.
63 M: قيتله.
64 T: اما يربون it invades, it dominates.
65 T: يتكون.
66 M: الخزف.
67 T: تكون بالقرب من.
68 T: وهو ثمنش.

Galen VI: The capacity of this plant[69] is composed of an earthy quintessence and a watery quintessence both of which are cold, and therefore its taste is astringent and it is cold, and if a bandage is made of it, it will be of clear benefit from all hot ailments.[70]

ح و[71] هذا النبات قوته مركبة[72] من جوهر ارضي وجوهر مائي وكلاهما بارد و لذلك ان طعمة قابض و هو برد[73] و اذا عمل منه ضماد نفع من جميع العلل الحارة نفعا بينا.[74]

---

69 Please note that in pre-modern taxonomy, algae are classified as plants.

70 According to what al-Ghafiqi cites, the entry on algae should be read at Book VI of Galen, however, when we investigate, we realize that it is at Liber VIII of De Simplicium Medicamentorum Temperamentis ac Facultatibus (*Γαληνου Περι Κρασεως Και Δυναμεωσ των Απλων Φαρμακων*, كتاب الادوية المفردة لجالينوس). The Greek and Latin texts are from Kühn (1826) and the Arabic text is from two different codices, Real Biblioteca del Monasterio de El Escorial No: 793 and No: 794.
Περὶ φακοῦ: Φακὸς ὁ ἐπὶ τῶν τελμάτων ὑγρᾶς καὶ ψυχρᾶς ἐστι κράσεως, ἐκ τῆς δευτέρας που τάξεως ὑπάρχων κατ' ἄμφω. De Phaco, Lenticula palustri: Lenticula palustris humidae frigidaque temperiei est, utrinque ex secundo quodammodo ordine (Kühn 1826: XII p. 149).

On Tuhlub: Its temperament is cold and humid and in both it is in the second degree (El Escorial No: 793, 137[b]).

ذكر الطحلب: مزاج هذا بارد رطب وهو فيهما كأنه في الدرجة الثانية.

On Tuhlub: Its temperament is cold and humid temperament and in all of these two traits it is in the second degree (El Escorial No: 794, 49[a])

ذكر الطحلب: مزاج هذا مزاج بارد رطب وهو في الخصلتين جميعا كأنه في الدرجة الثانية.

Περὶ φύκου: Φῦκος ὑγρὸν ἔτι καὶ χλωρὸν ἐξαιρούμενον τῆς θαλάττης, καὶ ψύχει καὶ ξηραίνει κατὰ τὴν δευτέραν τάξιν, ἔχει γάρ τι καὶ στρυφνὸν μετρίως. De fuco, alga: Phycos humens etiamnum et virens ex mari exemptus desiccat et refrigerat ordine secundo habet enim quiddam modice acerbum (Kühn 1826: XII p. 152) .

On Fuqus: It is a cochineal worm, if it is taken from the tree, it is considered humid and soft, it cools down and dries up in the second degree, because it contains something that is moderately astringent (El Escorial No: 794, 49[b]).

ذكر فوقس: وهو دود القرمز إذا اخذ هذا من الشجر وهو يعد رطب طري فهو يبرد ويجفف في الدرجة الثنية لان فيه شيا يقبض قبض معتدلا.

On Fuqus: It is a cochineal worm, if it is taken from the tree, it is considered humid and soft, it cools down and dries up in the second degree, when this is similar except that it contains something that is moderately astringent (El Escorial No: 793, 138[a]).

ذكر فوقس: وهو دود القرمز إذا اخذ هذا من الشجر وهو يعد رطب طري فهو يبرد ويجفف في الدرجة الثنية عند متشابها هذا الا ان فيه شيا يقبض قبض معتدلا.

Regarding the citation by al-Ghafiqi referring to Galen, we can conclude that it is the Φακὸς entry instead Φῦκος.

71 T: in this codex, ج is given instead ح و, which means Galen instead al-Hawi.

72 T: -.

73 T: مبرد.

74 T: -, -.

Dioscorides: It is very astringent, and it heals hot tumours that need cooling because of gout.
Other(s): The abdomen or any organ is imprisoned if / when coated with it, –especially the marine and riverine –. If / when it is boiled in old olive oil it softens the nerves.

د وهو قابض جدا ويصلح[75] للأورام الحارة المحتاجة الي التبريد من النقريس.
غيره يحبس البطن[76] من أي عضو[77] كان[78] إذا طلي به وبخاصة البحري[79] والنهري[80] إذا اغلي في الزيت العتيق لين العصب.

Since we have two codices of al-Ghafiqi's Kitab al-Adwiyyat al-Mufrada illustrated, we can find two pictures of طحلب – Tuhlub, Alga. Figure 7.12 shows the drawing of Tuhlub from the manuscript kept in Osler Library 7508 (Canada), folio 220^{a}, while Figure 7.13 shows that of Tuhlub from the manuscript in Library of Majlis-i Malek, 5958 (Iran), page 399. For a comparison I have added two illustrations from Dioscorides' Materia Medica. Figure 7.10 is φακός from BnF Gr. 2179, p. 109^{a} and Figure 7.14 is the illustration of Tuhlub from Harvard 1971.95.1,[81] f. 43^{b}. In this way I intend to show the similarities and differences among illustrations from different codices: Dioscorides in Greek, Dioscorides in Arabic, and finally al-Ghafiqi in Arabic.

5 Discussion

5.1 *Identification of Algae*

Identification of any plant species through ancient or medieval text is harder than ever, since in almost every case we do not have the specimens. In the case of algae, it is almost impossible at the species level, however, we can conclude on some possible genera through the textual apparatus as well as

75 M: يصلح و يصلح و same word repeats here probably due to a transcription error.

76 M: الدم البطن و the abdomen and the blood, T: -.

77 T: موضع.

78 T: مثال.

79 M: البحري من الطحلب the marine, of the alga.

80 M: النهري منه the riverine, of which.

81 The reference for this illustration should be cited as "*Struchnos Plants* (*painting with text, recto and verso*), *from a De Materia Medica of Dioscorides* (Unknown Artist), 1971.95.1," Harvard Art Museums' collections online, Aug 09, 2021, https://hvrd.art/o/216276. Special thanks to George Saliba, Prof. Emeritus of Arabic and Islamic Science, Columbia University and Founding Director of Farouk Jabre Center for Arabic & Islamic Science and Philosophy, American University of Beirut (AUB), who informed me about this image.

through the illustrative apparatus. Dioscoridean φακός should be a member of genus Lemna, namely *Lemna minor* L. of monocots. We find satisfactory evidence that Dioscorides (and others who refer to him) to mention 'lentil' in its resemblance. βρύον θαλάσσιον sea-moss should be something like hair, in short filament form. Beck (2005) identifies it as *Ulva lactuca* L. and *U. latissima* L. To our knowledge in today's taxonomy, *Chaetomorpha sp.* (unbranched filaments), or *Cladophora sp.* and *Bryopsis sp.*[82] are known as hair-like algae kinds.[83] When it comes to the identification of φῦκος θαλάσσιον, we can list more genera in this regard. The 'broad' one can be either *Ulva sp.* or *Porphyra sp.* while the 'long and reddish' one can be *Gelidium sp.* from Rhodophyta or some species like *Cystoseira sp.* or *Fucus sp.* from Phaeophyta.

5.2 *Textual Apparatus*

As we read Ibn Sina's *Canon of Medicine*, we understand that طحلب Tuhlub is classified among (and like other) Materia medica according to its qualities e.g., hot, cold, dry, and humid. It is also classified according to the habitat as طحلب النهري the *riverine* or طحلب البحري the *marine* kinds. These classifications which were inherited from the antique sources namely from Dioscorides and Galen, were thoroughly used in Islamic medicine and pharmacy. In the overall state, we understand that the medieval authors used a different taxonomy of the algae which – in my opinion – was a crucial part of their understanding or vision of the minerals, plants, animals, and humans as well. Moreover, we see that, the mediaeval authors – as well as the antique authors – grouped plants and alike according to their preference of habitats. This is somehow different in contemporary botany since we use the Linnaean system of classification, which is based mostly on the comparison of the reproductive organs. For instance, medieval authors grouped seaweed and Duckweed (*Lemna minor*) together because they grow in the water, in the same habitats. In today's taxonomy, Duckweed is classified in a totally different taxonomic group. This is very evident in the case of Ibn Sina, since in his account, Duckweed is counted as a kind of Tuhlub, again parallel to Dioscorides. However, Ibn Sina does not cite him (or any other sources) in this entry. We know that he often keeps silence to name his sources especially that of Dioscorides, regarding the Materia medica.

I think at least after the extensive use of and interaction with the modern biological sciences, the contemporary human vision and understanding

---

82 Bryopsis sp. are known as potential sources for a bioactive compound Khalalide F, which is well known for the therapeutic effects (Kan et al. 1999). Studies on such compounds may shed light on the antic and medieval uses of algae as Materia medica.

83 This morphological resemblance or analogy may be regarded as a sample of humanizing algae.

have changed a lot. What we find in the different taxonomic systems is the reflection of an idea of minerals, plants, animals, and probably every "thing" or "entity". Since in this chapter I investigated books on Materia medica, it is natural to see that algae are also classified among the very Materia medica, by means of their healing properties. Medieval physicians used algae for analgesic, anti-inflammatory purposes as Avicenna reports solely from Dioscorides. Regarding the therapeutic properties of algae, Ibn Sina's contemporary al-Biruni refers to three antique authors, namely Dioscorides, Galen, and Paul of Aegina.

I can certainly say that most antique authors were referred to by mediaeval physicians in the Islamic Geography. This shows us that their books were translated to Arabic, copied, distributed, and studied for centuries. Ibn Sina was the most famous figure of peripatetic philosophy in the eleventh century. His vision and understanding of the world built over the amalgamation of peripatetic tradition, Islamic creed, and local practices. On the other side, al-Biruni prefers to stand as a man of "science" rather than "philosophy" because he extensively used the mathematical and geometrical tools to learn and to understand the nature and the universe as he mentioned in his work entitled "*al-Athar al-Baqiya min Qurun al-Haliya – The Memoirs of Past Generations*". These two figures, Ibn Sina and al-Biruni perhaps represent two different world views of the eleventh century, the peripatetic and the mathematical and geometrical respectively. For instance, we see that in the same century, there are enormous efforts on algebra or better to say the attempts and studies of applying arithmetic to algebra in every respect (Baga 2017), as observable in works of al-Karaji.[84] Probably, after this century, mathematical sciences established their autonomous authority against Peripatetic Tradition. Al-Biruni, the great polymath of the tenth and eleventh centuries, presumed that this universe is geometrically structured, and it could be understood by following the rules of mathematics and geometry (Yavuz 2020). In doing so, he stands in the "scientific" side rather than the "philosophical" one, compared to Ibn Sina.

Al-Biruni, in his entry on algae, cites from five poets, a botanist, and a zoologist. What we read under the title al-Tuhlub, is more poetry, rather than pharmacy. Ibn Sina, nonetheless, does not say anything new after the reporting of what Dioscorides says. However, al-Biruni transmits to our consideration some famous medical poets, thus mentioning the sources of poetry, which we can then conclude based on a deeper look at his *Book of Pharmacy* that this is a part of al-Biruni's methodology. What al-Biruni cites from medieval authors are either descriptions or literal uses of algae in poetry. These literal uses may

84 Abu Bakr Muḥammad ibn al-Hasan al-Karaji (d. *c.*1029), Persian mathematician.

inform us about an imagery of algae, how much area they occupied in minds of the medieval authors and poets in the Islamic Geography.

When we examine the text supplied by al-Ghafiqi, we see that in the entry of Tuhlub, he refers to Dioscorides' Materia medica IV. 87 the duckweed, and then IV. 98 the sea-moss. Although he cites from Dioscorides in three different sentences, these are the additive combination of the texts from IV. 87 and IV. 98. The knowledge on algae is apparently unified or amalgamated, since in Arabic there is only one word to denote types of algae in spite of three Greek terms from the antiquity. Besides, what he quotes from Dioscorides exactly fits our citations which shows us al-Ghafiqi's precision or fairness in transmission of knowledge from his sources. Because al-Ghafiqi meticulously refers to his sources like Dioscorides, Galen and al-Razi (Rhazes).

We see that the authors in the Islamic Geography differ in their understanding and vision of the world, this diversity corresponds to their production and presentation of the knowledge on algae.

5.3 *Illustrative Apparatus*

Regarding the illustrative apparatus in this essay, we can compare *Dioscoridean* and *al-Ghafiqian* illustrations separately and together. First of all, it is obvious that Figures 7.2, 7.3, 7.5, 7.6, 7.8, 7.9, and 7.11 resemble branched kinds of algae. Among these, Figures 7.3, 7.6 and 7.9 are drawn as on water, while 7.2, 7.5, 7.8 and 7.11 are hairy or they have secondary branches and are drawn out of water. Figures 7.4 and 7.7 may depict a biofilm layer on water, or a thick cover of algae, while Figure 7.10 shows a pond or lake full of water and possible population of Lemna genus, probably *Lemna minor*. Among these illustrations from the Dioscoridean tradition, we can conclude that the two codices Or 3366 and Bodleian d 138 have a direct relationship and then A 3702 relates to these two. All of which may originate from an unknown X codex.

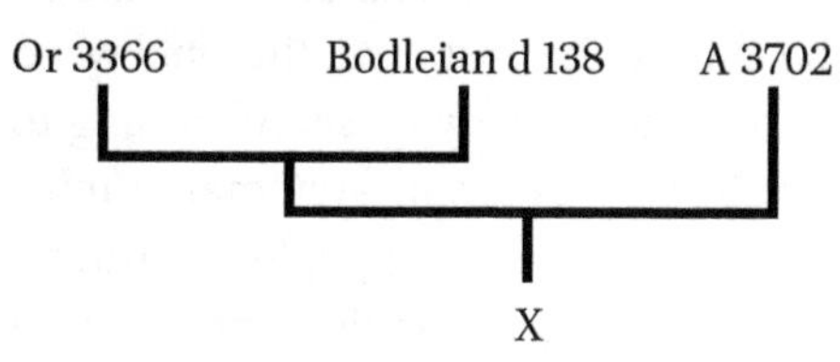

If we compare illustrations in al-Ghafiqi's opera, we find that Figure 7.12 again shows a pond or lake and duckweed or seaweed inside it. Figure 7.13 is similar to mosses on a stone rather than algae in a pond, however it may be a pond full of thick algal layer as well. If we make a comparison between Dioscoridean Figures (7.10 and 7.14) and al-Ghafiqian ones (7.12 and 7.13), we recognize that Figure 7.12 may be copied or derived from Figure 7.10.

The illustrative apparatus is the comparison of the very drawings of algae in different manuscripts from Greek and Arabic. The Dioscoridean tradition (in BnF 2179, Greek copy) seems to be represented in al-Ghafiqi's opera (in Osler 7508, Arabic copy) at least in the entries of φακός and Tuhlub (طحلب). Last, but not least, on the topic of algae, we find a sample transmission of knowledge from Dioscorides (1st c.) to al-Ghafiqi (13th c.), from Greek to Arabic, from the antiquity to the Middle Ages.

6 Conclusions

In order to prospect from a perspective of the future, it is necessary to decide and deduct in the present. Moreover, in order to act in the present; we have to know more about the past. Due to the temporal connection between cause and effect, we need to understand the historical origins of things and the background of processes. Therefore, historical studies in the sciences make sense in the 21st century. In this chapter, I try to shed light on the medieval uses of algae in the Islamic Civilization with an investigation through Materia medica books of some of the authors like al-Biruni (Aliboron), Ibn Sina (Avicenna), and al-Ghafiqi. The purpose of this chapter is to decipher the human-algae relationships in the Islamic Middle Ages. In doing so, this chapter supplies an informative base in order to question the imaginaries and narratives in the past. Moreover, this chapter tries to identify the algae in genus-level, according to the textual data derived from the manuscripts. It also uses an illustrative apparatus to compare the medieval-artist-reception of algae in the illuminated manuscripts. Consequently, this chapter investigates two contemporary figures Ibn Sina and al-Biruni in their reception and perception of algae as a kind of Materia medica, like many plants.

To conclude, I would like to reiterate that pre-modern scholars had a different taxonomy of the sciences and of the things. For instance, the duckweed and the seaweed were mentioned either together or very close. According to the current botanical taxonomy, they are in very different systematic units. From poems in literature, to drugs in medicine, medieval people "naturally" incorporated the algae in their daily or social lives. Perhaps this is the thing we miss in our contemporary, discrete, artificial views of natural life.

Figures from Various Manuscripts

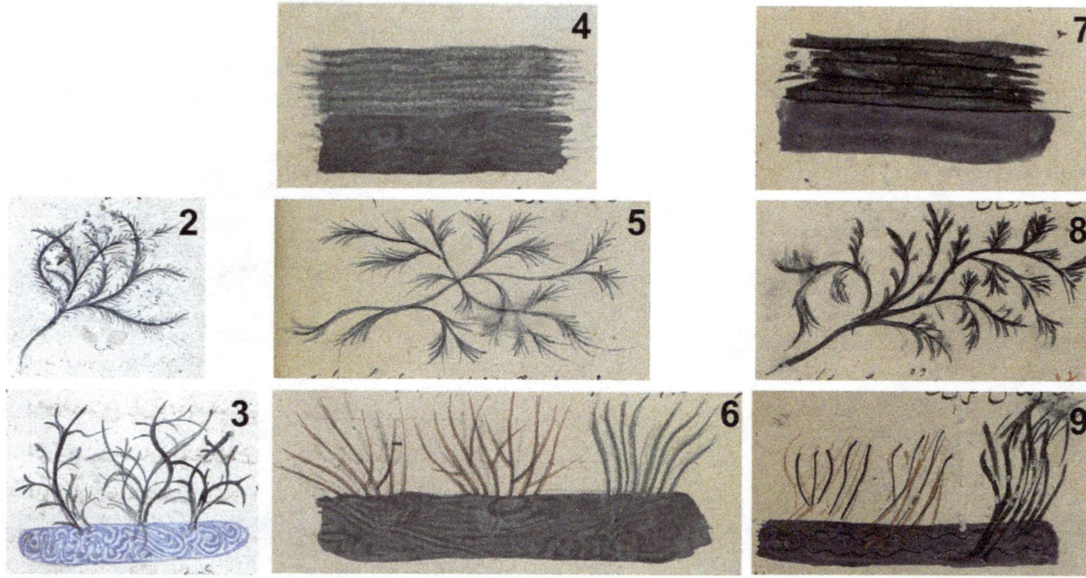

FIGURE 7.2 βρύον θαλάσσιον of Dioscorides (A 3702)

FIGURE 7.3 φῦκος θαλάσσιον of Dioscorides (A 3702)

FIGURE 7.4 φακός of Dioscorides (Or 3366)

FIGURE 7.5 βρύον θαλάσσιον of Dioscorides (Or 3366)

FIGURE 7.6 φῦκος θαλάσσιον of Dioscorides (Or 3366)

FIGURE 7.7 φακός of Dioscorides (Bodleian d 138)

FIGURE 7.8 βρύον θαλάσσιον of Dioscorides (Bodleian d 138)

FIGURE 7.9 φῦκος θαλάσσιον of Dioscorides (Bodleian d 138)

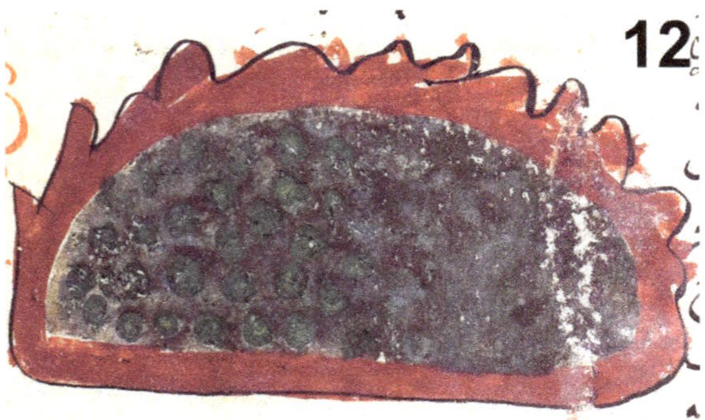

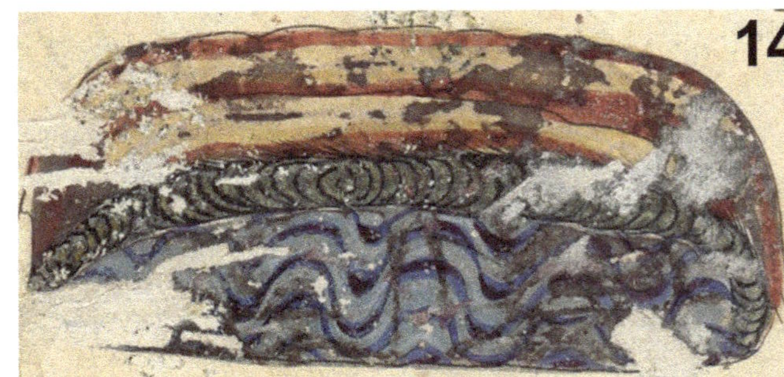

FIGURE 7.10 φακός of Dioscorides (BnF Gr. 2179)

FIGURE 7.11 βρύον θαλάσσιον of Dioscorides (BnF Gr. 2179)

FIGURE 7.12 Tuhlub of al-Ghafiqi's (Osler Library 7508)

FIGURE 7.13 Tuhlub of al-Ghafiqi's (Majlis-i Malek 5958)

FIGURE 7.14 Tuhlub of Dioscorides (Harvard 1971.95.1)

References

Abbott, N. 1972. Studies in Arabic Literary Papyri III Language and Literature, Chicago: University of Chicago Press.

al-Dinnawi, M.A. 1999. Ibn Sina, al-Qanun fi al-Tibb. Beirut: Dar al-Kutub al-llmiyya.

Baga, E. 2017. "Arithmetical Algebra in the Islamic History of Mathematics and Its Peak in the 9th/15th Century: Ibn al-Ha'im's *al-Mumti*", Nazariyat Journal for the History of Islamic Philosophy and Sciences 3 (2): 69–123. doi: 10.12658/Nazariyat.3.2.M0009en.

Beck, L.Y. 2005. Pedanius Dioscorides of Anazarbus De materia medica. Zürich and New York: Olms-Weidmann.

Borbone, P.G., M. Farina. 2014. "New Documents concerning Patriarch Ignatius Na'matallah (Mardin, ca. 1515 – Bracciano, near Rome, 1587) 1. Elias, the 'Nestorian' Bishop" EVO XXXVII: 179–189. doi: 10.12871/97886741501411.

Brunfels, O. 1531. Pauli Aeginetae Pharmaca simplicia. Parisis: Excudebat Georgius Ulricher Andlanus.

Casari, M. 2016. "Raimondi, Giovanni Battista". Dizionario Biografico degli Italiani. Torino: Istituto della Enciclopedia Italiana.

Casari, M. 2017. "'This language is more universal than any other' Values of Arabic in early modern Italy" In the City, Court, Academy: Language choice in Early Modern Italy, edited by Andrea Rizzi, and Eva Del Soldato, 173–198. Oxford: Routledge.

Di Vincenzo, E. 2009. Kitab al-adwiya al-Mufrada di Abu Ga'far Ahmad b. Muhammad b. Ahmad b. Sayyid Al Gafiqi Edizione del Capitolo Alif con Indici e Apparato Critico in Nota. Roma: La Sapienza University.

Dubler, C.E. 1991. "Diyuskuridis". Encyclopaedia of Islam. (EI²) Leiden: Brill.

Farina, M., S. Fani. 2016. "The Typographia Medicea and the Humanistic Perspective of Renaissance Rome". In The Grand Ducal Medici and the Levant: Material Culture, Diplomacy and Imagery in Early Modern Mediterranean, edited by Marta Carosci & Maurizio Arfaioli, 169–177. Brepols Publishers.

Gacek, A. 2015. "The Palaeographical and Codicological Features of the Osler Manuscript in the Context of the Manuscript Transmission of al-Ghafiqi's Herbal". In The Herbal of al-Ghafiqi (A Facsimile Edition with Critical Essays) edited by F. Jamil Ragep and Faith Wallis, Pamela Miller, Adam Gacek, 18–34. Montreal: McGill Queen's University Press.

Gunther, R.T. 1934. The Greek Herbal of Dioscorides. Illustrated by a Byzantine AD 512, Englished by John Goodyer AD 1655, edited and first printed AD 1933. New York: Hafner Publishing.

Ibn Sina. 1593. El Kanun fi't Tıbb. Romae: Typographie Medicea.

Kan, Y., T. Fujita, B. Sakamoto, Y. Hokama, H. Nagai. 1999. "Kahalalide K: A new cyclic depsipeptide from the Hawaiian green alga *Bryopsis* species", J Nat Prod 62: 1169–1172.

Kennedy, E.S. 1970. "Al-Bīrūnī". Dictionary of Scientific Biography. New York: Charles Scribner's Sons.

McGinnis, J. 2010. Avicenna. Oxford: Oxford University Press.

Meyer, E.H.F. 1841. Nicolai Damasceni Libri Duo Aristoteli Vulgo Adscripti. Lipsiae: Sumtibus Leopoldi Voss.

Osbaldeston, T.A. 2000. Dioscorides: De Materia Medica: A New Indexed Version in Modern English. Johannesburg: Ibidis Press.

Overduin, F. 2014. Nicander of Colophon's Theriaca: A Literary Commentary (Mnemosyne, Supplements). Leiden: Brill.

Ragep, J.F., F. Wallis, P. Miller, A. Gacek. 2015. The Herbal of al-Ghafiqi (A Facsimile Edition with Critical Essays. Montreal: McGill Queen's University Press.

Riddle, J.M. 1985. Dioscorides on Pharmacy and Medicine. Austin: University of Texas Press.

Said, H.M. 1973. Al-Biruni's Book on Pharmacy and Materia Medica. Karachi, Hamdard Edition.

Siraisi, N.G. 1987. Avicenna in Renaissance Italy: The Canon and Medical Teaching in Italian Universities after 1500. New Jersey: Princeton University Press.

Thatcher, G.W. 1911. "Abu Tammam". Encyclopædia Britannica. Cambridge: Cambridge University Press.

Toomer, G.J. 1996. Eastern Wisedome and Learning: The Study of Arabic in Seventeenth-century England. Oxford: Clarendon Press.

Touwaide, A. 1999. "La botanique entre science et culture au Ier siècle de notre ère". In the Geschichte der Mathematik und der Naturwissenschaften in der Antike, 1: Biologie edited by Georg Wöhrle and Jochen Althoff, 219–52. Stuttgart: F. Steiner.

Touwaide, A. 2009. "Translation and Transliteration of Plant Names in Hunayn b. Ishaq's and Istifan B. Basil's Arabic Version of Dioscorides, De materia medica". Al-Qantara. XXX (2): 557–580.

Witcombe, C.L.C.E. 2004. Copyright in the Renaissance: Prints and the Privilegio in Sixteenth-Century Venice and Rome. Leiden: Brill.

Yavuz, M. 2018. "A Comparative Study of Epiphytic Lichens Mentioned by Ibn Sina and Ibn al-Baytar", Medicina nei Secoli-Arte e Scienza 30 (2): 617–640.

Yavuz, M. 2020. "Lichens in al-Biruni's Kitab al-Saydanah fi al-Tibb", Early Science and Medicine 25 (2): 152–172. doi: 10.1163/15733823-00252P03.

Yavuz, M., P. Herraiz Oliva. 2020. "Botany as a New Field of Knowledge in the Thirteenth Century: On the Genesis of the Specialized Sciences", Teorie vědy / Theory of Science 42 (1): 51–75. doi:10.46938/tv.2020.478.

CHAPTER 8

Microalgae and Human Affairs: Massive Increase in Knowledge Drives Changes in Perceptions of Good and Bad Blooms

Gustaaf Hallegraeff

Abstract

Algal blooms have been utilised by humans as nutritious food, inspired naturalists with bioluminescent dinoflagellate spectacles, but remain much undervalued by the public as the source of fossil fuels, every second breath of oxygen we inhale, and even the origin of eukaryotic life on our planet. Instead, algal blooms have been increasingly associated by human society with beach health danger and seafood poisoning, signs of nutrient pollution or impending climate change. While the impacts of harmful algal blooms on human society have been increasing, this trend is largely driven by intensified monitoring associated with enhanced aquaculture. Satellite imagery has visualised the global scale of algal bloom phenomena, and newly highlighted their role in driving climate. We demonstrate how our massive increase in knowledge of microalgae has been driving ever changing perceptions of good and bad algal blooms, and recognition of the central role microalgae play on our planet, as well as for our human future.

Keywords

harmful algal blooms – red tides – fish kills – algal biotechnology – climate change – origin of life – artforms in nature

1 Introduction

While macroscopic seaweeds have been celebrated as sea vegetables for more than 2,000 years in Asian countries such as China and Japan (see Chapter 5), the detection of their microscopic counterparts had to wait until Van Leeuwenhoek (1673) used his first newly designed microscopes to study microalgae in Dutch pond water. Van Leeuwenhoek excitedly wrote to the

© GUSTAAF HALLEGRAEFF, 2024 | DOI:10.1163/9789004683310_010
This is an open access chapter distributed under the terms of the CC BY-NC 4.0 license.

Royal Society of the United Kingdom: "*Some of these [animalcules] are so exceedingly small that millions might be contained in a single drop of water. I was much surprised at this wonderful spectacle, having never seen any living creature comparable to these for smallness; nor could I indeed imagine that nature had afforded instances of so exceedingly minute animal proportions*". It was the German scientist Haeckel (1866) who first recognised that algae (without roots, flowers, and stems) were distinct from the better known kingdoms of Plantae and Animalia, and to accommodate them in his tree of life created the separate kingdom Protista. It had long been recognised by mariners that the land is green and the open ocean is blue. With massive increases in knowledge using algal culturing techniques, electron microscopes, molecular genetics and satellite imagery, we now know how this difference is underpinned by the existence of very distinct ecosystems (sometimes referred to as the paradox of the plankton; Hutchinson 1961) and varying abundances of phylogenetically divergent forms of plant life on the land and in the water. It is not surprising that human relationships with macroscopic land plants, aquatic seaweeds and microscopic algae are strikingly different. While seaweed farming practices have existed in East Asia for thousands of years, cultivation or harvesting of microalgae has been rare in the past (e.g. the Aztecs in Mexico). Water discolorations by microalgal blooms at various times instead have inspired awe and wonder, but more commonly in ancient times have generated fear of the unknown (e.g. the river Nile turning to blood; Exodus 1,000 BC). Starting in the 1990s with public concerns of our planet running out of fossil fuels, and in the 2000s with human-made climate change and the critical role of plants and algae in carbon sequestration being recognised, we started to see an increased public awareness of the role of microscopic algae for planetary health (e.g. James Lovelock's Gaia hypothesis inspired by the microalga *Emiliania huxleyi*; Lovelock 1991). Current public interests in the gut microbiome for human health and the science of viral pandemics similarly are not something that early explorers of microscopic life, such as Van Leeuwenhoek (1632–1723), Pasteur (1822–1895), Haeckel (1834–1919), could have foreseen. Similarly, the continuing exploration of life forms elsewhere in the universe has a strong microbial focus, which at the same time draws us back to the microbial origin of life on our own planet. I here review examples of microalgal bloom phenomena, both good and bad, which have been part of the microbial engine of our blue planet for billions of years, but which are either newly recognised or subject to significantly changing public perceptions. From a human-centric perspective, moral judgements of which algae are good or bad are continuously shifting over time, but also vary according to whether one desires clean water for tourism or algal rich water for fisheries productivity or aquaculture feed, biofuel production, or climate mitigation and carbon sequestration. From an algal-centric perspective, the question of whether humans are good or bad for algae (e.g. by fertilising oceans with waste nutrients, global ship ballast water transport, climate warming) will also be raised.

2 Algal Blooms: The Bad

FIGURE 8.1 Harmful algal blooms. 1a. *Noctiluca* dinoflagellate "red tide" responsible for beach closures in Sydney, Australia; 1b. *Noctiluca* slicks threatening salmon farms in Tasmania, Australia; 1c. Toxic *Microcystis* cyanobacterial surface bloom in a Dutch canal; 1d. Harmful bloom of *Chattonella marina* golden-brown flagellates responsible for USD71M mortality of finfish aquaculture in the Seto Inland Sea, Japan
PHOTO CREDITS: 1A. DAILY TELEGRAPH, SYDNEY; HTTPS://WWW.DAILYTELEGRAPH.COM.AU; 1B. JUDY MARSHALL, UNIVERSITY OF TASMANIA; 1C. G. HALLEGRAEFF, ORIGINAL; 1D. COURTESY PROF. T. OKAICHI, TAKAMATSU UNIVERSITY, JAPAN

2.1 *The River Nile Turned to Blood*

One of the first written references (1,000 years BC) to a harmful algal bloom appears in the Christian Bible: '*all the waters that were in the river were turned to blood. And the fish that was in the river died; and the river stank, and the Egyptians could not drink of the water of the river*' (Exodus 7: 20–21). There have been various suggestions as to the identity of the causative organism, including the freshwater flagellates *Euglena sanguinea* or *Haematococcus pluvialis* being carried downstream by torrential waters to overflow onto the flood plain of the Nile Delta (Hort 1957). In that case, a non-toxic bloom-forming alga became so densely concentrated that it generated anoxic conditions resulting in indiscriminate kills of both fish and invertebrates. Oxygen depletion can be due to high respiration by the algae (at night or in dim light during the day) but more commonly is caused by bacterial respiration during decay of the bloom. A more insidious red-water forming organism known from the area is the brackish water toxigenic dinoflagellate *Alexandrium minutum* (described by Halim 1960 from Egypt), which kills fish and can contaminate seafood with paralytic shellfish toxins dangerous to human beings.

2.2 Noctiluca *Dinoflagellate Red Tides*

Japanese archives reported coastal 'red waters' ('akashiwo') as early as the year 731, putatively thought to be due by the dinoflagellate *Noctiluca scintillans*. These single celled, bubble-shaped forms, 0.5 to 1 mm in diameter, carry a sticky tentacle used to snare away other plankton that come too close. When this organism was first described (as *Medusa scintillans* Macartney 1810) this caused them to be mistaken for small jellyfish. *Noctiluca* have no means of locomotion but can adjust their buoyancy to forage through the water column. When the food is finished, billions float to the surface in the hope of being carried by wind or currents to richer waters. If they end up washed ashore by onshore winds, the death throes of billions of stranded cells become visible as pink or tomato-soup red slicks but unlike other types of algal blooms this organism is essentially harmless to humans. As a precaution in November 2012 this organism triggered the temporary closure of ten Sydney beaches including iconic Bondi Beach (Fig. 8.1a). When *Noctiluca* slicks move through salmon farm pens in Southern Tasmania, Australia, this irritates the fishes' gills and causes them to stop surfacing and feeding (Hallegraeff et al. 2019; Fig. 8.1b).

2.3 *Cattle Deaths from Contaminated Drinking Water*

Toxic microalgae equally occur in marine and freshwater environments, and the latter can have serious implications for human drinking water reservoirs. The earliest written record dates back to 1878 when cattle deaths were associated with cyanobacterial toxins from the brackish water *Nodularia spumigena* from Lake Alexandrina in Australia (see Chapter 2 for *Nodularia* in the Baltic). Francis (1878) reported that "*a 'conferva' that is indigenous and confined to the lakes has been produced in excessive quantities, so much as to render the water unwholesome. It is swallowed by cattle when drinking. This acts poisonously and rapidly causes death. Sheep from 1 to 6 or 8 hours, horses 8 to 24 hours, dogs 4 to 5 hours, pigs 3 or 4 hours*". A much more serious problem occurred in central Australia in November–December 1991 when a bloom of the cyanobacterium (blue-green alga) *Anabaena* (*Dolichospermum*) *circinalis* covered 1000 km of the Darling-Barwon river system and killed an estimated 10,000 live stock and required emergency water supplies for several towns. The cyanobacterium *Microcystis* (Fig. 8.1c) causes widespread surface scums tainted with liver-damaging toxins, responsible for 55 human deaths in a dialysis clinic in Brasil that used water from a neighbouring freshwater lake (Jochimsen et al. 1998).

2.4 *Harmful Algal Blooms and Finfish Farms*

Algal blooms can cause devastating mortalities to finfish aquaculture operations. Finfish held captive in intensive aquaculture are extremely vulnerable to algal blooms (e.g. USD71M loss in Japan in 1972 (Fig.8.1d), USD800M in Chile in 2016; Mardones et al. 2021). Fish mortality can be caused by mechanical damage or clogging of fishes' gills, oxygen depletion during algal bloom collapse, but more commonly is caused by exudate chemicals (reactive oxygen radicals, free fatty acids) which usually are of no human health significance (Hallegraeff et al. 2017). On a global scale, the economic impacts of algal blooms on finfish aquaculture far exceed impacts on shellfish farms or tourism. While wild fish can avoid and swim away from algal bloom, captive fish cannot, hence this is largely a human generated problem but on the increase.

2.5 *Human Seafood Poisonings from Toxic Dinoflagellates*

One of the first recorded fatal cases of human poisoning after eating shellfish contaminated with dinoflagellate toxins happened in 1793, when Captain George Vancouver and his crew landed in British Columbia in an area now known as Poison Cove. He noted that for local Indian tribes it was taboo to eat shellfish when the seawater became bioluminescent due to algal blooms (Dale and Yentsch 1978). The identification of the causative organism, the dinoflagellate *Alexandrium catenella*, was not achieved until 1936 and the alkaloid toxins, now called paralytic shellfish poisons (PSP), were not chemically characterised until 1975. These toxins are so potent that a pinhead-size quantity (about 500 microgram), which can easily accumulate in just one 100 g serving of shellfish, could be fatal to humans. During the 1960s the United States Central Intelligence Agency (CIA) used saxitoxin in suicide capsules (Tester et al. 2020). To date, every year globally some 2000 human shellfish poisonings are reported, with approximately 15% fatalities, but ever increasing monitoring efforts to test water samples for the causative organisms and seafood for toxins are reducing human illnesses. Another precautionary approach is to instigate blank shellfish harvesting bans in particular areas (e.g. Alaska), or instigate seasonal bans (Fig. 8.2a). In the Netherlands, historically it was recommended to only eat mussels in months with the letter "r" in it (now known to be due to diarrhetic shellfish poisoning by *Dinophysis* dinoflagellates; a syndrome newly described by Yasumoto et al. 1978).

Even more insidiously, in 1774 Captain Cook and his crew suffered from the symptoms of Ciguatera Poisoning after consumption of red bass from the New Hebrides (Vanuatu). It was not until 1979 that the link with causative species of the dinoflagellate *Gambierdiscus* was established (Adachi and Fukuyo 1979). Ciguatera poisoning is a worldwide problem associated with ingestion of certain toxin-containing fish. The disease was initially named after a turban-shelled snail known as "cigua" in the Spanish Antilles, which was thought to be the cause of this disease. More than 200 species of fish have been implicated, the most common of which include grouper, red snapper, barracuda, amberjack, and less commonly mackerel, surgeonfish, and sea bass. In ciguatera endemic areas, including the Caribbean and the South Pacific islands, the incidence of ciguatera poisoning ranges from 50–500 cases annually per 10,000 population, making it one of the commonest illnesses in those areas. The causative microalgae are bottom-dwelling species attached to seaweeds and coral rubble but never cause water discolorations. Unfortunately, because of the migratory behaviour of fish and the lack analytical ciguatoxin standards no global monitoring programs are in place for these seafood toxins. Global warming is predicted to drive a range expansion of ciguatera into new areas (Hallegraeff 2010).

2.6 *Toxic* Pseudo-nitzschia *Diatoms and Hitchcock's "The Birds"*

On August 18, 1961, a local California newspaper reported that thousands of crazed seabirds were sighted on the shores of North Monterey Bay. The sooty shearwaters did regurgitate anchovies, flew into objects and died on the streets. This helped inspire Hitchcock's 1963 thriller "The Birds" (Fig. 8.2b). Then, 30 years later, disorientation and death struck brown pelicans and sealions in the same area. But this time, scientists were able to demonstrate that the birds had ingested a toxin, domoic acid, produced by the diatom species of *Pseudo-nitzschia australis* (Garrison et al. 1992; see also Chapter 1, section 2.1). Humans can also be impacted when shellfish ingested the diatoms and can

FIGURE 8.2 Harmful algal blooms. 2a. Warning sign posts in Tasmania, Australia, not to collect shellfish because of contamination with paralytic shellfish poisons from seasonal dinoflagellate blooms; 2b. Hitchcock's movie "The Birds" was based on a real event whereby toxic diatom blooms led to contaminated anchovies and seabird mortalities in California; 2c,d. "Rocksnot" in the Buller River, New Zealand, caused by the invasive diatom *Didymosphenia geminata*
PHOTO CREDITS: 2A. TASMANIAN GOVERNMENT DEPARTMENT OF HEALTH AND HUMAN SERVICES. 2B. THEATRICAL POSTER RELEASED BY UNIVERSAL PICTURES; 2 C, D. G.HALLEGRAEFF

suffer from a toxin syndrome called amnesic shellfish poisoning, first coined in 1987 when in Prince Edwards Island, Canada, 105 people consuming blue mussels had to be hospitalised, causing 3 fatalities. It should be noted that the harmful effects from algal toxins almost exclusively impact on mammals and birds with highly evolved nervous systems, while shellfish themselves can in fact benefit from having access to more food.

2.7 *"Rock Snot" and Ship Hull Fouling*

Whenever one immerses a boat, anchor chain, buoy or pylon in the water, it rapidly attracts a coat of fouling bacteria and microalgae, notably diatoms (see Chapter 3 for more on diatoms). Ship hull fouling costs the shipping industry billions of dollars every year in excess fuel bills. Similarly, once crystal clear trout fishing streams in the South Island of New Zealand since 2004 have been suffering from unsightly "rock snot" (Fig. 8.2c, d). The causative diatom *Didymosphenia geminata* is thought to have been accidentally introduced from the Northern Hemisphere via moist fishing waders or equipment. These blooms that smother a stream or lake bed may adversely affect water quality, ecology and fish stocks, and also can become a hazard for hydro-electric generation, agricultural irrigation and recreational pursuits (Killroy et al. 2009). The introduction of dedicated custom quarantine regulations to prevent further spreading of *Didymosphenia* as well as new International Maritime Organisation (IMO) protocols to stop spreading of microorganisms via ship's ballast water are driven by new scientific awareness (www.imo.org; Hallegraeff and Bolch 1992).

3 Algal Blooms: The Good

3.1 *Microalgal Scums as Human Food:* Spirulina *and the Aztecs*

The Aztecs living along the shores of Lake Texcoco in Mexico in the 13th century collected *Spirulina* cyanobacterial scums, called 'techuitlatl', by draining the water through bags of cloth (Fig. 8.3a) and spreading out the algae on the sandy shore for sun drying. The semi-dried algae were then cut into small squares and taken to the villages, where drying was completed. Women took these algal cakes, called 'dihé', for sale to the local market. Dihé was crumbled and mixed with tomatoes and peppers, and poured over millet, beans, fish or meat and eaten in 70% of meals (www.spirulinasource.com). Pregnant women ate dihé cakes because they believed its dark colour would protect their unborn baby from the eyes of sorcerers. It was not until 1967 that the nutritional properties of *Spirulina* were fully recognised by western science (60% of dry weight is protein) and this triggered many research projects for industrial purposes to establish large-scale production plants. Today, *Spirulina* (now reclassified as *Arthrospira*) is being produced in more than 22 countries and used in over 77 countries in nutritional supplements, high end cosmetics, for water purification but newly also algal biofuels (Fig. 8.3b).

Ever since Beyerinck (1890) first succeeded in culturing from Delft pond water microalgae such as *Chlorella* and *Scenedesmus*, people have proposed algal biomass production for animal and human food and biofuel. To cite from

FIGURE 8.3 Good algal blooms. 3a. Aztecs in Mexico harvesting *Spirulina* to produce algal cakes; 3b. Massive outdoor pond culture of *Spirulina* in the New Mexico desert for biofuel production; 3c. Lagoon production of *Dunaliella salina* in Western Australia for ß-carotene biotechnology.
PHOTO CREDITS: 1A. DRAWING IN HUMAN NATURE, MARCH 1978, BY PETER T. FURST; WWW.SPIRULINASOURCE.COM; 3B. A2BE CARBON CAPTURE LLC, USA. HTTPS://WWW.ALGAEATWORK.COM/INDEX.HTML; 3C. BETATENE (COGNIS), W. AUSTRALIA; HTTPS://ALCHEPHARMA.COM/

early attempts of algal mass cultivation by Burlew (1953) "*In regions of the world where population is especially dense, and fertile land is limited, it is entirely possible that process-industry methods of producing food may furnish a respite from the threat of famine and so contribute toward more salutary conditions for civilized living. If algal culture can serve such a purpose, it is well worth development for that reason alone*". This potential was rediscovered in the late 1980s and experienced a revival in the 2000, largely triggered by the prediction of fossil fuel peak oil production. While efforts using land plant biomass (sugar cane, maize, rape seed) are competing with using agricultural land for human food, mass production of microalgae in outdoor ponds or race ways (e.g. *Chlorella, Botryococcus, Arthrospira*; up to 70% oil by weight) can produce 13x higher biomass (liters of oil per hectare) from a significantly (49–132 times) smaller area (Benemann 2013). Harvesting microalgae and extracting and refining algal biofuels currently render such ventures uneconomical, unless biofuels are a by-product of high-value pharmaceuticals or nutraceuticals (Borowitzka et al. 1984).

3.2 *Microalgal Biomass Culture for Biotechnology:* Dunaliella salina *for β-carotene*

The green alga *Dunaliella salina* was named after its original discoverer, Michel Felix Dunal, who first reported in 1838 sighting the organism in saltern evaporation ponds in Montpellier, France. Under high light and hypersaline conditions, this green alga produces up to 5% of its dry weight as β-carotene (market price is $600/kg) to protect the organism from long-term UV radiation. From a first pilot plant established in the USSR in 1966, the commercial cultivation of *D. salina* has been pursued in Australia (since 1980), Israel and the USA. Grown in 5 ha open air ponds (Fig. 8.3c), Western Biotechnology in Australia produces $5M/yr. Similar ventures in the US focus on astaxanthin production from *Haematococcus* (valued at $2000–$3000/kg) to be used as a supplement in feed pellets for salmon aquaculture (Borowitzka et al. 1984) to maintain their pink flesh.

3.3 *Microalgal Blooms: The Green Pastures of the Sea*

Usually we think of the macroscopic algae or seaweeds, as the 'grass' of the sea, but microscopic diatoms, and more generally, phytoplankton, are the equivalent of green pastures on the land. There are diatoms in soil and other places, but diatoms represent the bulk of oceanic primary production. In the ocean, the microscopic organisms convert sunlight and nutrients through photosynthesis into the bottom of the food pyramid. These organisms form the basis of all marine foodwebs leading to edible fish. Only since 1978 satellites have been able to detect the amount of chlorophyll in the oceans. The SeaWIFS image in Fig. 8.4a represents the northern hemisphere spring with diatom spring blooms in the North Sea, supporting e.g. the rich herring and cod fisheries. Other phytoplankton rich areas are upwelling areas such as off Peru supporting sardine and anchovy fisheries, and polar regions supporting dense krill stocks being fed on by whales. As the ocean waters heat up, tropical waters tend to become more nutrient and plankton poor, but there are signs that polar waters may become more productive in near future (Doney 2006).

3.4 *The Breathing Lungs of Our Planet: The View from Space*

Land plants and algae help the planet breathe by turning carbon dioxide into oxygen. More and more, we are now becoming aware of the quantitative contribution by marine phytoplankton, because they account for 50% of global primary productivity (Longhurst et al. 1995). That means that every first breath of oxygen we humans take in derives from microscopic algae, and the second breath from land plants, which after all are only newcomers dating back to 400M years ago. The power of this microbial algal engine is well demonstrated by Fig. 8.4b. All the microalgal cells in the world oceans could be packed in a plank, 386,000 km long, 7 cm thick, and 30 cm wide, that is, stretching from the earth to the moon (Andersen 2005). On average this plank divides once per

FIGURE 8.4 Good algal blooms. 4a. SeaWIFS chlorophyll satellite image showing Northern Hemisphere spring diatom blooms (red orange); 4b. The power of the microbial algal engine; 4c. Coccolithophorid bloom of *Gephyrocapsa oceanica* in Jervis Bay, Australia; 4d. White Cliffs of Dover made up of coccolithophorid chalk; 4e. *Noctiluca* bioluminescent spectacle in Tasmania, Australia

PHOTO CREDITS: 4A. NASA HTTPS://OCEANCOLOR.GSFC.NASA.GOV/ATBD/CHLOR_A/; 4B. ORIGINAL, BASED ON DATA IN ANDERSON 2005; 4C. COURTESY FORD KRISTO, AUSTRALIAN NATIONAL PARKS AND WILDLIFE SERVICE; 4D. GEOLOGICAL SOCIETY OF THE UNITED KINGDOM; 4E. COURTESY KELLI MILLER, TASMANIA day, but the reason the oceans are not becoming greener or browner, is because of rapid consumption by zooplankton and larval fish. Large milky-white coccolithophorid (*Emiliania huxleyi*) blooms, visible from space, once a source of concern (e.g. Fig. 8.4c, mistaken for pollution from the milk industry in Jervis Bay, Australia; Blackburn and Cresswell 1993) are now be viewed as carbon sequestering blooms which are good for planetary health (Lovelock 1991). Realization of this crucial ecosystem activity also translated in a revived interest in understanding coccolithophorid calcareous deposits (150M years old) such as the white cliffs of Dover in England (Fig. 8.4d). In the past, massive algal blooms removing the greenhouse gas carbon dioxide from the atmosphere have been speculated to have triggered cooling periods leading to an ice age (Martin 1990). The nutrient input from both climate-driven dust storms and bushfires has been demonstrated to generate massive ocean algal blooms (Tang et al. 2021). Such scenarios have inspired ill-conceived geo-engineering plans to fertilise significant parts of our oceans to generate dense algal-blooms with the dual benefits of carbon sequestration (climate mitigation) as well as increased food production (Glibert et al. 2008).

3.5 *Algal Photosynthesis: The Source of Iron Ore*

The advent of cyanobacterial photosynthesis on our planet 3.5B years ago is prominently carved in the geological record in the form of banded iron formations deposited around 2.4B years ago. Early oceans contained high concentrations of dissolved iron which oxidised (the equivalent of rust formation) and then precipitated in parallel with the rise in oxygen concentrations in seawater. Proposals to name our current epoch "Anthropocene" to mark the time of humans having a substantial impact on the planet, using fossil fingerprints such as plastic or nuclear debris, fade in comparison with the huge impact of algal photosynthesis. Oxygen is highly toxic to life forms preceding cyanobacteria such as methane and sulphur bacteria which therefore were driven to exile in anoxic sediments or the deep-sea. Few people stand still today to realise that the success of the Australian, South African and Brasilian economies is almost entirely built upon the mining and export of iron ore deposits that mark this "microalgal" event.

3.6 *Bioluminescent Spectacles*

Charles Darwin (1839) already described bioluminescence in the sea:

> *While sailing in these latitudes on one very dark night, the sea presented a wonderful and most beautiful spectacle. There was a fresh breeze, and every part of the surface, which during the day is seen as foam, now glowed with a pale light. The vessel drove before her bows two billows of liquid phosphorus,*

and in her wake she was followed by a milky train. As far as the eye reached, the crest of every wave was bright, and the sky above the horizon, from the reflected glare of these livid flames, was not so utterly obscure, as over the rest of the heavens." Darwin almost certainly observed a dinoflagellate bloom such as Noctiluca (meaning "night-light") (Fig. 8.4e).

These fireworks have nothing to do with the chemical element phosphorus but use a molecule luciferin, an enzyme called luciferase, which combined with oxygen generates neon blue light. This biochemical strategy was first invented by the bacteria (Aristotle already described how damp wood can glow), but from there was borrowed by numerous other creatures to serve very different purposes such as with mushrooms to attract flies that spread spores, for glow worms to attract prey, and the well-known fireflies that use it to attract a mate. For sea creatures such as *Noctiluca*, the use of bioluminescence is thought to act as a burglar alarm to scare off predators Ironically, almost everything we know about bioluminescence dates from Navy funded research during World War 1 when a stealth German submarine tried to sneak through the Strait of Gibraltar, but lit up like a Christmas tree and was sunk by Allied forces. Despite all the research, these neon blue spectacles are tough to forecast with precision. Climate-driven bioluminescent blooms in southern Australian waters are turning into a tourist attraction (Hallegraeff et al. 2019). Similarly, a small tourism industry based on *Pyrodinium* bioluminescence in Baha Phosphorescente is based in Puerto Rico (Zahl 1960).

3.7 *Coral Reefs or Dinoflagellate Reefs*

Coral reefs are large underwater structures composed of the calcareous skeletons of colonial marine invertebrates called coral polyps belonging to the phylum cnidaria that includes jellyfish and anemones. Globally coral reefs generate tourism worth $36B per year. It was not until the 1960s that it was discovered that most corals have a symbiotic relationship with zooxanthellae in their tissues. These mostly dinoflagellate microalgae (the first genus to be described was aptly named *Symbiodinium*; Freudenthal 1962) live inside the coral polyp's body where they photosynthesize to produce energy for themselves and the polyps. The polyps, in turn, provide a home and carbon dioxide for the algae. Additionally, the zooxanthellae provide the coral with their lively colors – most coral polyp bodies are clear and colorless without zooxanthellae. Essentially, the entire productivity of coral reef systems can thus be credited to microalgae, but early suggestions to rename "coral reefs" to be termed "dinoflagellate reefs" were never taken seriously. A key threat to coral reefs from climate change is coral bleaching, when high water temperatures (as little as 1°C above the long term average) causes the corals to expel their zooxanthellae.

If this lasts too long they will not return and corals die. Different species of zooxanthellae impart lesser or better protection for their coral hosts, which is now being explored as a new bioengineering approach for coral conservation (Van Oppen et al. 2017).

3.8 *Algae: Living Fossils of the Origin of Multicellular Life*

The recognition of the importance of water in the genesis of life, and the fact that life started with microbial forms has never been doubted. "Where did life come from?" remains one the biggest questions in biology, that has preoccupied scientists for hundreds of years. How did a cosmic soup of inanimate matter eventually give rise to a diverse multitude of lifeforms? Charles Darwin

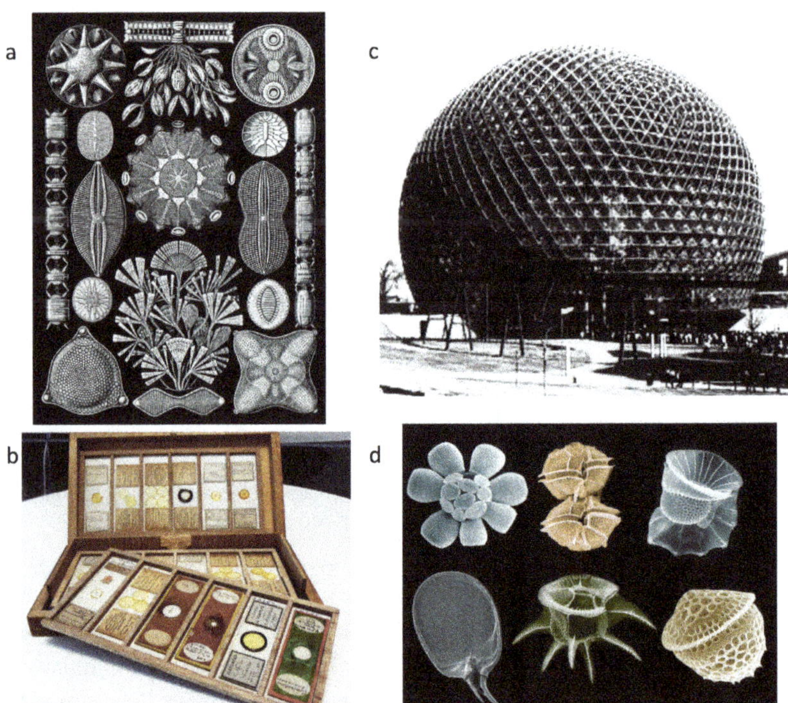

FIGURE 8.5 5a. Diatom drawings by Ernest Haeckel from his "Artforms of Nature" 1889; 5b. Permanent diatom slide mounts from the Victorian early 1900s; 5c. The design of the geodesic dome by Buckminster Fuller bears a strong resemblance to a scanning electron micrograph of the diatom *Pyxidicula* (the resting spore of *Stephanopyxis*); 5d. Tropical coccolithophorids and dinoflagellates from Australian waters

PHOTO CREDITS: 5A. HAECKEL 1889; 5B. ORIGINAL; 5C. MICHAEL ROUGIER, LIFE MAGAZINE; 5D. G. HALLEGRAEFF

(1871) already speculated "*But if. ... we could conceive in some warm little pond with all sorts of ammonia and phosphoric salts, light, heat, electricity etcetera present, that a protein compound was chemically formed, ready to undergo still more complex changes ...*". In the 1950's, Miller & Urey (1959) conducted an experiment which demonstrated that organic compounds (aminoacids) could be formed spontaneously by simulating the conditions of Earth's early atmosphere of methane, hydrogen and ammonia.

To date, Earth is the only place in the universe where we know for sure that life exists, but this may well change in the near future with spacecrafts busily attempting to retrieve soil samples from Mars and meteorites and return them to Earth. That life on Earth may have an extraterrestrial origin is an enticing idea, first promoted by British astrophysicist Fred Hoyle. How would organics have formed in space in the first place? And even if these organics could survive in space, would they be able to withstand entry through Earth's atmosphere?

The recent realisation that most of the water in the oceans of our planet is older the age of our immediate solar system (Cleeves et al. 2014) significantly raised the spectre of extraterrestrial watery life. Microscopic algae which evolved on our Planet around 1B years ago, with many forms essentially remaining unchanged, represent living fossils of our watery past. It perhaps is no coincidence that the salt content of the blood running through human veins matches that of seawater, and that our red and white blood cells bear an uncanny resemblance to unicellular algae?

4 Conclusions

Starting with Van Leeuwenhoek (1673), who first observed microalgae in pond water, early observers of microalgae were fascinated by their intricate designs and the possibilities these newly discovered creatures could offer. The German zoologist Ernst Haeckel (1866) enthusiastically exclaimed "*Nature has created an inexhaustable wealth of wondrous forms whose beauty and diversity way exceed anything that has been created by man*". "*Never will* I *forget the delight with which* I (...) *first observed* (...) *myriad animal forms,* (...) *and attempted to render with my paintbrush their splendid forms and colours*". His "Artforms in Nature" illustrations which include microscopic diatoms (Fig. 8.5a) and dinoflagellates continue to inspire artists even today. In Victorian times, it was a popular pastime to painstakingly mount cleaned diatom frustules in so-called "arranged" slides – made out of individually placed diatom shells (Fig. 8.5b). Even architects such as Buckminster Fuller, credited with the design of the geodesic dome (Fig. 8.5c), is thought to have used early scanning electron micrographs of a diatom for inspiration. Architects and diatoms struggle with the same problem to create the strongest possible structure using the least amount of building material.

Regrettably much of the early human perceptions of awe and wonder about microalgae gave way during the 1980s to 2000s to the predominantly negative perceptions of adverse impacts from harmful algal blooms (HABs). Carried over from historic beliefs (the red river Nile, cattle deaths from contaminated drinking water, Captain Vancouver and Captain Cook seafood poisonings), globally increasing investments in shellfish and finfish farming as well as tourism focused on the human-centric view of algal blooms as being harmful and unwanted. One of the most frequently asked questions about Harmful Algal Blooms has been whether they were increasing and expanding and what is the role of nutrient pollution, climate change, shipping and aquaculture. A statistical analysis on a global dataset extracted from the Harmful Algae Event Database (9,500 events) and Ocean Biodiversity Information System (7 M microalgal records) was recently conducted for the period 1985 to 2018 to investigate temporal trends in the frequency and distribution of marine harmful algal blooms (Hallegraeff et al. 2021). The authors found no uniform global trend in the number of harmful algal events and their distribution over time, once data were adjusted for regional variations in monitoring effort. Instead, the 16-fold global increase in aquaculture production was identified as one of the key drivers of increased concerns and reports of HABs, demonstrating how perceptions and moral judgements may eclipse scientific reality. An algal bloom commonly only becomes harmful to society once you put a shellfish farm or fish farm in the middle of it, hence this is largely a human generated problem.

Citing the algal poet Ralph Lewin (1987):

> *The biology of algae is a virtue, or a vice, that entails some tricky searching of the soul.*
>
> *It involves the growth of fishes, and the harvesting of rice, And pollution, and the origins of coal. It may get us into trouble; it may get us into space; Its dilemmas are as long as they are wide. It involves some moral judgements on the future of our race – And a little bit of science on the side.*

With massive increases in knowledge using algal culturing techniques, electron microscopes, molecular genetics and satellite imagery, we now recognise the critical role that microalgae play as the basis of all food chains leading to edible fish, the source of fossil fuels, iron ore deposits, the basis of coral reef tourism and planetary atmospheric health. Mass cultivation for aquaculture feeds, biotechnology and biofuels are attracting new interest. Bioengineering prospects of ocean fertilisation for increased fish production and carbon sequestration are on the horizon. Dedicated quarantine regulations to curb the spreading of unwanted species, for example via ship's ballast water, could never have been foreseen. It is now abundantly clear that most microalgal blooms are beneficial for humans. In turn, microalgae are stimulated by waste nutrients from human activities, the creation of artificial structures for attachment, the geographic spreading via ship's ballast water, enhanced growth by increased carbon dioxide and warming seawaters, and even nutrient inputs from climate-driven dust storms and bushfire ash. These fast-growing microorganisms represent a tremendous genetic diversity. In general they tend to be highly adaptable and, with the exception perhaps of tropical ocean plankton and coral zooxanthellae, are remarkably tolerant towards environmental change. The microbial engine of plankton algae plays a key role in our planet's ability to adapt and survive. It is perilous for our own human survival to ignore this critical creation.

References

Adachi R., Y. Fukuyo. 1979. "The thecal structure of a marine toxic dinoflagellate *Gambierdiscus toxicus* gen. et sp. nov. collected in a ciguatera endemic area". Bulletin Japanese Society Scientific Fisheries 45: 67–71.

Andersen R.A. (ed.). 2005. Algal Culturing Techniques, 596 pp. Elsevier, New York.

Benemann J. 2013. "Microalgae for Biofuels and Animal Feeds". Energies 6 (11): 5869–5886.

Beyerinck M.W. 1890. "Culturversuche mit Zoochlorellen, Lichengonidien und anderen niederen Algen". Botanische Zeitung 48, nr.45.

Blackburn S.I., G. Cresswell. 1993. "A coccolithophorid bloom in Jervis Bay, Australia". Australian Journal of Marine and Freshwater Research 44(2): 253–260.

Borowitzka L.J., M.A. Borowitzka, T.P. Moulton. 1984. "The mass culture of *Dunaliella* for fine chemicals: from laboratory to pilot plant". Hydrobiologia 116/117: 115–121.

Burlew J.S. (ed.). 1953. Algal Culture from Laboratory to Pilot Plant. Carnegie Institution of Washington publication no.600, 357 pp.

Cleeves, L.I., E.A. Bergin, C.M.O'D. Alexander, F. Du, D. Graninger, K.I. Oberg, T.J. Harries. 2014. "The ancient heritage of water ice in the solar system". Science 345: 1590–1593.

Dale, B., C.M. Yentsch. 1978. "Red tide and paralytic shellfish poisoning". Oceanus 21: 41–49.

Darwin, C.R. 1839. "Narrative of the surveying voyages of His Majesty's Ships Adventure and Beagle between the years 1826 and 1836, describing their examination of the southern shores of South America, and the Beagle's circumnavigation of the globe". Journal and remarks. 1832–1836. Henry Colburn. pp. 190–192.

Doney, S.C. 2006. "Plankton in a warmer world". Nature 444: 695–6.

Exodus, book of. In: "A Dictionary of the Bible". edited by W. R. F. Browning. Oxford Biblical Studies Online, http://www.oxfordbiblicalcstudies.com/article/opr/t94/e647 (accessed Jun 22, 2021).

Francis, G. 1878. "Poisonous Australian lake". Nature 18: 11–12.

Freudenthal, H.D. 1962. "*Symbiodinium* gen. nov. and *Symbiodinium microadriaticum* sp. nov; a zooxanthella: taxonomy, life cycle and morphology". Journal of Protozoology 9: 45–52.

Garrison, D.L., S.M. Conrad, P.P. Ellers, M. Waldron. 1992. "Confirmation of domoic acid production by Pseudonitzschia australis (Bacillariophyceae) cultures". Journal of Phycology 28: 604–607.

Glibert, P.M., R. Azanza, M. Burford, K. Furuya, E. Abal, A. Al-Azri, F. Al-Yamani, P. Andersen, D.M. Anderson, J. Beardall, G. Mine Berg, L. Brand, D. Bronk, J. Brookes, JA.M. Burkholder, A. Cembella, W.P. Cochlan, J.L. Collier, Y. Collos, R. Diaz, M. Doblin, T. Drennen, S. Dyhrman, Y. Fukuyo, M. Furnas, J. Galloway, E. Graneli, D. Viet Ha, G. Hallegraeff, J. Harrison, P.J. Harrison, C.A. Heil, K. Heimann, R. Howarth, C. Jauzein, A.A. Kana, T.M. Kana, H. Kim, R. Kudela, C. Legrand, M. Mallin, M. Mulholland, S. Murray, J. O'Neil, G. Pitcher, Y. Qi, N. Rabalais, R. Raine, S. Seitzinger, P.S. Salomon, C. Solomon, D.K. Stoecker, G. Usup, J. Wilson, K. Yin, M. Zhou, M. Zhu. 2008. "Ocean urea fertilization for carbon credits poses high ecological risks". Marine Pollution Bulletin 56: 1049–1056.

Haeckel, E. 1866. *Generelle Morphologie der Organismen*. Reimer, Berlin.

Haeckel, E. 1899. "*Kunstformen der Natur*". Bibliographisches Institut, Leipzig, Germany.

Halim, Y. 1960. "*Alexandrium minutum*, n. gen. n. sp. dinoflagellé provocant des 'eaux rouges'". Vie et Milieu 11: 102–105.

Hallegraeff, G.M. 2010. "Ocean climate change, phytoplankton community responses and harmful algal blooms: a formidable predictive challenge". Journal of Phycology 46: 220–235.

Hallegraeff, G.M., L. Albinsson, J. Dowdney, A. K. Holmes, M.P. Mansour, A. Seger. 2019. "Prey preference, environmental tolerances and ichthyotoxicity by the red-tide dinoflagellate *Noctiluca scintillans* cultured from Tasmanian waters". Journal of Plankton Research 41: 407–418.

Hallegraeff, G.M., D.M. Anderson, C. Belin, M.-Y. Bottein, E. Bresnan, M. Chinain, H. Enevoldsen, M. Iwataki, B. Karlson, C.H. McKenzie, I. Sunesen, G.C. Pitcher, P. Provoost, A. Richardson, L. Schweibold, P.A. Tester, V.L. Trainer, A.T. Yñiguez, A. Zingone. 2021. "Perceived global increase in algal blooms is attributable to intensified monitoring and emerging bloom impacts". Nature Communications Earth & Environment 2: 117.

Hallegraeff, G.M., C.J. Bolch. 1992. "Transport of diatom and dinoflagellate resting spores in ships' ballast water: Implications for plankton biogeography and aquaculture". Journal of Plankton Research 14: 1067–1084.

Hallegraeff, G.M., J.J. Dorantes-Aranda, J. Mardones, A. Seger. 2017. Review of Progress in our Understanding of Fish-Killing Microalgae: Implications for Management and Mitigation. In: Proença, L.A.O. and Hallegraeff, G.M. (eds). Marine and Fresh-Water Harmful Algae. Proceedings of the 17th International Conference on Harmful Algae, pp. 148–153. International Society for the Study of Harmful Algae and Intergovernmental Oceanographic Commission of UNESCO.

Hort, G. 1957. "The Plagues of Egypt". Zeitschrift fur die Alttestamentliche Wissenschaft 69: 84–103.

Hutchinson, G.E. 1961. "The paradox of the plankton". American Naturalist 95: 137–145.

Jochimsen, E.M., W.W. Carmichael, J. An, D.M. Cardo, S.T. Cookson, C.E.M. Holmes, M.B. Antunes, D.A. de Melo Filho, T. M. Lyra, V. Spinelli, T. Barreto, S.M.F.O. Azevedo, W.R. Jarvis. 1998. "Liver Failure and Death after Exposure to Microcystins at a Hemodialysis Center in Brazil". The New England Journal of Medicine 338: 873–878.

Kilroy, C., S.T. Larned, B.J.F. Biggs. 2009. "The non-indigenous diatom *Didymosphenia geminata* alters benthic communities in New Zealand rivers". Freshwater Biology 54: 1990–2002.

Lewin, R.A. 1987. *The Biology of Algae and diverse other verses*. The Boxwood Press, California, 191 p.

Longhurst, A., S. Sathyendranath, T. Platt, C. Caverhill. 1995. "An estimate of global primary production in the ocean from satellite radiometer data". Journal of Plankton Research 17: 1245–71.

Lovelock, J.E. 1991. "Gaia: The practical science of planetary medicine". Oxford University Press, 192 pp.

Macartney, J. 1810. "Observation upon luminous animals". Philosophical Transactions Royal Society of London, part 1, pp. 238–293.

Mardones, J.I., J. Paredes, M. Godoy, R. Suarez, L. Norambuena, V. Vargas, G. Fuenzalida, E. Pinilla, X. Rojas, J.J. Dorantes-Aranda, K.J. Lee Chang, D.M. Anderson, G.M. Hallegraeff 2021. "Disentangling the environmental processes responsible for the world's largest farmed fish-killing harmful algal bloom: Chile, 2016". Science of the Total Environment 766: 144383.

Martin, J.H. 1990. "Glacial-interglacial CO_2 change: The Iron Hypothesis". Paleoceanography and Palaeoclimatology 5: 1–13.

Miller, S.L., H.C. Urey. 1959. "Organic Compound Synthesis on the Primitive Earth". Science 130 (3370): 245–251.

Tang, W., J. Llort, J. Weis, M.M.G. Perron, S. Basart, Z. Li, S. Sathyendranath, T. Jackson, E.S. Rodriguez, B.C. Proemse, A.R. Bowie, C. Schallenberg, P.G. Strutton, R. Matear,

N. Cassar. 2021. "Widespread phytoplankton blooms triggered by 2019–2020 Australian wildfires". Nature 597: 370–375.

Tester, P.A., J. Matweyou, B. Himelbloom, B. Wright, S.R. Kibler, R.W. Litaker. 2020. "Saxitoxin and the Cold War". In: P. Hess (ed.) Proceedings of the 18th International Conference on Harmful Algae, Nantes, 191–194.

van Oppen, M., Gates, R.D., Blackall, L.L., Cantin, N., Chakravarti, L.J., Chan, W.Y., Cormick, C., Crean, A., Damjanovic, K., Epstein, H., Harrison, P.L., Jones, T.A., Miller, M., Pears, R.J., Peplow, L.M., Raftos, D.A., Schaffelke, B., Stewart, K., Torda, G., Wachenfeld, D., Putnam, H.M. 2017. Shifting paradigms in restoration of the world's coral reefs. Global change biology, 23(9), 3437–3448.

www.imo.org (accessed 22 June 2021).

www.spirulinasource.com (accessed 22 June 2021).

Yasumoto, T., Y. Oshima, W. Sugawara, Y. Fukuyo, H. Oguri, T. Igarashi, N. Fujita. 1978. "Identification of *Dinophysis fortii* as the causative organism of diarrhetic shellfish poisoning". Bulletin Japanese Society Scientific Fisheries 46: 1405–1411.

Zahl, P.A. 1960. "Sailing a Sea of Fire". National Geographic Magazine 119: 120–128.

CHAPTER 9

Becoming Marimo: The Curious Case of a Charismatic Algae and Imagined Indigeneity

Jon L. Pitt

Abstract

Lake Akan, located on Japan's northernmost island of Hokkaido, is home to world famous algae: spherical-shaped growths of *Aegagropila linnaei* known in Japanese as *marimo* (and sometimes called "moss balls" in English). Marimo have become a charismatic species internationally, a rare feat for an alga. Lake Akan hosts an annual three-day Marimo Festival, which began in 1950 as a way of drawing attention to the endangered species and as a means of celebrating the culture of the Indigenous Ainu community in the wake of settler colonialism. A key element to marimo's popularization is a widely-circulated story – a purported Ainu folktale that tells of two young, star-crossed lovers who jump into Lake Akan and become marimo. This tale has been used for decades to promote tourism to the Lake Akan area. However, in 2017, algae researcher Wakana Isamu traced the origins of the tale back to a Japanese writer named Nagata Kōsaku, who invented the tale and falsely claimed its Ainu origins in 1924. This chapter asks what it means for this tale of imagined indigeneity to have been embraced and reclaimed by the Ainu community of Kushiro in the name of conservation twice-over – that of rare algae and Ainu identity itself.

Keywords

Ainu – Japan – indigeneity – settler colonialism – marimo – folklore – tourism – algae – conservation

Lake Akan, in the Kushiro region of Japan's northernmost main island Hokkaido, is home to world famous algae: free-floating, spherical-shaped verdant growths of *Aegagropila linnaei* known in Japanese as *marimo* (Figure 9.1). Marimo have accomplished a rare feat for an alga. They have become a charismatic species and are sold as "pets" in Japan and abroad. A popular website promoting Japanese culture puts it this way: "If you're tired of fish but can't find

© JON L. PITT, 2024 | DOI:10.1163/9789004683310_011
This is an open access chapter distributed under the terms of the CC BY-NC 4.0 license.

FIGURE 9.1
Marimo of various sizes
PHOTO COURTESY OF MUSICEXPRESSION

another pet (besides plants … which aren't pets) to take care of, why don't you try marimo?" (Tolentino 2014). In 2005, a Japanese souvenir company created Marimokkori, an anthropomorphized marimo intended to promote tourism to Hokkaido. (See Figure 9.2) The name is a portmanteau, combining the word marimo with the word *mokkori*, which refers to a "bulge" produced by male genitalia under clothing (Marimokkori's round, green head is attached to a body with two arms and two legs with a conspicuous round bulge between them). Despite (or perhaps because of) this suggestive trait, Marimokkori remains a popular character that has appeared on a variety of souvenir goods, including a keychain in which he assumes the figure of a Buddha.

I first learned of marimo in 2001, when a friend gifted me one in a small glass bottle as a souvenir from a trip to Hokkaido. It sat on my desk in Tokyo, where I was studying Japanese language and literature as an undergraduate exchange student. I passed that marimo on to another friend when I returned to the United States, but my interest in the alga stayed with me. Years later, when I conducted my doctoral research on plant life in modern Japanese literature and cinema, I thought about marimo often. I was writing about authors and filmmakers who tried to rethink what it means to be human through a close engagement with plant life. Their novels, poems, and films imaginatively presented plants like mosses, vines, and trees as more similar to humans than is conventionally believed. I was tracking what I called a "botanical subjectivity" in these works, wherein humans attempted to experience the world in a plant-like manner. I was interested in how a botanical subjectivity was always presented as a multiple subjectivity – a subjective experience of a community rather than an individual, a forest rather than an individual tree. It struck me

FIGURE 9.2 Marimokkori
PHOTO COURTESY OF T. CSH

that marimo was a multiplicity in this way: a decentralized collection of filaments that seemed to constitute a singular being. I realized that this particular alga exists in popular consciousness somewhere on the spectrum between plant and animal (plants, after all, "aren't pets"). While it could be said that all algae exist in this liminal space between plant and animal, marimo's popularity marks it as a uniquely curious case. How, I wanted to know, did it become a household name in Japan in the first place?

Knowing that Lake Akan is home to the annual Marimo Festival, I began looking into the history of the region and quickly discovered that marimo's own history is thoroughly entangled within Japan's settler colonial history on Hokkaido. Central to Japan's "development" of Hokkaido, which officially became a part of Japan in 1869, was its forced assimilation and cultural erasure of the island's Indigenous Ainu population – settler colonial efforts to which the Marimo Festival seemed to be responding by celebrating Ainu culture. The festival features Ainu dance performances and culminates in a "sending off" ceremony in which a group of Ainu men row a hand-carved canoe out onto Lake Akan and return marimo to its waters. There is a dual nature to the festival that can be glimpsed in this latter ceremony, as it is intended to address both the suppression of Ainu culture and the endangered status of the marimo. One of the stated objectives of the festival is "protecting *aegagropilas* from poaching and the lowering of the water level due to hydroelectric power generation" (Irimoto 2004: 11).

Visitors to Lake Akan are generally led to believe that there is a long history of Ainu cultural affinity for the marimo, as the festival would suggest. Tourists are regularly greeted with a famous story long purported to be an Ainu folktale, one in which two young, star-crossed lovers jump into the lake and become a single marimo. This story piqued my interest, as it seemed to hint at a multiple subjectivity. Just as many filaments make up a single marimo, the two young lovers become one by becoming marimo. Conventionally called "The Legend of Marimo Love" (*Koi marimo densetsu*), the story is well-known throughout Japan and has taken form in numerous versions told across a wide range of media. But "The Legend of Marimo Love" is not a true Ainu folktale. It is, to borrow Richard M. Dorson's term, "fakelore" – a tale falsely ascribed to a folkloric tradition, having been written by a non-Ainu Japanese author (Dorson 1976). And while it draws from Ainu ceremonial practices, the Marimo Festival itself is not a "traditional" ceremony, having been created in 1950 by parties both Ainu and non-Ainu. And yet, members of the Ainu community around Lake Akan – which is home to the Akanko Ainu Kotan settlement that serves as a tourist destination and site for Ainu cultural promotion – have embraced both the story and the festival for their role in promoting tourism to the region. This chapter maps how marimo became a charismatic symbol of imagined indigeneity, beginning around Lake Akan and spreading throughout Japan (and ultimately the world at large). It considers what it means for the alga to have been embraced by the Lake Akan Ainu community in the name of preservation twice-over: of the endangered alga and of Ainu identity itself.

1 A Brief Algal History of Lake Akan's Marimo

The name "marimo" is comprised of two words: *mari*, which refers to a ball used for sports or play, and *mo*, which refers to aquatic plants like seaweed or algae. Despite its playfully evocative meaning, "marimo" is nevertheless a name that bears the legacy of Japanese settler colonialism. *Aegagropila linnaei* was given the name marimo by Kawakami Takiya (1871–1915), whose mention of the alga in 1898 in *The Journal of Japanese Botany* has been referred to as its "discovery" in Japan – a claim that effectively erases Ainu presence from the Lake Akan landscape (Nakazawa 1989: 14). Ainu ancestral lands once stretched from the northern end of Japan's main island, Honshū, up through Hokkaido, and out across the Kuril Islands and southern Sakhalin island (lewallen 2016). By the turn of the twentieth century, however, the Ainu were conventionally believed to be a "dying race" and were effectively written out of Japanese history by the passage of the 1899 Hokkaido Former Natives Protection Act (Hudson, lewallen, and Watson 2014: 1). It would take until the year 2008 for the Japanese government to officially recognize the Ainu as indigenous to northern Japan.

When Kawakami first wrote of marimo, he was a student at Sapporo Agricultural College, which was originally founded in 1872 in Tokyo as the Kaitakushi Tentative School, with "Kaitakushi" being the name of the governmental agency tasked with the "development" of Hokkaido (Wakana 2013). The school's purpose was to educate students to this end. It was relocated to Sapporo and renamed Hokkaido Imperial University in 1918, later becoming Hokkaido University, its current name, in 1947. While still a student, Kawakami was tasked by the local government in Hokkaido to survey both the weather and flora of the Lake Akan region. Such surveys were indispensable to a fledgling colonial government looking to transform the island of Hokkaido into the agricultural center it has become today.

But Kawakami's connection to marimo ties the alga further into Japan's colonial history, beyond the borders of Hokkaido, through imperial botany. Kawakami went on to name and document botanical species in Japanese-occupied Taiwan, where he helped establish the National Taiwan Museum and served as its inaugural director. In 1915, Kawakami published a travelogue/botanical field guide based on his trips throughout the South Pacific titled *In the Shade of Palm Trees* (*Yashi no hakage*). In his preface to the work, Kawakami writes that his botanical research is an important contribution to Japan's imperial expansion: "In my estimation, the South Pacific is a place we Japanese must certainly develop in the future, and we must exhaustively survey it from every aspect. Right now, it is of the utmost importance that we provide full knowledge of the true state of affairs to our nation" (Kawakami 1915: 11).[1] One can imagine Kawakami felt similarly about his work cataloguing the plant life around Lake Akan, including marimo.

A little over twenty years after Kawakami first published on marimo, the Japanese government officially recognized the alga as worthy of conservation. Following the 1919 passing of the Historic Sites, Scenic Beauty, and Natural Monuments Preservation Law (*Shiseki Meishō Tennenkinenbutsu Hozonho*), the government declared Lake Akan marimo a "Natural Monument" in 1921. One of the key factors that lead to this declaration was the uniqueness of the freshwater alga's biology. Of particular note was marimo's spherical shape and limited distribution; while *Aegagropila linnaei* can be found growing attached

1 Translation by the Author.

to rocks and as unattached filaments, the environment of Lake Akan resulted in unusually large ball-shaped marimo. (To date, the largest on record was 34 cm in diameter.) While *Aegagropila linnaei* have been found in freshwater sources throughout Japan and Europe, Lake Akan's marimo are renowned for their size. Another key factor leading to the designation of natural monument was the growing concern that human actions such as poaching could have adverse effects on the lake's ecosystem, leading to the loss of this unique species (Wakana 2020).

Such fears were justified. Around the time of marimo's designation as Natural Monument, a hydroelectric plant was established on the banks of the Akan River. As the river was used for electricity production, the water level of Lake Akan was lowered in areas in which marimo had accumulated. This led to marimo being exposed to the open air, causing many to dry out and die. In response to the increased energy demand occasioned by Japan's postwar recovery effort, the situation at Lake Akan continued to worsen after the end of World War II. In the spring of 1950, the water level dropped below 60 centimeters in some areas, exposing and killing many marimo. This incident, which entered public discourse as the "Electricity or Marimo?" debate, served as the impetus for the upgrading of Lake Akan's marimo to the status of "Special Natural Monument" in 1952 (Wakana 2020).

The "Electricity or Marimo?" debate also served as the background from which the Marimo Festival was created in 1950, the organizing guidelines of which highlighted its intention to promote conservation: "Various kinds of events are to unfold under the name of the 'Marimo Festival' to awaken people's love, which (has) faded for the marimo, among people within the (Kushiro region), and inspire them to protect them in order to preserve the fresh green marimo which lives at the bottom of Lake Akan, famous for its grand scenic beauty, as an eternal cultural property" (Irimoto 2004: 12). Yet due to the continued harvesting of marimo for souvenir use and further polluting of Lake Akan, the marimo population continued to decline after the 1950s, and by 1997 they were deemed critically endangered (Soejima et al. 2009). They continue to hold this status to this day (Boedeker et al. 2010).

2 On the Origins of "The Legend of Marimo Love"

One year after Lake Akan marimo were given the status of "Natural Monument," an unknown non-Ainu Japanese writer named Nagata Kōsaku published the story that would become "The Legend of Marimo Love" – the fakelore that would end up causing many to believe that there was a strong connection between the marimo and traditional Ainu culture. Nagata's story, "The Sad Sound of the Reed Flute on the Wind Blowing Down from Mt. Akan" (*Akan-oroshi ni kanashiki ashibue*), first appeared in the collection *Mountain Legends and Love Stories*, published by the Osaka office of the Asahi Shimbun newspaper. The book's introduction lauds the "expansive" collection of stories contained within, which include "mysterious ones, ones that are dramatizations of historical fact, and ones passed down by way of oral tradition" (Wakana 2019a).[2] The preface does not specify which stories belong to which category. But "The Sad Sound of the Reed Flute …" opens on the figure of an "elderly Ainu" who proceeds to tell the rest of the narrative, introducing it as "something that happened five-hundred years ago" (Nagata 1922: 173).[3]

This apparent oral folktale tells the tragic story of two young Ainu: Setona and Manibe (later, as the story took on a life of its own and spread through popular media, would his name change to Mani*pe*). Setona is the daughter of the village chieftain. Manibe is the son of a servant who works for Setona's family, and as such, is an inappropriate love match for her. To her dismay, Setona is instead promised to a vulgar man named Mekani. Near the end of the story, Manibe rescues Setona from Mekani's forceful advances, upon which she confesses her love for Manibe. Duty-bound to his position as a servant to her father, Manibe tells Setona that they cannot be together. Setona never recovers from this refusal. She ends the story confined to bed, "crying as if she had lost her mind" (Nagata 1922: 180). Manibe, meanwhile, is confronted by a knife-wielding Mekani late at night. A scuffle ensues, and Manibe manages to get a hold of Mekani's knife. He stabs Mekani, killing him on the spot. In fear, Manibe approaches the shore of Lake Akan and begins rowing his wooden canoe out onto the lake while blowing into his reed flute. He hears the sound of villagers rowing out after him. The story concludes:

> After that night, the figure of Manibe was never to be seen again. A few days later, Setona died while continuing to call out the name of her beloved. And so, it is said that the two have become one, within a single marimo that lives in the depths of the lake, and that the sad sounds of a reed flute come blowing from Mt. Akan down to the lake, mixed with the sound of a woman's crying (Nagata 1922: 181).

It is here that marimo makes its only appearance in Nagata's story. Like the legend it would go on to become, Nagata's story finds its two Ainu protagonists

2 Translation by the Author.
3 Translation by the Author.

becoming a single marimo after dying tragic deaths. But whereas "The Legend of Marimo Love" conventionally ends with "Manipe" and Setona jumping into Lake Akan in an act of lovers' suicide, Nagata's source material does not. Thus, part of the story's metamorphosis into "The Legend of Marimo Love" is its changed ending, in addition, of course, to its false attribution as an Ainu folktale.

How, then, did Nagata's story become what it became: a tale of lovers' suicide falsely ascribed to the tradition of Ainu mythology? This question weighed heavily on the mind of Wakana Isamu, the head researcher at the Kushiro International Wetland Center and one of the world's foremost marimo experts. It was Wakana (who is known by his nickname "Dr. Marimo") who first published on the connection between "The Legend of Marimo Love" and Nagata's forgotten source material. In a series of columns written for the Kushiro Shimbun newspaper between June 2018 and March 2019 (and later collected on his website), Wakana explains that over the thirty some-odd years that he has spent researching marimo, he has come across several texts that convinced him that "The Legend of Marimo Love" was not an Ainu story, but rather a creation of a Japanese writer (or *wajin*, which is the term often used to distinguish non-Ainu Japanese from Ainu).

While Wakana was the first to publish about Nagata's story being the source material for "The Legend of Marimo Love," he was not the first to doubt its authenticity. He mentions a specific essay by Satō Naotarō titled "On Marimo" (*Marimo ni tsuite*), which was included in a 1961 three-volume collection chronicling Satō's research on his native Kushiro region. In this essay, Satō claims that "The Legend of Marimo Love" is not an authentic Ainu folktale and that it began as a story included in the 1926 volume *Ainu Folklore* (*Ainu no densetsu*), compiled by Aoki Junji. The story Satō mentions, titled "The Sad Sound of a Reed Flute" (*Kanashiki ashibue*), is nearly identical to Nagata's version, but with a few key differences (Wakana 2019a). The most obvious difference is that the 1926 version is not ascribed to any particular author, even though it is clearly based on Nagata's story. It cuts out the character of the Ainu storyteller from Nagata's original version and renders Nagata's classically inflected literary prose into a more colloquial form of the Japanese language. These changes, bolstered by the fact that Nagata's name was not attached to the story, surely helped give it an aura of authenticity as "Ainu folklore," as the book's title suggests and its introduction outright claims when it says that "the stories included here are ones that I searched for in old documents, read thoroughly in exhaustive research on folklore, and furthermore, heard personally from the elderly inhabitants of Ainu villages" (Aoki 1924: 1).[4] And yet this 1926

4 Translation by the Author.

version of the story ends with a note that reads "From *Mountain Legends and Love Stories*" – the name of the 1922 book that featured Nagata's tale.

Having seen this note, Wakana decided to track down a copy of *Mountain Legends and Love Stories*, and thereupon developed his own suspicions that it was Nagata's version that was the true original. Then, out of the blue in 2017, he received a letter from Nagata's son, Natsuo. The letter explained how Natsuo wanted to clear up the origins of "The Legend of Marimo Love" for posterity's sake. Natsuo writes that his father had spent some time working in Kushiro, and while there, heard a story from an Ainu elder with whom he become acquainted. The story concerned a pair of young lovers from rival Ainu settlements who eventually drowned themselves in a lovers' suicide (Wakana 2019b). It was this story that apparently served as Nagata's inspiration when he later wrote "The Sad Sound of the Reed Flute on the Wind Blowing Down from Mt. Akan." However, given the fact that Nagata's story does not actually end with a lovers' suicide, one may wonder whether this anecdote is not a last-minute attempt to recuperate some amount of authenticity by linking the fakelore to an Ainu source in some small way. In any case, it is clear that marimo were added to the story at Nagata's discretion, likely owing to its having been granted "Natural Monument" status the previous year.

In the shift from Nagata's version to Aoki's modification of the story, the tale began its process of "becoming Ainu." I use this term in reference to a critical intervention into global indigenous studies laid out by ann-elise lewallen in her monograph *The Fabric of Indigeneity: Ainu Identity, Gender, and Settler Colonialism in Japan*.[5] In her study, lewallen calls the self-fashioning of identity that she examines in relation to clothwork "becoming Ainu," and argues that the term marks an important break from a biological view of indigeneity that "shifts the focus from a predetermined 'innate Ainuness' to an Ainuness selectively forged by each individual, thereby displacing the centrality of blood in regulating ethnicity and recentering individual agency and the process of self-determination" (lewallen 2016, 1). The transformation that begins with Aoki's version of the story and ends with the conventional assumption that "The Legend of Marimo Love" is an Ainu folktale did not come through a "process of self-determination," but rather through a long process of *interpolation*, understood here as a tool of assimilation in which *wajin* (non-Ainu Japanese) actors crafted an imagined indigenous cosmology from the outside, and imposed a cultural significance on the tale (and consequently the marimo itself) that had no historical basis. Marimo's "becoming Ainu" was thus initially a matter of "being made Ainu."

5 lewallen chooses to write her name in the lower case and I follow suit throughout.

As a rare species that aroused public imagination, marimo served as a kind of botanical terra nullius (or planta nullius) that was ripe for such inscription within a settler colonial system. Hokkaido had been deemed a terra nullius of its own, portrayed as a vast, empty frontier that, in Michele Mason's words, "served the state's goals by acting as a foil to confirm Japan's superior status and rationalize the colonial project" wherein its configuration as "a purely natural space, devoid of human habitation, history, and culture" not only rendered its Indigenous inhabitants an ahistorical blank slate, but also found newly "discovered" species like marimo free of any pre-existing cultural connotations (Mason 2012: 58). Trees, flowers, and grasses all had centuries of well-worn cultural associations in Japan, but marimo had no such deep aesthetic history.[6] Likewise, other species of algae had long culinary histories in Japan (as detailed in Ole G. Mouritsen and J. Lucas Pérez-Lloréns' contribution to this volume), but no such history existed for marimo on the Japanese mainland. Nor was there any use of the alga akin to the production of agar (the subject of Melody Jue's chapter in this book). This lack of Japanese knowledge and use of the alga made it easy for someone like Nagata or Aoki to dream up fakelore that linked the marimo to Ainu culture. As a planta nullis, marimo could be "made Ainu" from the outside – its aesthetic "otherness" bolstered by its new association with the colonial "other" of the Ainu. It was on this botanical blank slate that "The Legend of Marimo Love" was inscribed as fakelore, over and over again.

3 Singing Marimo

Nagata's short story metamorphosized in the years following its publication in 1922. By 1931, it had taken on enough of a life of its own that the Kushiro Shimbun newspaper ran an article promoting an upcoming radio broadcast devoted to "local Ainu folklore" that included a tale in which two Ainu lovers jump into the lake and become a single marimo through death – which, again, was not the original ending of the story (Wakana 2019c). But this version of the tale gained traction in the Lake Akan region as a means to promote tourism (especially after the establishment of the Marimo Festival in 1950), leading to a song inspired by the story to become a nation-wide hit in 1953. "Song of the Marimo" (*Marimo no uta*), written by lyricist Iwase Hiroshi and composer Yashima Hideaki, was performed by popular singer Andō Mariko, who was

6 In classical Japanese poetry, for example, plants were codified by lexicons called *saijiki*, which were essentially dictionaries of "seasonal words" (*kigo*) that regulated how and when plant names could be used.

born near Lake Akan, in Kitami. The song has since been covered numerous times and, like "The Legend of Marimo Love," it can often be heard in tourist establishments around the lake.

"Song of the Marimo" has three verses. The first mentions Lake Akan by name and speaks of the "lonely wind that crosses the surface of the water." It addresses the "floating marimo" directly, asking them what they are thinking. The second verse continues:

> Floating to the water's surface when clear
> Sinking to the water's depths when cloudy
> Love becomes one with sorrow
> Oh marimo, oh marimo
> Tearful marimo

Biologist Sakai Yoshio begins his 1991 book *The Science of Marimo* (*Marimo no kagaku*) by disputing this verse's claim that Lake Akan's marimo float and sink depending on the weather. "To be sure," he writes, "marimo in a laboratory aquarium with good light will float and sink, but in Lake Akan, they don't move the way the song claims" (Sakai 1991: iv).[7] Clearly, the "Song of the Marimo" was still popular enough some four decades after it was released that Sakai felt the need to dispel the misinformation it promoted.

But what Sakai does not feel the need to address is the song's invocation of the "The Legend of Marimo Love." For what is hinted at in the second verse ("love becomes one with sorrow") becomes explicit in the third verse:

> Even now, in the Ainu village
> Of a lasting romance it sings, sorrowfully
> Marimo, the shadow it casts is lonely
> Oh marimo, oh marimo
> Green marimo

While it does not mention Manipe and Setona by name, the song references their doomed romance. It would seem that by 1953, the fakelore was known well enough that it could be referenced in such an indirect way and still connect with a national audience that understood the referent. Today, a large stone memorial bearing the song's lyrics sits on the shore of Lake Akan (Figure 9.3). The story it tells, a secondhand product of imagined indigeneity, stands literally engraved into the landscape.

7 Translation by the Author.

FIGURE 9.3 The lyrics of "Song of the Marimo" engraved into rock on the shores of Lake Akan
PHOTO COURTESY OF 663HIGHLAND

4 Dancing Marimo

Said landscape would go on to serve as the setting for an international collaboration between a Japanese composer and a Russian ballet dancer during the height of the Cold War that once again suggested "The Legend of Marimo Love" was an Ainu tale. In 1962, composer Ishii Kan debuted his ballet *Marimo* in Tokyo. Four years earlier, he had won a prize for his composition titled *Symphonia Ainu*. Ishii was invited to compose the music for *Marimo* by the Tchaikovsky Commemorative Ballet School of Japan, which also invited A.A. Warlarmov from the Bolshoi Ballet School in Russia to help develop the production.

According to the liner notes included in the soundtrack album for *Marimo*, Warlarmov "became very interested in the legend about the 'marimo'" and thus "carried out research on the customs of the Ainu" (*Marimo Ballet Suite*, n.d.). Warlarmov reportedly visited the Ainu community at Lake Akan and made recordings of Ainu music and 8mm videos of Ainu dance. He also consulted with Japanese artists and intellectuals in order to avoid having his ballet turn into an "inappropriate work made by a foreigner" (Saitō 2018). Perhaps most notable among Warlarmov's consultants was author Takeda Taijun, whose 1955 novel *Forest and Lake Festival* (*Mori to mizu'umi no matsuri*) is set around Lake Akan and features, in its opening a chapter, a tour guide who "must recite the tragic love story" of Setona and Manipe "countless times each day" (Takeda 1962: 17).

The fact that "The Legend of Marimo Love" was fakelore seems not to have been communicated to Warlarmov, and so Manipe and Setona became the focus of the ballet he helped author even as he reportedly strove for authenticity. The story once again took on a new iteration. In the ballet, Setona is no longer the village chief's daughter. In this version, she is poor, and meets Manipe as he saves her from a bear attack. The two fall in love, but it is Manipe who ends up being forced to marry another. In order to nullify Manipe's engagement to this other woman, Setona must face a series of challenges to appease the god of Lake Akan. She has her eyes plucked out and loses her ability to hear and speak. Manipe recognizes Setona despite these changes when she arrives at his wedding ceremony. The two leave the ceremony together. Manipe then faces three challenges of his own: he is attacked by a murder of crows, forced to struggle against a massive windstorm, and is told that he must leave his life behind and jump into the lake, where he will then find his beloved. Having completed these tasks, the two become marimo, and dance "a pas de deux of love" (*Marimo Ballet Suite*, n.d.).

5 Another Legend of Becoming Marimo

Through song and dance, the fakelore that is "The Legend of Marimo Love" took on new forms and found new life as inspiration for variations on a theme that was still believed by many to be an Ainu folktale. But not everyone believed the story's supposed provenance. It was Satō Naotarō's essay that convinced Wakana Isamu that "The Legend of Marimo Love" was not a true Ainu folktale, but Satō was not alone is doubting its origins. Poet, farmer, and Anarchist Sarashina Genzō (1904–1985), for example, raised doubts about the authenticity of "The Legend of Marimo Love" in the first volume of his three-volume account of Ainu ecology titled *A Kotan Wildlife* (*Kotan seibutsuki*), which was originally published in 1942 and then revised and republished in 1976. Sarashina, who was born in Hokkaido to first generation *wajin* settlers and both collected Ainu folklore and wrote about Ainu life in his poetry, devotes a section of this first volume of *A Kotan Wildlife* to marimo. He begins the section with a direct reference to "The Legend of Marimo Love," claiming that marimo is "the main character of (the story), which you are bound to hear somewhere if you visit Lake Akan" (Sarashina and Sarashina 2020: 257).[8] But, Sarashina points out, as beautiful as that story is, it is not to be found in any of the Ainu tales handed down from antiquity.

Sarashina writes that there *is*, however, a tale known among the elderly Ainu around Lake Akan concerning marimo, but that it is radically different in tone from "The Legend of Marimo Love." He recounts it as follows:

> Long ago, for some reason, the god of Lake Akan disliked water chestnuts (*pekanpe* in Ainu). Yet somehow or other, some water chestnuts found their way into the lake, and hoping for company, they asked the god to help them multiply. The god replied coldly, "Just by being here, you dirty up the lake. And if your numbers were to grow, humans would come to collect you, and further muck up this beautiful lake. By no means can I allow you to be here." This angered the water chestnuts, and they plucked up the grass along the lake shore and all the water plants and threw them all at the god of the lake They then left. The grass and water plants became entangled in refuse, thus forming marimo. And so, the Ainu word for marimo is *tokarip* (which means "that which lies in the lake") (Sarashina and Sarashina 2020: 258).

Sarashina notes that this folktale is in no way a "love story" and explains that Ainu feelings toward marimo were less than favorable. He recounts that he was taught by Ainu elders that as marimo numbers increase, it becomes difficult to catch fish. He lists a few other Ainu words for marimo that express this dislike of the alga: *torasanpe* (or "lake goblin") and *tosuruku* (or "lake poison"). Sarashina concludes that "If there is a story claiming to be an Ainu folktale that has a beautiful love story written into it, we can be sure, without a doubt, that it is a fictional tale written by a *wajin*" (Sarashina and Sarashina 2020: 258). The fact that his *Kotan Wildlife* would gain mainstream popularity some seventy years later due to its association with the highly popular and yet controversial manga *Golden Kamuy* (which began serialization in 2014) would likely have amused Sarashina. The story, which is set in Hokkaido around the turn of the 20th century, briefly features a scene in which a child moves to Lake Akan and is bullied into eating marimo (Noda 2020). It is a scene befitting the description of marimo as "lake goblin."

8 Translation by the Author.

6 On Becoming Ainu through Becoming Marimo

Yet even before Sarashina attempted to set the record straight, Yamamoto Tasuke (1904–1993), who would become a major figure in the Ainu independence movement after the war and the driving force behind the creation the Marimo Festival in 1950, was writing of the inauthenticity of "The Legend of Marimo Love." In 1940, Yamamoto published his own collection of Ainu folk tales titled *Akan National Park and Ainu Folklore* (*Akan kokuritsu kōen to Ainu no densetsu*) and included a version of "The Legend of Marimo Love" that he called "The Tale of Setona and the Marimo." Yamamoto summarizes the story and follows the conventional telling, including the lover's suicide at the end. But his framing of the story casts doubts on its origin. Yamamoto suggests the story is in fact fakelore: "In order to write this book, I conducted research for three years in the Akan area, but none of the Lake Akan Ainu knew this famous legend" (Yamamoto 1940: 16).[9]

Given that Yamamoto knew "The Legend of Marimo Love" was not a story told among the Ainu community at Lake Akan, his decision to include it in his collection can be seen as a first step toward reclaiming the fakelore. His inclusion of the story was an effort to "become Ainu" within lewallen's parameters of the term, in which "self-craft describes a process of forging an Ainu identity firmly rooted in ancestral values, worldview, and lifeways, but one that is sufficiently flexible to adapt these values to meet the needs of the present" (lewallen 2016, 2). Yamamoto seemed to understand that marimo was a planta nullis and decided to begin an inscription of his own. He saw the potential to reclaim the supposed "Ainu-ness" of marimo perpetuated by "The Legend of Marimo Love" and use the alga's imagined indigeneity to self-craft a new tradition ten years later in the form of The Marimo Festival – a tradition that better reflected Ainu "ancestral values, worldview, and lifeways." A brochure for the Akanko Ainu Kotan that promotes the festival elaborates on these values and lifeways:

> The Ainu people are an indigenous people to Hokkaido where they lived for countless generations before the arrival of the sisam (Japanese). Their ancestors referred to Hokkaido as Ainu Mosir (tranquil land of the people) and believed spirits, or kamuy, inhabited the natural world. They humbly paid homage to these kamuy and lived in thanks of the bounty of nature. Under the spirit of coexistence symbolized by the saying "I am where the kamuy are and the kamuy are where I am," the Ainu people have lived in

9 Translation by the Author.

> harmony with their surroundings without modifying, destroying or polluting the natural environment (Honda 2013: 2).

Created to help preserve both Ainu cultural heritage and the endangered alga itself, the Marimo Festival thus enacts a form of traditional ecological knowledge in which the sacrality of the natural world informs ecological conservation. As a first step toward self-fashioning a cultural relationship to marimo that would ultimately result in the creation of the Marimo Festival, Yamamoto's *Akan National Park and Ainu Folklore* ultimately becomes a site of resistance within a settler colonial logic that looked to define what is and what is not authentic Ainu tradition.

At the same time, however, the Marimo Festival has held (and continues to hold) a contested, ambiguous status among the Ainu. As Tessa Morris-Suzuki has argued, "the preservation and presentation of culture have been deeply contested, divisive, and problematic issues for Ainu society" (Morris-Suzuki 2014: 50). The Marimo Festival is no exception. In director Fukunaga Takeshi's 2020 film *Ainu Mosir*, which is a story about the Ainu community around Lake Akan that features actors from the local community, the Marimo Festival is presented as primarily public-facing and is contrasted with a controversial bear sending-off ceremony that the film's narrative portrays as a more authentic and private ritual.[10] Indeed, since its inception, the Marimo Festival has been criticized for exploiting Ainu culture for the sake of tourism. In the face of such criticism, Yamamoto has defended the creation of the event as a means "to regain our pride" (Ziomek 2019: 367). The Akanko Ainu Kotan continues to defend the festival as a way to "offer devout prayer to the kamuy (spirits) through the marimo" and thus turn Lake Akan "into a focal point for the Ainu people to pass down their spiritual heritage to future generations" (Honda 2013: 19).

Yamamoto recognized the flexibility that the ambiguous status of the marimo afforded him in regard to notions of authenticity. He used this ambiguity to reawaken an Ainu culture he felt had faded in the wake of World War II, claiming that "when the Ainu came to gather for the (Marimo) festival, I saw a forgotten people … become excited. It was planned as a cultural exchange. As a result, in each region the practice of Ainu dances began again. Each year the Ainu came together and the level of the arts continued to improve, and their spirit changed" (Ziomek 2019: 368). lewallen notes that tourism has been a "double-edged sword" for Ainu communities, as it "has packaged Ainu cultural knowledge and difference for mass consumption" while also "(serving)

10 While director Fukunaga was raised near Lake Akan, he does not identify as Ainu himself.

as an incubator to maintain this knowledge and link culture once again with economic livelihood" (lewallen 2016: 90). Yet she also stresses that tourist communities like the one at Akanko Ainu Kotan, "have enabled the preservation of *Ainupuri* (proper comportment based on ancestral protocols) and conservation of the natural resources necessary to sustain it" (lewallen 2016: 92).

Yamamoto's reclamation of "The Legend of Marimo Love" and subsequent creation of the Marimo Festival can thus be read as an act of "becoming Ainu" through an agential self-inscription onto the planta nullius that is the marimo. And now that it has been firmly established that "The Legend of Marimo Love" is fakelore, members of the Akanko Ainu Kotan appear willing to let the story continue to circulate in order to draw visitors to Lake Akan. A 2017 newspaper article titled "'The Legend of Marimo Love' was the Creation of an Ethnic Japanese (*Wajin*)" closes with the words of Nishida Masao, the head of the Ainu Craft Collective at the Akanko Ainu Kotan who can be seen participating in the Marimo Festival in Fukunaga's film: "The story's plot is interesting, and it encourages tourism to Lake Akan. It's fine to continue introducing this story to people from here on out, as long as we make it clear that it's not an Ainu legend" (*Hokkaidō Shimbun* 2017).[11]

7 Coda: Invasive Algae

In March 2021, the U.S. Fish and Wildlife Service issued an alert to anyone who had purchased "a moss ball aquatic plant product" after February 1st of the same year, asking them to "Destroy! Don't Dump!" their marimo. The charismatic pet algae, it was discovered, had been quietly smuggling an invasive species into the United States: the zebra mussel. The alert calls the zebra mussel "one of the most destructive invasive species in North America" as it "can quickly take over once they get established in a waterbody and cause significant damage" (U.S. Fish & Wildlife Service). To ward off this threat of invasion, the Fish and Wildlife Service mandated marimo be destroyed through freezing, boiling, or bleaching, and then be disposed of in sealed plastic bags.

Given marimo's place within the settler colonial history of Hokkaido, the concerns over its invasive potential and the systematic plan to eradicate its existence in North America take on an uncanny hue. Once again marimo finds itself being used by invasive actors, its charismatic charm hiding the threat of uncontrolled expansion hidden within its green "body." The invasive marimo in question, however, did not come from Lake Akan, but rather were imported

11 Translation by the Author.

into California from Ukraine (Montana Fish, Wildlife, and Parks). According to a 2010 article published in *BioScience*, Ukraine has been the major supplier of *Aegagropila linnaei* to the international aquarium trade, with the alga being harvested from the Shatsk lakes (Boedeker et al. 2010: 192).[12] Chain retailers such as Petco pulled these "lake goblins" off their shelves, although one can still purchase them from online retailers that continue to mistakenly promote "The Legend of Marimo Love" as an authentic Ainu folktale (Moss Ball Pets). And recently, marimo have begun appearing on pet shop shelves once again.

It is tempting to see this zebra mussel-infested chapter of marimo history as something of a conclusion to the long tale of becoming marimo that began with Nagata Kōsaku's decision to write fakelore in 1922. As he wrote of an imaginary indigenous cosmology in which Ainu lovers became forever united within a single marimo, Nagata set off a chain of events that would distance the marimo further and further from Ainu traditional ecological knowledge. As "The Legend of Marimo Love" helped fuel the commodification of the marimo, the alga eventually entered into an economic network of international distribution that would take it from Ukrainian lakes to the United States, where it would come to threaten the very waterways that would serve as its new expanded ecosystem.

And thus, while consumers who may have been lured in by the romantic tale of Manipe and Setona must now freeze, boil, and/or bleach their marimo in preparation for the unbecoming that is disposal, the community around Lake Akan prepare for another year of the perpetual becoming that is the Marimo Festival, where they will sing and dance and return marimo to the watery depths of the lake from whence they came. In the process, their actions will echo the words of Norway-based Ainu artist Uzawa Kanako, who describes her contemporary take on traditional Ainu dance as "a new way to emphasize" the "continuously developing" nature of culture. "As I am of a generation that is the result of strong assimilation", she continues, "I feel I have lost many aspects of traditional culture but I refuse to accept that I am not in touch with my culture when I choose to author it in my own way" (Uzawa 2014: 89).

12 The article explains: "Balls from Lake Svityaz are shipped even to aquatic plant-breeding facilities in Southeast Asia before they are returned to the European market. Japanese aquarium shops sell balls only of European origin. No other natural source of *A. linnaei* balls other than the Shatsk lakes (including Lake Svityaz) was ever mentioned by people in the aquarium plant-trade business" (Boedeker et al. 2010: 192).

References

Aoki, J. 1924. Ainu no densetsu to sono jōwa. Sapporo: Fūkidō Shobō.

Boedeker, C., A. Eggert, A. Immers, E. Smets. 2010. "Global Decline of and Threats to *Aegagropila linnaei*, with Special Reference to the Lake Ball Habit." BioScience 60 (3): 187–198. doi:10.1525/bio.2010.60.3.5.

Chiba, N. 2010. Marimo o mamoru: Wakana Isamu-san no kenkyū. Tokyo: Rironsha.

Dorson, R. 1976. Folklore and Fakelore: Essays Toward a Discipline of Folk Studies. Cambridge: Harvard University Press.

Fukunaga, T., dir. 2020. Ainu Mosir. AINU MOSIR LLC/Booster Project.

Hokkaidō Shimbun. 2017. "Marimo koi densetu, wajin no sōsaku datta – Ainu minzoku no hanashi kiki chakusō." August 22, 2017.

Honda, Y., ed. 2013. Tracing the History of the Land of Lakes and Forests: Akan Ainu Cultural Heritage. Kushiro City: Akanko Hot Springs Ainu Cultural Promotion Executive Committee.

Hudson, M., A. Lewallen, M. Watson. 2014. Beyond Ainu Studies: Changing Academic and Public Perspectives. Honolulu: University of Hawai'i Press.

Irimoto, T. 2004. "Creation of the Marimo Festival: Ainu Identity and Ethnic Symbiosis." Senri Ethnological Studies 66: 11–38. http://doi.org/10.15021/00002684.

Kawakami, T. 1915. Yashi no hakage. Tokyo: Rokumeikan.

lewallen, a. 2016. The Fabric of Indigeneity: Ainu Identity, Gender, and Settler Colonialism in Japan. Santa Fe: School for Advanced Research Press.

Marimo Ballet Suite: Ainu's Racial Ballet in Three Acts. n.d. Modern Japanese Composers Series No. 14, Kan Ishii. Toshiba Records TA7008. Vinyl record.

Mason, M. 2012. Dominant Narratives of Colonial Hokkaido and Imperial Japan: Envisioning the Periphery and the Modern Nation-State. New York: Palgrave Macmillan.

Montana Fish, Wildlife, and Parks. "Invasive Mussels Found in Aquarium Moss Balls Sold in Montana." Accessed May 26, 2021. https://fwp.mt.gov/homepage/news/2021/march/0304-invasive-mussels-moss-balls.

Moss Ball Pets. "A Love Story." Accessed May 20, 2021. https://mossballpets.com/pages/a-love-story.

Nagata, K. 1922. "Akan-oroshi ni kanashiki ashibue." In Yama no densetsu to jōwa. Osaka: Osaka Mainichi Shimbunshakan.

Nakazawa, S. 1989. Marimo wa naze marui – sono seitai to keitai. Tokyo: Chūō Kōron.

Noda, S. 2020. Golden Kamuy 18. San Francisco: Viz Media.

Saitō, K. 2020. "Baree 'Marimo' (1962nen) to shakaishugi riarizumu: Ainu ni matsuwaru sōsaku densetsu no soviet bareeka." Minzoku Geijutsu 34: 123–128.

Sakai, Y. 1991. Marimo no kagaku. Sapporo: Hokkaidō Daigaku Tosho Kankōkai.

Sarashina, G., Sarashina, K. 2020. Kotan seibutsuki 1: Jumoku, zasō-hen. Tokyo: Seidosha.

Soejima, A., N. Yamazaki, T. Nishino, I. Wakana. 2009. "Genetic variation and structure of the endangered freshwater benthic alga Marimo, *Aegagropila linnaei* (Ulvophyceae) in Japanese lakes." Aquatic Ecology 43 (2): 359–370. http://doi.org/10.1007/s1045020.

Takeda, T. 1962. Mori to mizu'umi no matsuri. Tokyo: Shinchōsha.

Tolentino, M. 2014. "Want a super low maintenance pet? Try marimo!" tsunagu Japan. https://www.tsunagujapan.com/want-a-super-low-maintenance-pet-try-marimo/.

U.S. Fish & Wildlife Service. "Invasive Zebra Mussels Found in Moss Balls." Accessed May 26, 2021. https://www.fws.gov/fisheries/ans/zebra-mussel-disposal.html.

Uzawa, K. 2014. "Charanke." In Beyond Ainu Studies: Changing Academic and Public Perspectives, 89–90. Honolulu: University of Hawai'i Press.

Wakana, I. 2013. "Marimo no tōsai o kangaeru." Hokkaidō no shizen 51: 64–75.

Wakana, I. 2019a. "Ainu no densetsu ka Wajin no sōsaku ka." Marimohakushi no kenkyū nikki. https://ameblo.jp/maromo-lab/entry12433135776.html?frm=theme.

Wakana, I. 2019b. "Ainu no hiren o moto ni sōsaku." Marimohakushi no kenkyū nikki. https://ameblo.jp/maromo-lab/entry-12435375245.html.

Wakana, I. 2019c. "Hōsōkaishi no rajio de kakusan." Marimohakushi no kenkyū nikki. https://ameblo.jp/maromo-lab/entry-12438776671.html.

Wakana, I. 2020. "Tokubetsu tennen kinenbutsu: Akanko no marim." Hokkaidō no shizen 58: 104–105.

Yamamoto, T. 1940. Akan kokuritsu kōen to Ainu no densetsu. Tokyo: Nihon Ryokō Kyōkai.

Ziomek, K. 2019. Lost Histories: Recovering the Lives of Japan's Colonial Peoples. Cambridge: Harvard University Asia Center.

CHAPTER 10

"A Seaweed Goes to War": Agar as a Thermal Medium in C.K. Tseng's Research at the Scripps Institution of Oceanography (1943–1946)

Melody Jue

Abstract

This chapter examines how C.K. Tseng's wartime scientific writings during 1943–1946 at the Scripps Institution of Oceanography tell a story about agar as a thermal medium of labor in transpacific contexts. This story is framed by war and agriculture through the role agar played as a bacteriological culturing medium, as well as a thermal medium – a "cold food" and a gel with unique thermal properties. By accounting for agar's thermal properties from both western scientific and East Asian perspectives, I argue that Tseng avoids the fallacy of "thermal objectivity" (Starosielski 2021). Tseng's comparison of thermal approaches to understanding agar is cross-cultural and transnational, requiring a consideration of what I call different "thermal epistemes," or systematic practices of knowing through temperature. I show how Tseng addressed both eastern and western thermal epistemes in the way he described agar as a thermal medium, a culturing medium, and finally a medium of labor. Tracing these three medial qualities enables me to show that Tseng's personification of Gelidium as a soldier was no mere literary flourish, but rather an extension of his thinking about his own conditions of labor in the laboratory, under the shadow of WWII.

Keywords

agar – temperature – labor – objectivity – science and technology studies – WWII – Asia – seaweed – Scripps Institution of Oceanography

© MELODY JUE, 2024 | DOI:10.1163/9789004683310_012
This is an open access chapter distributed under the terms of the CC BY-NC 4.0 license.

When the young phycologist C.K. Tseng[1] took a postdoctoral position at the Scripps Institution of Oceanography in San Diego in 1942, he had no idea that he would be unable to return to his home country of China until the conclusion of WWII. While he would later become a leading force in the development of Chinese aquaculture, as a postdoc at Scripps he was asked to take charge of experiments on agar. Agar is an important seaweed product used to culture and identify bacteria in water supplies, milk, and medical contexts. Since agar was supplied almost entirely by Japan prior to the war, this shortage created a security crisis for the home front – and Tseng was conscripted into scientific service.

Agar's connection to national security appears in surprising places. Among Tseng's many publications during this three-year period was his essay, "A Seaweed Goes to War," in *California Monthly*, which began:

> In 1917 The United States went to war; Private *Macrocystis* was drafted. In 1941 the United States went to war again: this time Private *Gelidium* answered the call. The macroystis is generally known as "giant kelp" and is specifically *Macrocystis pyrifera*. The gelidium is known as "agarweed" and is specifically *Gelidium cartilagincum*. Both are natives of California, also living in South Africa and a few other places. Both belong to the same group of plants which we used contemptuously to call "seaweeds," but from which we are now deriving millions of dollars worth of precious products (Tseng 1944, 10).

In a striking act of personification, Tseng describes two different seaweed species – *Macrocystis* and *Gelidium* – as soldiers or "privates" answering a call to war for the United States, a curiously patriotic statement for a Chinese citizen. While Tseng's metaphor of a seaweed going to war might conjure science fictional images of an aggressive plant fighting for a nationalist cause, seaweeds were less about fighting an enemy than about performing certain forms of agrarian and industrial labor on the home front. During WWI, Giant Kelp, or *Macrocystis pyrifera*, was a source of the fertilizer additive potash, previously imported from Germany; in WWII, *Gelidium* was important for making agar, used in bacteriology: "To determine the type of pathogenic bacteria in a

1 曾呈奎, or Zeng Chengkui in Pinyin, the current system for romanizing Chinese characters. C.K. Tseng alumni profile, https://sites.lsa.umich.edu/chinese-alumni/zeng-chengkui-%E6%9B%BE%E5%91%88%E5%A5%8E/, accessed June 29, 2021.

disease or to determine whether certain bacteria found in the water supply are dangerous," Tseng wrote, "they have to be cultivated" (Tseng 1944, 10).

Tseng explained agar's role in national security through describing its importance as a bacterial culturing medium. In later writings, Tseng elaborated:

> The most important service that agar renders to mankind, in war or in peace, is as a bacteriological culture medium. To determine the type of pathogenic bacteria in a disease or to determine whether certain bacteria found in the water supply are dangerous, they have to be cultivated. In the preparation of serum and anti-toxins for protection against diseases the particular types of bacteria involved have also to be cultivated (Tseng 1994, 10).

This bacterial cultivation was only possible because of agar's thermal properties as a medium, which also made it useful in the hot drawing of tungsten wire, in making dental impressions and prosthetics, in sealing marine storage batteries, and in stabilizing a variety of foods.[2] Whereas bacteria liquify gelatin (and eat it), agar remains solid and inert when heated to temperatures that promote bacterial growth. Culturing bacteria was a precondition for identifying it, more of a forensic process than a martial one. "Culturing" in this sense is related to the practice of growing bacteria (Landecker 2010), not unlike the aquacultural interests that initially drew Tseng to the study of seaweeds.

Tseng's personification of *Gelidium* as a soldier going to war also channels a story about labor, a category that need not apply only to humans, but can also extend to animals, microbes, and non-human others (Barua 2018; Paxson 2018). Yet it is notable that Tseng personifies the living seaweed, *Gelidium*, rather than the extract product of agar. In his framing, *Gelidium* is the soldier/worker in service of the US war effort, whereas agar gel is more like a microcosm of the factory – or laboratory, or soil – an engineered environment for the production and culturing of living bacteria. "A seaweed goes to war" might have been more accurately titled, "A seaweed goes to the laboratory," not to fight bacteria, but to grow it, and in doing so ensure the safety of water and food supplies. Tseng's writings situate Private Gelidium

2 Tseng's discussions of another economic seaweed in China, Gloiopeltis, also interweave reflections on its properties as a medium. One of the reasons why Gloiopeltis was so valuable in Amoy was because of its important use as a sizing medium (to coat and smooth threads) in the production of silk. It was also used as a varnish on paper lanterns, paper walls, and even "ghost money" (冥幣) – a symbolic form of currency, burned as an offering to ancestors. Tseng, *The Selected Writings of C.K. Tseng*, 929.

not so much in the heat of battle, as in the warmth of a culturing dish and the kitchen.

Across his writings, Tseng connected agar to non-Western understandings of temperature – not as an external measure, but as an intrinsic property of foodstuffs. Indeed, what is remarkable about Tseng's agar publications is that many of them include literature reviews about agar that address agar's culinary uses in Japan and China as well as its laboratory applications. This work of synthesizing was likely motivated by the martial context in which he was writing, since laying out what was known about harvesting *Gelidium* seaweed and producing agar was a key first step for the United States' goal of jump-starting its fledgling agar industry. Tseng's fluency in both Chinese and English was essential for navigating the global circulation of agar, considering that it had been produced in Japan successfully for over 350 years prior to the time of his writing. Subsequent American literature reviews on agar quoted Tseng, rather than reviewing East Asian sources in their original languages (Humm 1947).

Tseng's interest in seaweeds as food framed how he compared their uses and thermal properties. Sometimes called the "Father of Chinese Oceanography" for his formative work on Chinese aquaculture, Tseng consistently describes his motivation for studying seaweeds as part of a larger humanitarian goal: feeding people (Tseng 1994, 1).[3] *The Selected Writings of C.K. Tseng* (曾呈奎文选, 1994) not only discuss how temperature and thermal manipulation play a role in the making of agar from *Gelidium* and other seaweeds, but also contain substantial ethnobotanical detail about the ways that seaweeds were used as a source of food in East and Southeast Asia. As I will elaborate, Tseng regularly mentions how edible seaweeds like agar were seen considered part of a "cool diet," a perspective that constituted a different "thermal episteme" (my term) than Western perceptions of temperature. In an East Asian context, temperature can describe literal temperature, or characterize the essence of a food or medicinal substance – part of a system of thought predicated on maintaining a balance of yin and yang (cold and hot) in the body.

In this chapter, I trace how C.K. Tseng's wartime scientific writings during 1943–1946 at the Scripps Institution of Oceanography tell a story about agar as a thermal medium of labor in transpacific contexts. This story is framed by war and agriculture through the role agar played as a bacteriological culturing

3 C.K. Tseng alumni profile, https://sites.lsa.umich.edu/chinese-alumni/zeng-chengkui-%E6%9B%BE%E5%91%88%E5%A5%8E/. C.K. Tseng first developed an interest in seaweeds through coursework at Amoy University (now "Xiamen"), where he earned a B.Sc. in Biology in 1931. For several years after, Tseng conducted some of the first and most extensive taxonomic surveys of seaweeds on the Chinese coastline, before pursuing a PhD at the University of Michigan in 1940 to pursue studies in phycology.

medium, vital to national security through testing and ensuring safe water and food supplies. Yet in Tseng's writings, agar is not just a culturing medium, it is a thermal medium – a "cold food" and a gel with unique thermal properties, different across eastern and western thermal epistemes. By accounting for agar's thermal properties from *both* western scientific and east Asian perspectives, I argue that Tseng avoids the fallacy of "thermal objectivity," a term that Nicole Starosielski uses to name a sense of temperature as "independent of both culture and perception" that "offers a universal language – degrees Celsius and Fahrenheit (and Kelvin) – that can describe engines, people, and climate change alike," an epistemic position often taken for granted by scientists (Starosielski 2021, 2). In order to counter the ways that thermal objectivity masks the operations of thermopower, Starosielski aims to evoke "a multitude of thermocultures," defined as, "the cultural processes of thermal modulation and exposure that pattern sensitivities to heat and cold, construct normative ways of making sense of thermal stimuli, and set expectations for temperature" (Starosielski 2021, 5).

Tseng's comparison of thermal approaches to understanding agar is cross-cultural and transnational, but it requires a stronger term than thermocultures. It requires thinking with *thermal epistemes*, or systematic practices of knowing through temperature. In an eastern thermal episteme, agar is a "cold" food, where cold is its essential or ontological property (the agar *is* cold). By contrast, from a western scientific thermal episteme, agar exhibits interesting properties in the laboratory due to thermal changes (agar *responds to* coldness or hotness). In the following sections, I show how Tseng addressed both eastern and western thermal epistemes in the way he described agar as a thermal medium, a culturing medium, and finally a medium of labor. Tracing these three medial qualities enables me to show that Tseng's personification of *Gelidium* as a soldier was no mere literary flourish, but rather an extension of his thinking about his own conditions of labor in the laboratory, under the shadow of WWII.

1 Agar as a Thermal Medium

Tseng's allusions to an East Asian thermal episteme most clearly emerge in his descriptions of agar as a food that is part of a "cool diet." In an influential paper where he coined and defined the term "phycocolloid" – a gluey substance derived from seaweed cell walls – Tseng notes that, "For ages, sweetened and flavored agar-gel has been a favorite 'cool diet' in the summer in the Orient, especially in the hot tropical and subtropical countries. In Japan, agar jellies are mixed with bean paste and some coloring matter, and made in various shapes"

(Tseng 1994, 1048). He also alludes to the way a different seaweed, *Gloiopeltis* can work as an agar substitute,

> serving as a cool, sweet diet during hot summer weather. For making the dish, the seaweed is first cooked with water in order to extract its gelose content. After boiling from half an hour to an hour, by which time all the gelose will be dissolved out, a little vinegar is added to counteract the slightly unpleasant odor of the seaweed. [...] The resulting firm gel constitutes a very palatable and refreshing dish (Tseng 1994, 933).

Tseng's references to a "cool diet" in these passages is not about foods that are literally cold and refreshing, but also cold in their essence, according to their properties in traditional Chinese medicine.[4]

What does it mean to say something is cool in its essence, or nature? Understanding this requires familiarity with conceptions of dynamic balance (yin, yang) in traditional Chinese medicine, and their thermal connotations. Among other conceptual dyads, yin (陰) is associated with coolness, while yang (陽) is associated with heat.[5] Not everyone should eat a "cool diet" even when it is hot outside, and Chinese medicine and eating practices prescribe different kinds of eating regimens along the lines of gender and age, among other factors.[6] Medical anthropologist Cordia Ming-Yeuk Chu observes, "When they [Chinese people] complain that they are 'hot,' they do not necessarily mean hot in temperature, and they will take medicine or food of a cold nature (for instance, see below) in order to balance this condition. Conversely, when they feel too 'cold,' they will take in 'hot-natured' foods or medicines." (Chu 1980, 42).

4 Tseng isn't the only seaweed biologist to note the use of agar jelly as a "cold food." Bangmei Xia and Isabella A. Abbott write of agar jelly, "The gelled product, 'liang fen,' is cut into cubes about 12 cm on a side, which are sold in markets and along the street; in summer it is a common sight. Dried Gelidium and Pterocladia are also sold in street markets in summer. Zhejiang and Fujian inhabitants especially like sweets, so they add sugar and fruit juice to liquid before it gels. In summer, it is an especially popular 'cooling' food throughout coastal China" in "Edible Seaweeds of China and Their Place in the Chinese Diet," Bangmei Xia and Isabella A. Abbott, *Economic Botany* 41:3 (1987), 349.

5 There are thermal connotations in compound uses of these characters, such as in 陰天 (a cloudy day) and 太陽 (the sun). Yin and yang can also refer to the balance of other elements, for example, depletion and repletion. On Chinese Traditional Medicine, see Volker Scheid, *Chinese Medicine in Contemporary China: Plurality and Synthesis* (Durham: Duke Press, 2002); Mei Zhan, *Other-Worldly: Making Chinese Medicine through Transnational Frames* (Durham: Duke Press, 2009); Judith Farquhar, *Appedites: Food and Sex in Post-Socialist China* (Durham: Duke Press, 2002).

6 While eating dim sum, I have personally been advised by older Chinese women to avoid drinking ice water, even during the sweltering summer.

While everyone has some balance of yin and yang, women are seen as having more yin (陰) and men more yang (陽), leading to the cautionary advice that women should not eat too many "cold" foods. Drawing on her ethnographic conversations, Chu notes that this is especially true for menstruating women:

> During her menstrual period, a woman should not drink or eat cold food or have contact with cold water; she should not bathe, swim, wash her hair, feet, or face, or work in water. Coldness in food can refer to the temperature of a food or to its nature. Foods that are cold in temperature are ice, cold liquid drinks such as suan-mei t'ang (sour plum drink), soft drinks, ice cream, cold water, and so on. What is cold in nature varies from place to place, but some foods commonly agreed to be liang-hsing [cold in nature] are watermelon, bitter melon, fragrant melon, pears, turnips, oranges, seaweed, cabbage, mustard greens, and unboiled water taken directly from a well or river (Chu 1980, 43).

Although this advice reflects a form of gender essentialism, it presents a different way of thinking about temperature and embodiment in contrast to the thermal objectivity of Western science. In traditional Chinese medicine, the body is seen to be sensitive to any elements from the outside that are cold in either their physical qualities or their essence.[7] Finding relief from temperature (say, a hot day) is not the same for all people; while many might opt to eat the refreshing treat of agar jelly, menstruating women might not.[8] Chinese medicine and eating practices provide an example of how temperature not only "materializes in divergent ways across geographies and history," but also how cultural practices can give rise to different thermal epistemes (Starosielski 2021, 8). In the context of traditional Chinese medicine, the body is in a dynamic condition that can be affected by dietary choices. Eastern ways of understanding temperature are not based on an external technical measure alone – say, the reading of a thermometer – but instead, recognize the

7 In *Other-Worldly*, anthropologist Mei Zhan notes that other concepts in traditional Chinese medicine, like "organs" in the zangfu system, are an important "representation" that, despite names like heart and kidney, do not literally "correspond to anatomical parts of the body" (76). The tendency to look for Western equivalents in a Chinese medicinal context can be misleading, a barrier to learning.

8 Examples of traditional Chinese medicinal treatment can also be much more complicated than balancing just hot and cold; for example, Farquhar notes that a "bitter" drug or herb can help cool or lower a fever, but is often coupled with a "sweet" drug to counteract the impact of the bitter drug on the digestive system, where the sweet drug replenishes fluids and restores nourishment. See Farquhar, "Eating Chinese Medicine," *Cultural Anthropology* 9:4 (1994), 480.

gendered human body as a thermal medium that is affected by the conditions and changes of an environment.

Working with seasonal environmental temperature was also an important part of manufacturing agar, an elemental process dependent on negotiating patterns of warmth, rain, and coldness. Tseng recounts the following origin story of agar making in Japan, over 350 years before the time of his writing:

> Around 1658, the process of purifying and dehydrating agar in its present form was accidentally discovered by an inquisitive and ingenious Japanese innkeeper, Tarozaemon Minoya, according to a story quite commonly accepted by the Japanese agar manufacturers. Minoya happened to note one morning that some of the agar jellies which he had thrown outdoors and which had evidently frozen the night before, had thawed and dried in the sun to become translucent, membranous, porous flakes. Thus, the method of making agar utilizing natural freezing, thawing, and drying was evolved, which to date is still being used by the majority of the Japanese agar manufacturers (Tseng 1994, 972).

Tseng goes on to describe how traditional methods of producing agar in Japan took place in accordance with different seasons:

> Since the traditional Japanese method of agar manufacture depends almost entirely on atmospheric conditions, the time and the place for making agar are vitally important. The suitable season is generally from the latter part of November to February or March of the following year, when the temperature is cold enough for the freezing process. The selection of place depends on its winter temperature, the quality of its water, and the topography of its ground. The most suitable temperature is between 25F and 35F. An ideal locality is bounded on the northwest by mountains or hills, with a stretch of meadow or plain to the southeast. The mountains or hills serve to intercept the moist north wind in the winter months, and the meadow or plain serves for spreading the agar gels to be frozen in the gold nights, and to be thawed and dried in the sun (Tseng 1994, 1024).

Unlike industrial techniques that could happen anytime (use of a freezer or desiccator), making agar in Japan relied on temperature and seasonal changes that naturally occurred over the course of a year. Seasonal timing was crucial: for example, the first step of bleaching agarophyte seaweeds (not just *Gelidium*,

FIGURE 10.1 Drying seaweed before processing to produce agar, Japan 1947
SCRIPPS INSTITUTION OF OCEANOGRAPHY, DIGITAL ARCHIVES

but several others) would take place around the end of August or beginning September, "when the summer heat has abated slightly. Rainy weather is to be avoided; otherwise rotting may take place" (Tseng 1994, 1023). Although the process normally took several days, "with exceptionally favorable weather 24 hours may be sufficient" (Tseng 1994, 1023). The pre-industrial Japanese techniques of making and purifying agar were labor intensive (Figure 10.1), timed to seasonal changes in temperature and sunshine. This seasonality is reflected in agar's naming; Tseng notes that agar is, "known in Japan as 'kanten', meaning 'cold weather', an allusion to its being made only in winter, and in China as 'tungfen', meaning 'frozen powder'" (Tseng 1994, 972).

Thus, even at the level of naming, agar was indexed to thermal considerations of seasonal change and weather. The eastern thermal episteme that becomes visible through agar involves considerations of both the body and environment – agar's somatic cooling effects, as well as its pre-industrial production during winter. As we will see, agar loses these connections to body and environment once it moves into the thermal episteme of Western science, which develops techniques of producing agar in from Gelidium at any time of year, and sees agar *responding to* changes in temperature (rather than *causing* changes in body temperature). With the laboratory, agar becomes articulated as a thermal medium for culturing bacteria – a story which, as I show below, involves both colonialist trade networks and women's labor in the kitchen.

2 Agar as a Culturing Medium

Tseng's writings on agar not only enable a comparative perspective on thermal epistemes, but also track the movement of agar within longer histories of resource use that predate European colonialism. For instance, Tseng notes that the name "agar" is not Japanese, but Malaysian, suggesting that Chinese immigrants imported the Japanese kanten (agar) to the East Indies for their own use, "as they usually do with other food articles when they settle in other countries" (Tseng 1994, 973). Tseng speculates that, "To avoid introducing a Japanese or Chinese name, they merely called it agar-agar," a term already used by Malayans for seaweed jellies made from a different species of seaweed, *Eucheuma muricatum* (Tseng 1994, 973). When the Dutch East India company later established trade routes between Indonesia and Europe, colonists bought what they thought was a Malaysian product for use in their own cooking, shared among friends and relatives: "Thus a Malayan term became permanently attached to a Japanese product" (Tseng 1994, 973). It was through this colonial trade network – and women's labor in the kitchen – that agar was, by chance, recognized to be an invaluable culture medium in microbiology by Frau Fanny Eilshemius Hesse.

Tseng recounts Frau Hesse's influence in suggesting agar as a bacteriological medium in a lengthy paper he wrote entitled, "Phycocolloids: Useful Seaweed Polysaccharides" (1946), and unlike most people recalling the story, Tseng uses her full name.[9] Early attempts at culturing bacteria on media like potato slices or gelatin ran into a number of problems – the bacteria would eat the medium, or if the temperature became too warm, the gelatin would liquidate (Hesse 1992). Bacteriology needed a medium that would stay firm at the high temperatures most ideal for bacterial growth. Frau Hesse – who was married to a technician in Robert Koch's laboratory – suggested that agar might be a better medium. Tseng writes,

> The idea of using agar as a medium for bacteriological culture was actually first conceived by a housewife, Frau Fanny Eilshemius Hesse. Frau Hesse had been using agar in her kitchen for years in the preparation of jellies, and was fully aware of its value as a gelatin-substitute. It was said that she received the recipe from her mother, who in turn had

9 Some contemporary versions of the story erroneously credit the use of agar to Frau Hesse's husband, or simply to Koch himself. See for example, https://www.saltwire.com/newfoundland-labrador/lifestyles/local-lifestyles/the-right-chemistry-seaweed-and-its-surprising-subplots-332209/, accessed July 8, 2021.

> obtained the formula from some Dutch friends formerly living in Java (Tseng 1994, 975).

Agar's journey from Japan to the Malaysia continued, then, through Dutch colonial networks to continental Europe through its usefulness as a food. It is significant that agar's "discovery" as a culturing medium was made possible through the social networks of women, since Koch is usually credited for inventing a novel method of culturing bacteria in agar. Tseng reminds us that Koch's original paper mentions agar only once in a "single, insignificant sentence."[10] Yet this was the sentence that revolutionized microbiology, the hint that resulted in agar's use in many other laboratories to establish agar as the ideal culturing medium – the jelly that wouldn't melt.

It is this ability of agar gel to remain solid at high temperatures that Tseng highlights in nearly all of his agar writings. As we can see in this passage, agar's thermal properties are more than just a high melting point – once melted, agar also remains *liquid* below its melting point.

> Agar is also remarkable for melting at about 95C and solidifying at 42C. Because it remains liquid when cooled to 42C, organisms may be thoroughly mixed with it at a temperature that does not injure them; on the other hand, once it gelifies, it will remain firm at about the same temperature, usually 37C, at which the pathogenic forms are generally incubated. The reversibility of the agar gel enables the operator to alternately warm it into the sol[id] state and cool it into the gel state (Tseng 1994, 942).

This lag in a material phenomenon is called "hysteresis," a word whose Greek origin is related to husterēsis (ὑστέρησις), a shortcoming, delay, or sense of being late or behind (and not, as one might erroneously guess, *hysteria*).[11] What Tseng calls agar's "reversibility," or ability to phase change from liquid to solid and back again at a delay, is an example of hysteresis, and one of the most important properties of agar as a culturing medium. Tseng also details other important qualities of agar gel as a medium. He writes, "its neutrality in reaction, its transparency, its ability to withstand high pressure for a certain length of time without undergoing hydrolysis, all contribute to make agar

10 "So wachsen sie beispielsweise auf einter mit Agar-agar bereiteten, be Blutwären hart bleibenden Gallerte, welche einen Zusatz von Fleischinfus und Pepton erhalten hat," quoted in Tseng, *The Selected Writings of C.K. Tseng*, 975.

11 OED, "Hysteresis", https://www.oed.com/view/Entry/90636?redirectedFrom=hysteresis#eid, accessed July 13, 2021.

gel the most valuable culture medium that science has ever known to date" (Tseng 1994, 35). It has an outstanding gelling power, and part of its value "as a medium for bacteriological culture derives from its low degree of syneresis," or sweating (Tseng 1994, 1015). Gelatin tends to sweat, producing a surface layer of water that tends to dislodge bacterial colonies. By contrast, agar's lesser degree of sweating means that bacterial colonies can grow relatively undisturbed, instead of sliding into one another.

Initially, producing agar as a culturing medium was a variable process. At the time that Tseng was writing, what was sold under the name "agar" from Japan was not well standardized, and was in fact produced from a varying ratio of species. Tseng notes that, "Japanese agar was so inconsistent in its quality that research workers found it very difficult to standardize the preparation of culture media. Hence in many laboratories the agar had to be purified before use" (Tseng 1994, 987). Agar had variable borax content, which Tseng speculates could have been from epiphytic organisms such as bryozoans, living on blades of the *Gelidium* and other harvested seaweeds (Tseng 1994, 989).

These two contexts – varying ratios of seaweed species, and the organisms that lived on those seaweeds – prevent the easy equation where *Gelidium* = agar. Tseng notes how the US opted for a definition of agar "not only on the basis of its source and nature, but also as to its contents of foreign organic matter, acid-insoluble ash, and moisture" (Tseng 1994, 987). Defining agar according to chemical composition was a way to standardize a substance that, in practice, could never be the extract of just one species hauled out of seawater. *Gelidium*, like all seaweeds growing in the ocean, is an appealing surface for other life forms to encrust and is thus often entangled with other kinds of seaweeds. Agar gel is thus an unavoidably multi-species product that, through a refining process, can approach the asymptote of purification.

So what defines agar, then, if not just *Gelidium*? This was a question that Tseng definitively answered during his research at Scripps. In one paper, Tseng defines agar as, "The dried amorphous, gelatin-like, non-nitrogenous extract from *Gelidium* and other agarophytes, being the sulfuric acid ester of a linear galactan, insoluble in cold but soluble in hot water, a one percent neutral solution of which sets at 35 to 50 C to a firm gel, melting at 80 to 100°C (Tseng 1994, 951). Tseng's definition expands an earlier and more rudimentary description of agar as simply, "the sulfuric ester of a linear polygalactose" (Tseng 1994, 993).[12] Tseng also identified agar as a type of phycocolloid – a term that he coined to describe the specific properties of seaweed extracts (mostly from their cell walls), from the Greek "phykos" (seaweed) and "kolla" (glue) (Tseng 1994, 971).

12 Tseng, *The Selected Writings of C.K. Tseng*, 993.

FIGURE 10.2A (LEFT) AND 10.2B (RIGHT) American Agar Company (Los Angeles, 1933); Close-up view of kelp slurry at American Agar Company (1940)
SCRIPPS INSTITUTE OF OCEANOGRAPHY, DIGITAL ARCHIVES

While seaweed extracts had previously been called referred to by a variety of names such as phycocolle, Japanese isinglass, seaweed glue, gelose, mousse de Chine, Tseng's term "phycocolloid" is the preferred term that is still used today.

The American process of purifying agar involved the application of thermal change but through a different process of labor than in Japan, freeing agar production from seasonal timing. At the start of WWII, the US barely had a domestic seaweed harvesting industry. Three factories extracted seaweed products: Kelco Co (San Diego) processed algin from Giant Kelp (*Macrocystis pyrifera*); American Agar Co. (Los Angeles) processed *Gelidium cartilagineum* for the making of agar (Figure 10.2a, 10.2b); and Krimko Co (Chicago) processed *Chondrus crispus*, for the extract carrageenin. Tseng recalls how the first agar factory in the United States was started by a Japanese man, Chokichi Matsuoka, but that the Japanese method of processing agar "involved too much labor for an American industry," or more accurately, *human* labor (Tseng 1994, 35). The factory was later bought by American John Becker, who "improved methods of cooking, filtering, bleaching and also invented the 'dewatering' equipment and the sizing and dehydrating of agar flakes" (Tseng 1994, 35). The American purification method involved a multi-step machine assisted process designed to remove impurities, a process that involved filtration, gelation, freezing, thawing, vacuum filtration, dehydration, bleaching, and grinding (Tseng 1994, 1029). Agar became useful as a thermal medium only through a long process of thermal and elemental manipulation – a machine-assisted process that no longer had to be timed to the seasons.

Agar's many uses in industrial, food, and medicinal products further elaborate its properties as a medium, not only through its unique chemical properties, but also through its technical usefulness in hot weather. Because bacteria do not eat agar (in microbiology, a nutrient has to be mixed in), it can be used as an effective seal in canning. Tseng discusses its usefulness for canning meat and fish, recommending it "for hot countries where refrigeration facilities are not generally available. The meat to be preserved is cooked with agar" which helps to "prevent the entrance of bacteria which will otherwise spoil the meat" (Tseng 1994, 1058). Agar's usefulness as a medium not only has to do with its own thermal properties, but also with how agar holds up in hot environments. Indeed, agar substitutes for refrigeration technologies through performing the same role that refrigeration sought to achieve: the inhibition of bacterial growth in food products. Beyond the work of culturing, agar is also an effective seal against the intrusions of bacteria, of value in *preventing* bacterial growth – a role closer to the kind of barrier or "guard" to bacterial contamination that Tseng implied in "A Seaweed Goes to War." It is to these questions of labor that I now turn, involving seaweeds as well as Tseng's own conditions of labor.

3 Agar as a Medium of Labor

In this chapter, I have shown how agar emerged as a thermal medium across Eastern and Western scientific contexts in Tseng's writings – from its usefulness in Chinese medicine, to its winter cultivation in Japan, to its purification for use in culturing bacteria in microbiology, to its applications in food safety in hot climates. Yet agar's many uses as a medium obscure a different type of story about labor: the labor that goes into collecting seaweeds, scientifically observing seaweeds, and the implications of seeing seaweeds themselves as laboring organisms for national security. In conclusion, I turn to the conditions of labor that made it possible for Tseng to imagine agar as a thermal medium and a culturing medium, conditions that were framed by the pressures of WWII and demand for domestic production of agar.

As part of his work at the Scripps Institution of Oceanography (SIO) during WWII, Tseng not only wrote literature reviews about agar; he worked on a number of experiments to study the growth of *Gelidium* seaweed both in the laboratory and *in situ*, and is recognized as the first scientific research diver at SIO.[13] Using a Japanese-made suit attached to hoses for surface-supplied air, Tseng regularly dove 10–30 feet below the surface in order to observe *Gelidium*

13 Scripps Scientific Diving Program History, https://scripps.ucsd.edu/scientific-diving-program/history, accessed July 13, 2021.

FIGURE 10.3 C.K. Tseng surfaces from observing seaweed during a helmet dive. Note the glasses. 1944
SCRIPPS INSTITUTE OF OCEANOGRAPHY, DIGITAL ARCHIVES

as it normally grows in the ocean (Figure 10.3). Tseng describes this early dive work in the following way:

> My experimental method was to dive to the sea bottom, to measure the growth of the *Gelidium*, record the salinity and temperature and bring up water sample[s] for analyzing the nitrate and phosphate contents of the seawater. This required that I had to learn diving with the help of a diving suit, a diving helmet, a pair of lead shoes and a life-line connected to a pump on the boat. I was, perhaps, the first botanist to dive and work underwater and it was somewhat dangerous. However, for the sake of Science I had no other choice (Tseng 1994, 10).

Helmet diving was a labor-intensive activity that required a second person to pump air to the diver from a boat or dock. San Diego has annual ocean temperatures that average 57F in the winter, and 68F in the summer.[14] To understand something about the growth of *Gelidium* in natural conditions, as opposed to the laboratory, involved the physical act of submergence in the cold ocean, only modestly insulated by a suit.

14 San Diego average water temperature, https://www.currentresults.com/Oceans/Temperature/san-diego-average-water-temperature.php, accessed July 13, 2021.

FIGURE 10.4
American Agar and Chemical Company kelp diver underwater, 1965
SCRIPPS INSTITUTION OF OCEANOGRAPHY, DIGITAL ARCHIVES

While Tseng's own diving labor was non-negligible, the work of divers harvesting seaweed was even more intensive. In his 119-page paper defining phycocolloids and their properties, Tseng describes the type of labor involved in *Gelidium* harvesting, which included harvests from Southern California but was primarily collected in Baja California (Tseng 1994, 1105). "An experienced diver generally works continuously for one to two hours under water. Then he comes up for a short rest. He may dive two or three times a day, working under water for four or five hours. The diving boat usually goes out in the early morning, since the ocean is as a rule quieter in the morning than in the late afternoon" (Tseng 1994, 1103). The conditions for diving for *Gelidium* were physically challenging:

> Since *Gelidium* grows most abundantly around the edges and on the slanting surfaces of rocks and boulders in places where the water is generally turbulent and the water movement fast, the agar diver usually has to crawl from one rock to another, and occasionally grasps the large kelps for support. With one hand he carries a basket made of small ropes; with the other he pulls a bush of the agarophyte from the rocks and puts it into the basket. When he has filled the basket he ties it to his life-line and pulls the rope twice. Upon receiving the signal, the life-line tender hauls up the basketful of agarweed, weighing between 60 to 75 pounds, and lowers an empty one to the diver (Tseng 1994, 1103).

This labor of harvesting seaweed (Figure 10.4) in cold water was the precondition for much microbiological work, supplying the organism that would become processed into a medium of bacterial culturing.

We are now in a position to read "A Seaweed Goes to War" differently, having moved through the ways that Tseng's other writings develop a picture of agar gel as entangled with questions of eastern and western thermal epistemes, culturing practices, and labor. It is to labor that Tseng turn to conclude "A Seaweed Goes to War" by wondering what would happen in the aftermath of the war, if his call to ramp up the agar industry was successful. Returning to his original metaphor of the seaweed as soldier, Tseng speculated,

> Just as the conscientious public is deeply concerned with the future of the returning service men when final victory comes, persons who have known Private *Gelidium cartilagineum* intimately are also worrying about his future. Under present conditions, most probably soon after the cessation of hostilities the agar industry, unable to compete with the cheaper Japanese agar, will have to fold up, and *Gelidium* will be "unemployed" (Tseng 1944, 36).

This gendered and tragicomic image of an "unemployed" seaweed evokes two registers of war – martial and economic – as Tseng imagines war with Japan becoming capitalist competition for increasingly cheap products. It is also the moment where Tseng slides between a personification of *Gelidium*, and objectification of agar. *Gelidium* is the soldier, but agar is the commodity. Indeed, *Gelidium*'s employment hinges on agar's successful commodification, in competition with the Japanese product.

This passage also shows how Tseng justifies further scientific research on agar through invoking patriotic care for the returned soldier, a spectre that haunts other descriptions of agar's applications. Extrapolating from the existing use of agar in dental molds, Tseng posited that, "An important use of agar as a result of war will perhaps be in taking molds for making artificial fingers, hands, ears, etc." (Tseng 1994, 1069). In other words, agar would be useful in the making of prostheses, a casting medium for the formation of replacement body parts lost during the course of war. Thus, even as Tseng worried that American agar might be replaced by cheaper Japanese agar, he also thought of it as a malleable medium for replacing body parts. In this way, agar was both like a soldier returning home, and a physical material that could augment the wounded bodies of soldiers – personification collapsing with its referent.

Looking back on Tseng's brief window of research from 1943–1946, before he returned to China to work on aquaculture, and before his research was put on hold by the anti-intellectual climate of Mao's Cultural Revolution in the 1960s, Tseng's fear that agar would be "out of work" did not come to pass. In fact, quite the opposite has been the case during a *Gelidium* shortage in 2016, due to overharvesting in Morocco.[15] *Gelidium* – despite being an ideal culturing medium – has not been possible to cultivate at a level that would satisfy global demand. As the journal *Nature* reported in 2016, *Gelidium* species "favour cool, turbulent waters that provide a steady supply of oxygen and other nutrients – a preference that makes industrial-scale farming impossible."[16] The oceanic conditions that make harvesting *Gelidium* a labor-intensive process are also those that prevent its culturing on a large scale, even when agar is more in demand than ever.

Yet what is perhaps most striking about Tseng's article "A Seaweed Goes to War" is how he positions the reader to imagine the living seaweed *Gelidium* as a fellow laboring body – albeit one that might be out of work by the end of the war. Seen in this way, Tseng's literary metaphor of the seaweed-as-soldier was no mere accident, or use of hyperbole. *Gelidium* is a double for Tseng himself – another being of East Asian origin conscripted into thermal service for the American war effort, in the laboratory rather than the front lines. Both seaweed and scientist were caught between colonialist and extractive networks, between eastern and western thermal epistemes, laboring under thermal variation.

References

Barua, M. 2018. "Animal Work: Metabolic, Ecological, Affective." Theorizing the Contemporary, *Fieldsights*, July 26. https://culanth.org/fieldsights/animal-work-metabolic-ecological-affective.

Chu, C.M.Y. 1980. "Menstrual Beliefs and Practices of Chinese Women." *Journal of the Folklore Institute* 17(1): 42.

Hesse, W. 1992. "Walther and Angelina Hesse: Early Contributors to Bacteriology." *American Society for Microbiology News* 58(8): 425–428.

Humm, H.J. 1947. "Agar – a pre-war monopoly." *Economic Botany* 1(3): 317–329.

Landecker, H. 2010. *Culturing Life: How Cells became Technologies*. Cambridge: Harvard University Press.

15 Where Japan was the leading supplier before WWII, most of the world's Gelidium is harvest from Morocco, followed by Spain and several other countries. "Lab staple agar hit by seaweed shortage," https://www.nature.com/articles/528171a, accessed July 11, 2021.

16 "Lab staple agar hit by seaweed shortage," https://www.nature.com/articles/528171a, accessed July 11, 2021.

Paxson, H. 2018. "The Naturalization of Nature as Working." Theorizing the Contemporary, *Fieldsights*, July 26. https://culanth.org/fieldsights/the-naturalization-of-nature-as-working.

Starosielski, N. 2021. *Media Hot and Cold*. Durham: Duke University Press.

Tseng, C.K. 1944. "A Seaweed Goes to War." California Monthly. May.

Tseng, C.K. 1994. *The Selected Writings of CK Tseng Vol. 1 & 2* (曾呈奎文选 (上 / 下)). Beijing: Ocean Publishing House.

CHAPTER 11

Augmented Polycultures: Scaling up Algal Ecosystems and Design of a Biofouling Aesthetic

Brenda Parker and Marcos Cruz

Abstract

To lay the foundations for the Biocene, a potential future era of our Anthropocene human habitat, the infrastructure of our built environment should play a more active role in carbon mitigation and reduction. Algae and cryptogrammic species will become important elements of bio-integrated "photosynthetic cities". However, to realise this, we will need to relinquish notions of monoculture and purity associated with highly maintained and controlled cultivation. This chapter will look back at the origins of contained microalgal culture in the realms of science and engineering to understand the basis for our current design language. We assume the position that in future, consortia-based approaches with direct exposure to the outdoor environment will be required in order to deliver the vision of algae for bioremediation or microbiome-inspired green infrastructure in a resilient way. Ultimately, our photosynthetic human habitat will embody a more provocative and disobedient condition. Reconciling with the abject nature of biofouling, overcoming disgust and ultimately reaching an acceptance of the sublime will be needed in order to form ecologically relevant and environmentally meaningful interventions. The role of design will be pivotal to introduce a new aesthetic which is based on how we embrace self-regenerative conditions while promoting heterogeneity and biodiversity in buildings.

Keywords

microalgae – polyculture – heterogeneity – symbiosis – biofouling – bioreceptivity – bioremediation – complexity – scale-up – aesthetics

1 Introduction

The concept of a "photosynthetic city" defines a radically new vision for the future human habitat whereby our spaces and structures become

© BRENDA PARKER AND MARCOS CRUZ, 2024 | DOI:10.1163/9789004683310_013
This is an open access chapter distributed under the terms of the CC BY-NC 4.0 license.

bio-integrated. It encapsulates two core, biologically-mediated, tenets: where buildings can become scaffolds for growth, or where construction materials are grown *ab initio*. The motivation for this is to reverse the devasting effects of climate change, and to create an environment for more ecologically-driven societies where we live in greater proximity to biodiversity. The process of photosynthesis underpins all ecosystems, and will be essential to the realisation of a shift from our current era of the Anthropocene to a future Biocene.

If we are to create such a photosynthetic city, how might it look? Unicellular photosynthetic organisms such as microalgae and cyanobacteria are frequently keystone species when transforming hostile landscapes. However, it is time to question what it might really mean to embed algae within our built environment and to look back critically at some of the underlying paradigms that have been established in previous design work. To understand how we might progress, it is worth reflecting on the methods of scaling up algal cultures for biotechnology, design and architectural purposes. Within the diversity of applications, this procedure has many common themes, including a focus on enclosed systems, monoculture and purity. While the appearance of algae as individual species has a greater social acceptability, this state is not reflective of natural ecosystems and as a result, considerable effort is required to maintain stability.

To move forward, this chapter will assume the position that in future, consortia-based approaches with direct exposure to the outdoor environment will be required in order to deliver the vision of algae for bioremediation or microbiome-inspired green infrastructure in a resilient way. Ultimately, our photosynthetic human habitat will embody a more provocative and disobedient condition. Reconciling with the abject nature of biofouling, overcoming disgust and ultimately reaching an acceptance of the sublime will be needed in order to form ecologically relevant and environmentally meaningful interventions. The role of design will be pivotal to introduce a new aesthetic which is based on how we embrace self-regenerative conditions while promoting heterogeneity and biodiversity in buildings.

2 An Initiation into Algal Biotechnology in the Built Environment – the B.I.Q. House

Following the construction boom in the late 20th century, the start of the 21st century brought about an awareness of the devastating carbon footprint of the built environment and the urgent need to find new sustainable solutions for our future human habitat. This coincided with rising oil prices, which created the drive to produce algal biofuels. In fact, these endeavours were reviving work started in the 1970's by the Aquatic Species Program where the research effort sought to create bioenergy from non-land based crops (Sheehan et al. 1998).

The heady mixture of highly funded and much-disseminated research in algal technology, coupled with the promise of lowering carbon emissions, gave architects and engineers a sense of opportunity to integrate such novel systems in buildings and urban infrastructures. Numerous projects were envisioned, but the first large functioning façade was designed by ARUP Engineering in collaboration with the Strategic Science Consult and Colt International in 2013 for the International Building Exhibition in Hamburg (Figure 11.1). Known as the BIQ House (Bio Intelligent Quotient), the building became the first of its kind to host a 200m^2 area of flat panel *SolarLeaf* photobioreactors (Wurm and Pauli 2016). The vision was the creation of a "bioreactive" façade that was capable of responding to light by the increased growth of microalgae. In addition, the panels acted as solar thermal collectors, and used to generate hot water. The design included the control and processing equipment needed to separate the algal cells in a facility located under the building (Figure 11.2). Biomass collected could be used to generate methane for electricity generation following a secondary conversion step. For the inhabitants the façade creates a self-cleaning and bioresponsive screen to filter light into the building. The initial sketches were formally more expressive, however budget constraints and practical considerations meant the engineering realisation of flat panels was the most effective means of construction. The building remains an important precedent in the history of applying living systems within our built environment due to the scale and ambition. Nonetheless, the project had a number of challenges. User experience of residents indicated that the noise from the airlift pumping system creates disturbance, and steps were taken to reduce the frequency. Materials of construction when handling algal cultures are important, even though algae are seen as benign. During certain phases of growth pH can rise significantly, resulting in corrosion of aluminium components. Also, incident sunlight within such thin film bioreactors can result in temperatures that challenge the physiological limits of algal cells. Despite this, the preliminary data from mass balances indicates that from the façade over 600 kg of microalgal biomass was produced in a year, and the net energy balance was favourable (Fraunhofer 2016). If this system was to be applied to many surrounding buildings, it could produce a significant amount of biomass in specific urban environments in a distributed manner.

The use of flat panel façades for culture of other, valuable microalgal species has been further explored in an architectural context by Jérémy Pruvost and

FIGURE 11.1 *Solarleaf* façade of *B.I.Q. House* with algae monoculture in flat bioreactors built on the occasion of the International Building Exhibition in Hamburg, 2013. Team: ARUP Engineering in collaboration with the Strategic Science Consult and Colt International
PHOTO CREDIT: MARCOS CRUZ

French architects from XTU (Pruvost et al. 2016; Todisco 2019). The engineering environment has been characterised and the patterns of growth examined in the photobioreactors designed. However, it highlights an interesting gap in our present thinking, and a dichotomy between two possible approaches. Should a reactor be designed for the species in a bottom-up manner, or if the species and its medium should be selected and designed to be able to deal with the conditions of the bioreactor as per a top-down approach? The comparison of both scenarios has not been exhaustively examined. Applied phycology has focussed on the phenomenon of algal growth within a pre-defined design, with the objective of understanding how it behaves in the vessel. For instance, scale up of species such as *Tetraselmis* for carbon dioxide sequestration has been evaluated in this manner (Pereira et al. 2018) using flat panels and tubular reactors. When considering an application within the urban landscape, the complexity of industrial photobioreactors (Figure 11.2) on the other hand offers little possibility for customisation or participatory design, as per the vision of an open source architecture (Ratti and Claudel 2015). Nonetheless, embarking

FIGURE 11.2 *B.I.Q. House* plant room to service and maintain bioreactor façade
PHOTO CREDIT: MARCOS CRUZ

on a design process for a photobioreactor to cultivate a single species could be perceived as a high risk endeavour relying on both an exhaustive knowledge of cell physiology and metabolism, as well as suitability for the environmental condition.

3 Photobioreactors and Engineering Cell Environments

To understand the challenge of photobioreactor design is worth going back to the origins of contemporary algal bioreactors, which echoes the development of mass cultivation of microbial cells in other forms of biomanufacturing. It has roots in the engineering of fermentation vessels and the evolution of industrial biotechnology since the 1940s. Geometries of stirred tanks have typically focussed on cylindrical models with modifications to the interior conditions to facilitate oxygen demand or decrease mass transfer limitations. The addition of spargers, impellers or baffles introduces mixing and turbulence, and through these interventions creates an environment amenable to cell growth. It has also established a conventional mode of manufacture based on standardised equipment and engineering heuristics in order to minimise risk. Largely driven by the pharmaceutical industry, the typologies of upstream cultivation vessels are inherently conservative. The strict procedures of validation leading to product approval have further cemented the approaches accepted for scale up of biological cultivation.

To scale up algal culture in closed systems requires a photobioreactor (PBR) – that is, a vessel capable of facilitating gas exchange, holding liquid and permitting light transmission. As noted previously (Pulz and Scheibenbogen 2007), light is the main limiting factor in scale up of algal cultures. The distance light must travel through the medium (path length) and exposure to sufficient solar energy to conduct photosynthesis is a primary concern. To this end, there are families of geometric forms based on two mechanisms: either bubble columns or airlift reactors, which comprise a specific compartment for the injection of gas. Both typologies may be cylindrical in nature.

Alternatively, scale up can also be achieved by maximising surface area to volume ratio. One approach for this in regards to contained photobioreactor systems was to move towards flat "plate" systems, a format first explored in the 1950s (Burlew 1953). The small path length and flat geometry has led to their inclusion in façade projects. While the internal mechanisms of flat panel PBRs may function as either bubble columns or airlift reactors, they crucially begin to break with biotechnology tradition as the format is rectangular. In microbiology, vessels are typically designed to have smooth junctions and rounded boundaries (cf. the petri dish) to minimise accumulation of matter or accommodate contaminating microbes. The principal advantage of the flat panel reactor is the high areal productivity. However, this is at the expense of capital expenditure and material usage given the low volumes they contain.

In the examples of bioreactors given above, the emphasis has been on cultivation of a single species. All forms of biotechnological manufacturing, from food to fine chemicals and pharmaceuticals, rely on reproducibility. Introduction of other organisms creates variable yields due to competition for carbon, nutrients and light; or can be lethal to the "desired" organism. Therefore, control mechanisms and sterile techniques are put in place to eliminate the potential for contamination. Interestingly, this approach does not apply when working with outdoor systems where biological systems are used to remove nutrients from wastewater. Water treatment facilities use membrane reactors and shallow raceways known as high rate ponds that work with mixtures of microorganisms, and even encourage flocs of algae and bacteria. While the exact species may vary with temperature or composition of wastewater, the functional structure is preserved despite taxonomic variability (Craggs et al. 2014; Louca et al. 2016). In this respect there is a divergence between biomass production (single species) and remediation (consortia). Herein lies the challenge for the built environment that so far inherited the aesthetics of the former, but for its wider use and application will have to benefit from the properties of the latter.

4 Monoculture and Purity

Monocultures in agriculture or biotechnology imply the growth of a single species, usually with either containment or preventative actions taken to prevent contamination with other species. Within algal cultivation these invaders may be benign, growing alongside the algal cells and feasting on organic carbon shed into the media. In some cases, these other organisms can be deleterious to algal growth. Viruses, bacteria, grazers in the form of ciliates or other zooplankton can rapidly decimate cultures and cause a "crash" (Day et al. 2017). This terminology hints at the precarious thermodynamic state of a monoculture. Energy is expended in the form of additional apparatus and precautions to maintain a single organism in a pure culture. A state of equilibrium is achieved through intensive human intervention, which could be interpreted as "Organisation [is] maintained by extracting order from the environment" (Schrödinger 1948).

The requirement for algal monoculture can be seen to stem from two origins which reflects the different communities who have been working within the phycology domain. Firstly, within the mindset of process engineers from biotechnology and pharmaceuticals where control over the organism is paramount to ensure safety of the product. Use of antibiotics, aseptic technique and genetic manipulation are cornerstones of this approach. While this can be successfully scaled up for valuable therapeutic products in other host organisms, there is an inherent challenge in trying to apply this mindset to outdoor culture in the volumes required (Carney and Lane 2014). Algal cultures have much lower cell density and much more extended elaboration times than bacteria or yeast typically employed in biotechnology processes. Maintaining the strain of interest has been important for yielding triacylglycerols for biofuel or astaxanthin for antioxidants in mass culture, and much of the tacit knowledge of scale up derives from these two applications. The second influential domain has been the need for pure culture to study the physiology of photosynthesis. Certain species, such as *Chlorella*, *Chlamydomonas* and *Phaeodactylum* have become model organisms thanks to the relatively facile cultivation. The choice to study a few model species intensively to gain translatable knowledge that can be applied to other species (Bolker 2019) has led to many advances in

FIGURE 11.3 *Algaegarden* – installation of steel and wooden frames, with hanging pods containing algal cultures. Exhibited at the Reford Gardens for the Metis International Garden Festival, Quebec, CA, 2011–13. Team: Wayward Plants / Heather Ring in collaboration with Brenda Parker and Synnøve Fredericks. Partners: Canadian Phycological Culture Centre, NutrOcean Rimouski, Martin Cooper (Engineer)
PHOTO CREDIT: LOUISE TANGUAY

our understanding of light harvesting and metabolism in algal species. Model organisms are typically easy to grow in monoculture, or have been demonstrated to be resilient within a laboratory environment. Nonetheless these algal species can be a source of anchoring bias whereby many phenomena observed in these experimentally tractable organisms elevates them to an enshrined status within applied phycology. In the 1930's at Harvard University graduate students formed "Chlorella club" – the members included Charles Stacey French and Peisong Tang who would go on to make enormous contributions to our understanding of photosynthetic processes (Zallen 1993). The vivid green of *Chlorella* is often a synonymous with the imagery of microalgal culture, and the study of this organism for biodiesel production has vastly expanded the expertise in mass cultivation.

In the contemporary sense, designers working with biology have inherited a legacy from these schools of thought regarding monoculture. There may also be ideas that to be accepted within an interdisciplinary field it might be advantageous to work with recognised organisms. In addition, there is a further aesthetic dimension to the use of monocultures. Separation of algal species enables a consistency and purity of pigment. In the case of green algae such as *Chlorella* this represents chlorophyll, or a blue-green mixture of chlorophyll and chromoproteins in cyanobacteria (Jeffrey et al. 2011). Diatoms are a rich brown because of carotenoids such as fucoxanthin (Kuczynska et al. 2015). Other pigmental changes are induced by stresses on the culture system, for instance beta carotene production is triggered in hypersaline conditions in *Dunaliella* creating a vibrant pink hue, and astaxanthin produced by high light or nitrate deprivation of *Haematococcous pluvialis* turns the culture crimson. Beside functional advantages of monocultures, an underlying, perhaps more unconscious, aspect to this lies in the expectation and drive towards sterility and cleanliness when producing any type of growth outside its original environment. The homogeneity of a monoculture instinctively implies a level of control – it feels safer than the heterogenous and potential hybrid biofilm where pathogens may breed.

5 The *Algaegarden*

The application of monoculture in a design context was tested in the art installation *Algaegarden* (Figure 11.3) created by Wayward Plants led by Heather Ring in collaboration with Brenda Parker and Synnøve Fredericks for the Métis International Garden Festival in Quebec in 2011 (Reford Gardens 2011). *Algaegarden* celebrated the beauty and productive potential of algae through a design that underlines its diversity and meaning. It deliberately presented individual species in hanging pods in order to invite a closer dialogue and understanding. The project aimed to create a narrative environment where visitors could experience and engage with algal cultures through the air pumps that enabled bubbles to race through the hanging tubes (Figure 11.4). Algae, often considered a nuisance, become an object of secret beauty and curiosity. The installation chose to amplify a selected number of individual species of algae through their presentation in rows to create an immersive sensation of being surrounded. Referencing an aquatic edge, the garden was lined with pond grasses to reference the types of ecosystems that one might encounter planktonic organisms such as algae. The interactivity invited the visitor to consider how to be more aware of the ecological roles of these organisms and their significance as well as evocative colouration.

The organisation of the garden by species taps into the human need for systemising nature – grouping by taxonomy. Botanic gardens have a long legacy

FIGURE 11.4 *Algaegarden* – detail of bioreactor tubes containing *Chlorella* cultures
PHOTO CREDIT: LOUISE TANGUAY

of systematic beds where plants are presented according to their evolutionary relationships. Here, in this case the colour spectrum was used as a proxy for biodiversity within the installation. Algal cultures collections enable us to understand an organism by the provenance: the location it was isolated from, by whom and what year. After the peak of the algal biofuel era, it felt urgent to reconnect humans to the vital role of algae in our ecosystems. Each algal species presented within the garden had a story illustrating its importance within ecology, commerce or our understanding of plant physiology. Golden brown *Pavlova lutheri* was supplied by a local aquaculture company who used it to feed oysters at a hatchery; a pink *Porphyridium purpureum* because of the pigment phycoerythrin, and CPCC 90 – the same species of *Chlorella vulgaris* used by Emerson to calculate the quantum yield of photosynthesis. Algae pods were prepared with media and nutrients to sustain the cells for a short season as the installation was a temporary one created for the festival without the goal of long-term culture maintenance. Therefore, preserving the organisms as single species within the pods was vital to the readability of the design and the message of the garden. Contamination was reduced through the positive pressure created from pumping the bags with the aeration system, but it was not controlled for.

6 Challenging the Format of the Bioreactor: INDUS

Taking the B.I.Q. House and the *Algaegarden* as valuable departure points for scale-up, the project INDUS, designed by Shneel Bhayana, Brenda Parker and Marcos Cruz for the A/D/O Water Design Futures competition in New York in 2019 (Figure 11.5), sought to question the format of a bioreactor entirely. By placing algal cells inside a matrix and circulating water over the surface it enables a departure from the mere design of a "vessel". In addition it offered the opportunity from a technical standpoint to understand how immobilisation might be used to reduce energy consumption and challenges of separating algal cultures from the liquid media. Building upon previous work on additive manufacturing of viscous membranes (Bhayana et al. 2020), hydrogels were used to maintain the algal cells in a favourable environment. Rather than printing the hydrogel layer-by-layer into three-dimensional morphologies, a ceramic scaffold was designed and moulded to host the hydrogel with immobilised algae. As an application, the issue of scalability in the outdoors was addressed in the context of water pollution. Bio-integrated design was used as a method (Cruz and Parker 2021) that offered a dual advantage wherein not only the material but also the scaffold for its growth was designed to embed this new living system.

The INDUS project was conceived as a tile-based, modular bioreactor wall with the goal of cleaning water through bioremediation. The design of INDUS reflected an interdisciplinary approach to the vast and nebulous issue of water pollution that poses a significant threat to human health (Landrigan et al. 2018). By embedding ethnographic research, scientific research, computational simulation and material studies it used a holistic approach to the design (Parker et al. 2022). The project evolved from a series of site visits, case studies and interviews between 2014 and 2018, including a community of artisans, a panchayat, who manufacture bangles in a region outside Kolkata, India. A case study site within the Howrah Domjur informal bangle and ornament letting industries, a two-room building with a soil pit for waste, became the departure point for subsequent design renders. The community reiterated the need for a simple, scalable and a sustainable system to treat heavy metal contaminated wastewater on a local level due to changing regulation from the Central Pollution Control Board. In response, the design made use of vernacular materials and traditional clay making methods. Inspired by the architecture of a leaf, water was envisioned to flow over a series of vein-like channels containing algae prepared in a seaweed-based hydrogel developed previously (Bhayana et al. 2020). The main objective was that pollutants such as cadmium were

FIGURE 11.5 *INDUS 1.0* – ceramic tiles with extruded algae-laden hydrogel, 2018–19. Team: Shneel Bhayana, Brenda Parker, Marcos Cruz and the Bio-Integrated Design Lab at UCL
PHOTO CREDIT: SHNEEL BHAYANA

sequestered by the algae and the hydrogel could then be processed to recover heavy metals safely.

In parallel to the design evolution, experimental validation of performance with a range of algal species and studied uptake kinetics of heavy metals such as cadmium was carried out. Building upon previous work (Scarano and Morelli 2003; Torres et al. 1998), laboratory tests (Stoffels et al. *unpublished data*) confirmed the uptake of cadmium by microalgal cells encapsulated in hydrogels as well as the production of cadmium nanoparticles by actively growing cells of *Phaeodactylum*. The aim was to enable the rural community of artisans to regenerate water for reuse within their manufacturing processes. The modularity of the system was deliberate as this enables the communities to construct this wall depending on the site availability along with the amount of water to be treated. INDUS was designed to be integrated within the existing community, which can be customised and also reflect the cultural identity of the region. Unlike conventional water treatment facilities, the leaf-like tiles complete with their "veins" when assembled into a wall add ornamental richness (Figure 11.6). This aspect of the decorative sought to instill pride and encourage visibility to what otherwise what could be a very alienating apparatus in a very rural context. Its morphological expression was aimed to generate a sense of empathy of users when engaging with its technology, while adding to the traditional pattern-like motifs of India. It could be supported on free-standing or existing walls of densely populated industrial clusters.

The long-term vision for such a bioremediation system is to understand the value of materials recovered. In the case of cadmium, metal nanoparticles synthesised by the algae can present a source of revenue (Cueva and Horsfall 2017). This can potentially create an incentive-based closed loop system, removing pollutants from the cycle entirely. INDUS has potential to become a precursor to existing constructed wetlands (Scholz and Lee 2007) or work in tandem with technologies such a microbial fuel cells or biophotovoltaics obtaining a higher efficiency of wastewater treatment. The visibility of the system encourages communities to engage with and consider the impacts of water contamination and environmental clean-up. Further, the design and fabrication of the wall gives a new dimension to traditional clay making practices.

A modular demonstrator wall of INDUS 1.0 was built outdoors for the Brompton Design District at London Design Festival in 2019. Tiles were computationally designed based on a generative algorithm which recreates biological patterns. From this, a mould was cast and tiles produced by hand pressing. A drip irrigation system was used to maintain moisture of the hydrogel component. It was apparent that while the deep channels and geometry of the tile enabled stable residency of the hydrogel, there was a need to decrease

FIGURE 11.6 *INDUS 1.0* – installation in exhibition at the Brompton Design District, London Design Festival, 2019. Team: Shneel Bhayana, Brenda Parker, Marcos Cruz. Collaboration: Bio-Integrated Design Lab at UCL. Ceramic casting: Richard Miller, Froyle Tiles, UK
PHOTO CREDIT: SHNEEL BHAYANA

the speed at which water flowed over the surface. A further design iteration of INDUS 2.0 considered the idea of *lentic* and *lotic* zones (Figure 11.7). Borrowing from riverine ecology, these terms define regions of still and flowing water respectively. In the lentic zone of INDUS, water flows more slowly and the contact with the hydrogel is longer. The lotic zones are faster, and more turbulent. To facilitate uptake of cadmium by the immobilised algae retention times would be extended through the lateral striations, increasing the proportion of lentic zonation to slow water flow over the surface.

In terms of architectural precedents that have explored the cultivation of a mixture of species, the concept of immobilisation had been explored in the *Alga(e)zebo* installation by the team marcosandmarjan – lead by Marcos Cruz and Marjan Colletti – for the London Olympics in 2012 (Figure 11.8) The design of the bioreactor vessels permitted the ingress of aerial organisms. Using a matrix of semi-solid agar, wild cultures could mix with laboratory strains of algae previously inoculated and viewed inside the transparent columns. The *Alga(e)zebo* intertwined human artifice with natural surroundings, working

FIGURE 11.7 *INDUS 2.0* – ceramic tiles with extruded algae-laden hydrogel, 2019. Team: Shneel Bhayana, Brenda Parker, Marcos Cruz. Collaboration: Dali Alnaeb and Aurora Tairan Li. Ceramic casting: Richard Miller, Froyle Tiles, UK
PHOTO CREDIT: SHNEEL BHAYANA

FIGURE 11.8 *Alga(e)zebo* – photobiorector with multiple algae species from the surrounding embedded in double curved steel structure. Installation located at Euston Square Gardens built on the occasion of the London Olympics, 2012. Team: marcosandmarjan (Marcos Cruz and Marjan Colletti). Manufacturer: Formstaal / CSI, Stralsund Germany. Engineering: Bollinger, Grohmann und Schneider, Vienna Austria. Photobioreactor: Richard Beckett / DMC London with UCL Algae
IMAGE CREDIT: MARCOS CRUZ

from the microscopic scale to the interaction with vegetation and trees, examining the complex boundary negotiations that take place between architecture and nature in contemporary cities (Cruz and Coletti 2016). The provocation was to suggest architecture is behaving and looking more like a biological construct, whilst nature is manipulated via human interference. In one sense, the use of a viscous matrix for the immobilisation of algae unites both INDUS and the *Alga(e)zebo*. However, there is an important difference in terms of ambition and functionality. While the remit of INDUS was directed at the needs of a community facing water pollution in a resource-constrained site, the *Alga(e)zebo* was a highly elaborate construction located in central London. The challenge, yet also the paradox of INDUS, however, was that the design is celebrated for an aesthetic that relies on the installation staying pristine and clean despite the goal of treating polluted water. Although ultimately aimed for an outdoor environment, it did not encompass the effects created by the natural weathering process in which chemical changes of the material, dusts, the colonisation of new species or any ecological disturbances trigger profound changes of the overall surface. INDUS was conceived and designed with an immobilised single species relating to a biotechnological agenda of bioremediation performance that was supposed to maintain a permanent sense of newness, avoiding the unpredictable murky impacts of time.

7 Towards Polyculture and Interkingdom Interactions

> No claim for novelty can be made for mixed cultures: They form the basis of the most ancient fermentation processes. With the exploitation of monocultures having been pushed to its limits it is perhaps time to reappraise the potential of mixed culture systems. They provide a means of combining the genetic properties of species without the expense and dangers inherent in genetic engineering which, in general terms, aims at the same effect. (Harrison 1978).

In nature, monoculture is seldom found. While organisms may dominate a particular ecological niche, for instance extremophiles, this is usually down to a selection pressure. For instance, in the case of algae adapted to high pH levels they have relatively few competitors due to the niche environment. Availability of carbon, and other key nutrients is a limiting factor on the diversity a system can support. Mutualism in algal-bacterial systems and the formation of stable consortia is based on inter-kingdom exchange of resources (Rawat et al. 2021). Our understanding of the constant communication between trophic levels of the food chain is still nascent. Take for example, microalgae and bacteria trading Vitamin B_{12}, using a currency of carbon (Croft et al. 2005). We are beginning to unravel the metabolic exchanges – albeit painfully slowly via a process of eavesdropping on conversation and trade between organisms. The relationship between algae and bacteria facilitates tasks that either organism alone could not accomplish. For instance, in most aquatic environments, including wastewater, iron is scarcely bioavailable due to poor solubility. Bacteria have evolved pathways to secrete metal-chelating compounds called siderophores that can bind to iron and enable transportation within the cell. These public goods are the basis of a complex eco-evolutionary dynamic between heterotrophic bacteria and autotrophic organisms (Amin et al. 2009). Capitalising on fixed carbon from photosynthesis, algae shed polysaccharides and monosaccharides into the phycosphere (Seymour et al. 2017), an important region in the microscale relationship between cells. This dynamic proximity enables food security for the bacterial partner, while resourcing trace elements and vitamins for the phototroph.

This alliance may be invisible to the naked eye in a planktonic state, but frequently this takes on a physical presence between bacteria and algae in form of a biofilm. Biofilms are defined as 'aggregates of microorganisms in which cells are frequently embedded in a self-produced matrix of extracellular polymeric substances (EPS) that are adherent to each other and/or a surface' (Vert et al. 2012). EPS, or slime, is composed of various extracellular biopolymers which suspends the community, and enables the system to embody properties distinct from individual members (Flemming et al. 2016). Elaboration of biofilm-based structures requires an understanding of their formation and ecology, distinct from the single organism as described previously. Algal biofilms operate as intricate societies, with cooperation between some microbes and conflict with others (Queller and Strassmann 2009) while maintaining a resilience to environmental perturbances.

In the context of a future iteration of an INDUS bioremediation wall system based on consortia, cooperation exists at the microscopic level in order to maintain the overall community structure via mutualism, cross-feeding or syntrophy (Cavaliere et al. 2017) or through the formation of eco-evolutionary partnerships from environmental isolates (Borchert et al. 2021). Meanwhile at the macroscale, we must simultaneously consider how to design for a pattern of activity that is materially, morphologically and contextually determined. Autotrophy powers the biofilm community, and exchanges of nutrients within the slime manifold enable the system to create a very robust matrix, counteracting the forces of precocious decline or deterioration in fluctuating conditions. We are still in the process of understanding the myriad mechanisms

FIGURE 11.9 Algal biofilm formation on concrete scaffolds
IMAGE CREDIT: MARCOS CRUZ

of microbial cooperation (Buhmann et al. 2016; Nowak 2006; Villa et al. 2015) facilitating the emergent properties of the system. But what we do know is that these biodiverse algal biofilm communities create nutrient-rich substrates that in time are vital to the long-term and self-sustained nature of other forms of life that are integrated in large-scale constructions.

Design of algal ecosystems therefore navigates the boundary between the artificial and the natural. It has been previously stated that one should not confuse "biological" with "natural" when considering artefacts of our ingenuity (Simon 2019). Humans are profligate ecosystem engineers, through the modification of biotic and abiotic materials. In the previous case of INDUS we acted as allogenic engineers, as we modify materials such as clay to become ceramic and alginate to become hydrogel. Ultimately as we move to creating materials capable of bioreceptivity or biocolonisation we seek to create structural scaffolds in our built environment (Figure 11.9) to scale-up while augmenting biological performance, making us ultimately become "autogenic designers" (Jones et al. 1994). That is to say, we are capable of modifying our environment and resource exchanges through our own physical structures. This has significant implications for our methods of design and fabrication. While the overall ecosystem service of a building or an urban infrastructure may be agreed upon, the architecture will vary when considering designing in a microbially-centred manner vs a more traditional programme, space and form driven approach. We predict that here lies the tension, but also complementarity, of the bio-integrated design method (Cruz and Parker 2021), which negotiates the territory on an equal weighting.

8 Urban Abject Biofilms: The Preston Road Wall

In our aim to animate the built environment by promoting the biocolonisation of building surfaces, there is, however, a "germ of disequilibrium" to borrow a phrase attributed to MacNeice (Kearney 1979) where, despite the rationalisation of the biofilm as a highly sophisticated organisational unit that should appeal to our innate biophilia (Wilson 1984), it can invoke for most people very negative qualities, especially when grown into larger and more complex substrates. To understand this, it is worth looking at it from an aesthetic and cultural perspective. The biological hybridity and material viscosity of such humic strata create dark and slimy conditions, which potentially place it into the category of the abject (Kristeva 1982). Visually, they have formal associations with conditions that trigger our response to what is culturally mostly repressed in our body – excreta such as snot, pus, sperm, etc. Biofilms are filmy, scummy, curdly, mucky and viscid and therefore imbued with a sense of lowness due to their defiling appearance. This sits in close proximity to our boundaries of revulsion at the indeterminate and unresolved. The relationship of nature to body and mind creates ultimately a link to decay as a proxy for death.

Our built environment is exposed to a natural weathering process that makes all its outdoor surfaces prone to be receptive to biofilm formation. Encounters with such emerging biofilms are often signifiers of the unkempt and neglected areas of a city. Disrepair creates areas that are so common in our cities where pipework creates uncontrolled flows of water over brickwork, as for example seen at Preston Road Station in London (Figures 11.10 and 11.11). The presence of green, brown and black slime is a *memento mori* for the human observer. Leaking surfaces in city walls accumulate lots of growth, forming what could be considered highly unpleasant features that are associated with damp and decaying places. The presence of moisture enables the establishment of a complex biological infrastructure to form an ever-evolving "life soup" (Miller 1997). But in reality, these areas are saturated with vitality and the potential for growth. Successful establishment of the initial colonists by a process of adsorption triggers production of the EPS bioadhesive and an irreversible binding of a collective of cells (Bixler and Bhushan 2012). The final stage of biofilm colonisation involves dispersion and propagules colonise new habitats.

Microalgae are the most recognisable component of outdoor biofilms due to their abundance and pigmentation (Gaylarde and Gaylarde 2005). They form highly rich microbial, phycological or cryptogamic environments. This results in biodiverse polycultures full of excreting matter – soggy, clammy, slithery, sticky, tacky and dank. Soon, a synthetic biotope – an ecosystem created by

FIGURE 11.10 Synthetic biotope – leaking wall with biofouling at Preston Road Station, London, 2021
PHOTO CREDIT: MARCOS CRUZ

anthropogenic activity – is generated that is teeming with life, attracting inorganic sediments of dust particles, fibres and hair. Fed by humic matter, bacteria, cyanobacteria and algae, mosses and fungi proliferate to eventually form a rich microbiome that feeds the ecological succession of plants. For common architecture, however, this natural and highly site-specific phenomenon represents a fundamental challenge: the predictive dimension of design feels undermined and redundant when biology generates unpredictable growth conditions that are hard to forecast. At the same time, biofilm and biotope formations imply ugly states that are difficult to accept in a discipline that had for long as its ultimate goal the creation of beauty and purity. Our society has developed an intolerance towards 'dirty' and stained surfaces, especially when produced by nature's growth: it means that any visible biofilms in our immediate surroundings have to be removed – the result of a sanitation impulse that in our Western culture has symbolised for long a 'higher' status of civilization and cultural differentiation.

Design, in particular, has followed all-encompassing concepts of cleanliness (Vigarello 2008) and promoted hygiene aesthetics (Forty 1986) that has avoided heterogeneous, rough and viscous conditions in favour of more homogenous, smooth and matt finishes, a phenomenon epitomised with architecture's 'Modern whitewash' (Wigley 1996). In an ultimate instance, architecture has repressed any evidence of the abject in pro of clean-looking, clear and flat morphologies. In fact, our societal expectation has been constructed around an aesthetic and ecologic baseline principle in which biodiverse walls filled with messy growth are considered problematic, with too many irregular and patchy patterns. They contradict established norms and categorisations. As such, the scientific vocabulary switches to describing the biofilm as biofouling, with anthropocentric characteristics of stubbornness (Monroe 2007) and recalcitrance. The abject is here not an objective condition but instead a culturally constructed judgement of dirt and pollution where species and growth are mostly considered what the anthropologist Mary Douglas so famously defined as "matter out of place" (Douglas 1966).

At the same time, there is an inherent relationship with the uncanny (*unheimlich*) that has to be acknowledged when these biofilms occur in an overtly exposed context. Would they be inert or inorganic, they may not create too much of a problem, but anything that is alive in this form triggers a potential sense of disgust, and ultimately our innate fear that it all could overgrow and expand. Abject biofilms trigger an instinctive sense of defence from what could be polluting or contaminating matter that inflicts us harm. Ultimately, it is our visual perception and any possible smells that instigate our subconscious that is fed by our primal experiences of touch and taste. Geosmin and 2-methylisoborneol are compounds with musty and muddy aromas produced by members of the actinomycetes, cyanobacteria and green algae that can be detected by humans at low thresholds (Jüttner and Watson 2007). Viewing biofouling matter suggests the prospect of "unnerving touches, nauseating smells, foul odours" (Miller 1997). The real problem of putrefaction is not so much the biological content, but the simple phenomenon of being 'potential' which makes it all feel disturbing. One does not know what to expect and how much these biofilms and substrates could take over in time. There is also a problem with our difficulty to identify the filmy goo that will in time become overgrown with vegetation. There is a lack of terminology to objectively describe such conditions that seem to defy our orders of classification, challenging the aesthetic hegemony of a "specific model" of beauty (Eco 2011). Newly formed biotopes can have undefined and ambiguous contours and colouration. The more life they have the more they could feel like being too cluttered, disordered and uncontrolled. The biotope's lush ends up being uncanny because it is anomalous nature – disturbing, but highly vital.

FIGURE 11.11 Synthetic biotope – abject biofilms at Preston Road Station, London, 2021
PHOTO CREDIT: MARCOS CRUZ

9 Biofouling and the Regenerative Power of Decay

Key to the formation of synthetic biotopes is the occurrence of biofouling in which biofilm formation and bacterial adhesion on a building surface produces a first phase of microfouling. This is then followed by an attachment of larger organisms which defines a second phase of macrofouling. This results in the decay of organic and inert matter which in turn has a vital regenerative power for the system to be kept alive. What was once fully growing will at some point fade away. Species and matter will age and degrade and in time shift to a yellowish, brownish or darkish hue; it may wilt or rot and gradually decompose, producing a nutrient-rich substrate to feed other forms of life. At the same time, alterations to physico-chemical parameters or temporal changes within the environment can promote heterogeneity within a biofilm (Wimpenny et al. 2000) In terms of aesthetics, on the other hand, the integration and scaling-up of polycultures and their life cycles in the built environment allows for a new type of design method where abject features define a basic "creative formlessness" (Douglas 1966) for growth and decay to become essential performative and formal parameters.

We need to assume the impacts of algal biofouling and the epibiotic accumulation of matter on building surfaces. Ultimately, it is important that we design for what is aging as opposed to our common expectation that buildings will look for ever new (Cruz 2021). Even more so does this apply to the homogeneous ever-green appearance of common green roofs and walls that, similar to monocultures in bioreactors, have been conceived to remain pure, therefore requiring a costly and unsustainable apparatus to be kept unchanged for long periods of time. On the contrary, we need to embrace a new approach to design that rejects two key aspects: architecture's obsessions with the eternally stable, clean and geometrically controlled, and horticulture's demands for a biological efficiency that is purely functional and mechanistic. We need to delve into a more spontaneous and surprising, also far more provocative and even disobedient aesthetic.

Historically, elements of the fecundity of crytogramic takeover of structures may be seen in the Gothic (Hughes et al. 2015), and mostly in Romanticism where the effects of time and weathering produced an existential cycle between the natural and synthetic. From the medieval fascination with the monstrous to nineteenth century obsession with imperfection, incompleteness and cyclic entropy, the return to the darkness and dankness of growth on buildings is abundant in these periods. The imagery is nostalgic, yet also exuberant and transgressive, invoking sex and death. It also speaks of unravelling, madness and the esoteric – characteristics later exorcised by the clinical treatment of twentieth century Modernism where stained, deformed and degenerated conditions became aesthetically unacceptable.

10 Phases of Bio-Integrated Design for Heterogeneity

How do we reconcile incongruous values of biofouling, the abject and ugly with the functional in our future cities, especially one with aspirations to be photosynthetic? To examine this, the phases of design in a biologically integrated manner has to be explored. Echoing the three stages of growth in cell culture – lag phase, logarithmic phase and stationary phase – the evolution of an outdoor bioreactor might be anticipated. The role of autogenic design is here to synthesise computation, fabrication and biology in a manner that can be augmented and "read", reflecting different stages of its lifecycle.

Considerations include the role of morphological variation and the intersection of environmental parameters, including moisture, light, nutrient level, etc. all of which are essential for the long-term creation of a synthetic biotope. This designed area offers environmental conditions providing ultimately a place for a specific assemblage of microbes, plants and animals. By developing a nomenclature to describe the phases of colonisation and development of sub-aerial biofilms on a designed substrate we may begin to understand how to interpret the lifespan of placing this outdoors. Here, we suggest three phases that can be actively considered and designed for, and a fourth phase where control is relinquished to the succession of organisms.

1. Inchoate phase of the design is an incipient, nascent stage, prior to establishment of visible algal biofilm growth. Here the material scaffold in form of a building roof or façade is presented. Depending on the context, a design may remain in this state for several months or years. Bio-colonisation of bacteria and algae and a shift from this latency period has been modelled previously for common architectural substrates such as brick (Quagliarini et al. 2021).

The rate of progression into the next phase may be through deliberate design, material choice or manipulation of environmental parameters as potential interventions for augmentation of bioreceptivity.

2. *Burgeoning phase* of the design focuses on the proliferation of growth in regions where surface roughness or water activity is favourable for growth, which may be predicted by Avrami's Law (Tran et al. 2013) to predict the temporal evolution of attachment and colonisation to a surface. This may not be uniform. It may be seen as the formation of a heterogenous patina of rich algal ecologies on the surface, or an accumulation of layers. The role of the scaffold is to support this and create desirable conditions of hydrodynamics and light exposure. In this phase materiality has a role to play in zonation – defining areas where growth is facilitated and limits where the scaffold surface is exposed.

3. *Confervant phase* of the design explores a moment of great intensity where the material has an abundance of growth that is well established. In this phase the design and biology work in tandem. The biofilm will have an internal heterogeneity of pH or light penetration, resulting in a dynamic system that may undergo taxonomic shifts at the community level but maintains functional structure. Here, the emergent properties of the system (Flemming et al. 2016), for instance bioremediation or nutrient uptake, become apparent. The algal biofilm will also have an external heterogeneity of colouration and texture, dependent on the composition and species associated.

4. *Biocoenosis phase* is the last stage in which the functional bioremediation process shifts towards a more uncontrolled bioreceptive moment. In this phase interacting organisms live together forming a highly complex ecological community with inter-species interactions shaped by evolutionary forces (Lee and Ryu, 2021). Cryptogramic and vascular species are established and intertwined in what is the emergence of a synthetic biotope on a building scale. This phase recognises that with biodiversity will come a multiplicity of forms, and will transcend the microbial.

Revisiting Guillette's work (Guillitte 1995), the term *quaternary bioreceptivity* has been proposed, where a structure may be coated to become more hospitable to colonisation (Sanmartín et al. 2021). A secondary material, such as a water retaining hydrogel (Snoeck et al. 2022), is applied or impregnated into the surface in order to enhance the ability to attract biological life. In this phase the materiality is paramount. Within the lens of design this may be considered in two ways. It may take form of seeding of algal biofilms through deposition directly, or through an initial phase of artificial biofilm created in a hydrogel structure. Parameters such as surface roughness, porosity and permeability will determine the rate at which a biofilm adheres or establishes. Designing with this mechanism of enhanced bioreceptivity in mind may accelerate the transition between initial phases to realise a performative ecosystem service.

11 The Future Human Habitat Lab of St Andrew's: Polyculture Design and Embracing the Abject

These design stages define the underlying methodology of the future lab at the St Andrew's Botanic Garden in Scotland – an experimental project aimed to create the prototype of a fully grown building that is tested and understood and ultimately applied to create a forthcoming photosynthetic human habitat (Figure 11.12).

Speculating on the evolution of such a project over a matter of years, we may anticipate how the phases described above may manifest over the various transitory states. In the initial phase, such a construction may embed elements

FIGURE 11.12 *Future human habitat lab / Bio.HAB* – sketch of building with merging synthetic biotope. Proposed location: St Andrew's Botanic Garden, Scotland, 2021. Team: Brenda Parker and Marcos Cruz
IMAGE CREDIT: MARCOS CRUZ

of photosynthetically biomineralised engineered living materials (ELM) which will follow the progression from *inchoate* to *confervant*, becoming progressively more structural (Tamuli et al. 2021). Comprising of cyanobacteria and silica aggregates, the ELM will provide a tectonic presence. The gradient of semi-translucent components allows for light to be embedded in the building in a variable degree of intensities without the need for traditional windows or openings.

Biocolonisation of other parts of the living lab will take place on designed synthetic biotopic zones, created through the intersection of materiality, porosity and water flow. Selection pressures from the ambient environmental conditions and seasonality may favour aerial, or poikilohydric organisms, which later become enmeshed in the exopolymeric material of an algal biofilm. The pristine walls will become murky and pigmented as regions of growth establish and augment over time. A disobedient ecology, as the botanist Mark Spencer would call it, is unfolding that blends surrounding landscape with the building's surface morphology. It results in a somehow creepy, but also extraordinarily alluring scene of unprecedented dimensions – engaging and immersive, the project starts defining a radically new aesthetic. The gradual formation of the biofouling repels and even provokes initial revulsion, yet compulsion to touch when experienced from very close. In this way the building confronts the limits of social acceptability. It may even elicit outrage provoked by disgust and fear of disease and the complex relationship of humans to what they may perceive as contamination (Davey 2011).

As our conception of humans as holobionts becomes more prevalent, this building may become a fascination, especially when viewed as a provocation of how we may envision a microbiome-inspired green infrastructure (MIGI) within landscape design (Watkins et al. 2020). MIGI promotes interactions between humans and environmental microbiomes, with explicit considerations for sustaining microbially-mediated ecosystem functionality and resilience. In this way, the ecosystem services provided by such a building as a hub for beneficial organisms for immunoregulation or bioremediation are placed at the foreground. Pathogens co-exist and are kept in balance via the presence of far more pervasive pre- and probiotic communities, creating a microbial equilibrium in the system that is beneficial to us. MIGI seeks to apply the "Old Friends Hypothesis" (Bloomfield et al. 2016; Rook et al. 2003) whereby dysbiosis is prevented through the creation of a varied exposome (Robinson and Jorgensen 2020) which could be generated by the *confervant* and *biocoenosis* phases of the design. The result is a scenario that is defined by its moist viscosity and a rather obscene biological beauty as well as a growing sense of connection to our own health (Flandroy et al. 2018). Over time, algae and microorganisms are no longer the only species present. The building flourishes and development of the *biocoenosis* becomes apparent. Common species give way to more rare species in a visual and dynamic display of biodiversity (Goulden 1969)· triggering a process of ecological re-wilding that makes it feel all rather rambunctious (Marris 2011). The scientific opportunity to examine inter-kingdom interactions and succession within this bioreceptive environment creates a bridge between the empirical and experiential. Considering our place in a microbial world (McFall-Ngai 2015), and our relationship in terms of the climate (Cavicchioli et al. 2019) will require an empathy and understanding of the "other".

A key aspect to achieve this positive shift in perception is the aestheticization of biofouling and abject biofilm formations. Several compositional features are important to help us engage and appreciate this living spectacle. The most important parameter is the overall readability of the biotope that acquires objectual qualities that are nonetheless interpreted in a highly subjective manner. The clarity of the building or wall tectonic is therefore crucial which can be achieved when applying Gestalt Psychology principles such as integral dimensions, emergent features, configural superiority, global precedence, primacy of holistic properties, law of *Prägnanz* (Wagemans et al. 2012). Most important are legible contours (macro geometry) and/or specific reference spots that are recognisable amongst the spooky and heterogenous micro landscape of algal biofilm formation. No matter what, our eye will look for anchor points within the filmy and vegetative clutter, making sense of inherent relationships in search for a reassurance of scale and intention of what is proliferating on the walls. This can be done by strategically setting intelligible features that can be understood amidst the overall wall composition. Familiar geometric forms or identifiable species, for instance, will enable this. At the same time, figurative qualities of the underlying surface morphology as experienced in figural ornaments of our past heritage help to overcome the predominant abstract dimension of biofouling. Especially when the surfaces have patchy gradients of intensity and colouration along with what looks like randomly scattered growth, figurative elements help mitigating the unfamiliarity of the biotope, in fact reversing our initial scepticism towards a growing attraction. Repetitive and single orientated protrusions or infolds will create topological variation as well as specific shadow plays that can further help the readability and enhance our curiosity. Amidst the unpredictability and unfamiliarity of the biotope, we are now not only able to validate the system but finally indulge with it.

In the applied sense the building surface may be considered an exo-bioreactor, or in the ecosystem sense it is a synthetic biotope. Regardless, the role of design is to negotiate the precarity of how the formal and poetic expression can overcome accepted norms of aesthetic presentation while maintaining scientific validity. The ugliness of the project offers a contemporary form of beauty and splendour – a contemporary sublime. In this context, the Burkean concept of sublime is distinct from beauty (Burke 1757). The sublime delves in the awe of the extraordinary but also contains elements of fear of the uncanny. It is a painting in a murky palette which transcends death and scale. The notion of the sublime encompasses the terrible to elicit a deeper response. In this way the infinite and irregular nature of such designs can be understood. But the sublime is here also grotesque due to the exaggerated and overtly visceral qualities that it implies. Design in this context offers a means to interpret the complex phenomena and allows the public to overcome loathing, towards a new acceptance of biological uniqueness and ever-changing complexion.

12 Conclusion

In this chapter we explored how algal monocultures are a well-established and highly efficient production system, with defined outputs and an inherent beauty of pigmentation. Nevertheless, they are dependent on complex apparatus to support purity. The B.I.Q house and the *Algaegarden* are examples of how such bioreactors have been applied to larger-scale constructs. The energy expended in maintaining the monoculture through mechanical procedures is unsustainable in the outdoor context. To achieve the ambitions of a truly bio-integrated architecture, there is a need to rethink our practice. Polycultures are more hybrid and less explored by design. Due to their multiple species cooperation and when exposed without any mechanical protection, they are more capable to survive in external conditions. They create resilient consortia that are self-regenerative and exist in constant exchange with the surroundings. In this context, algal biofilm formation is essential to establish such highly enriched ecologies. As a product of the design, algal-bacterial interactions are fostered, which when established exhibit emergent properties of bioremediation and/or microbiome-inspired green infrastructures. Yet the design and form cannot seek to emulate the monoculture, it requires its own aesthetic to deal with heterogeneity. Biofilms are viscous and become murky in time. They have an adverse visual impact that is associated to the abject. These conditions have been historically rejected as they reflect a sense of neglect, decay and disease that is supposed to be threatening us. Yet this perception is finally shifting, not only because of a recognition how ecologically rich and diverse these biofilms are, but also due to the emergence of a new design protocol that accepts the material impact of biofouling and decay as generative qualities.

In order to provide a way forward we posit a methodology to design for heterogeneity. The overall macro geometry or tectonic should have a sense of readability and/or figurativeness (not necessarily of bodies) so that it can handle the patchy irregularity of what grows on it. Here we have presented the various phases to be designed for: *Inchoate*, *Burgeoning*, *Confervant* and *Biocoenosis*. This is a form of bio-integrated design that already considers intermediary and indeterminate stages as valid parts of the life cycle. The intersection of materiality and geometry is key to lending the uncontrolled surface of growth some contrasting order. Material properties of charge, conductivity, porosity, roughness, wettability will interact with the physical environment: light, water, laminar vs turbulent flow, friction, etc. Geometry and form will give the uncontrolled surface of growth a contrasting order.

But this mode of design poses an inherent challenge: to integrate polycultures in architecture there is not only a problem of an unpleasing look and an *a priori* lacking readability and order, but also the problem of scaling-up itself. What is acceptable in a petri dish in a controlled lab environment feels threatening and overwhelming when applied and exposed outdoors. The sticky, slithering, wriggly, oozing, or slimy conditions that are large have the potential to multiply and expand even further due to its unprotected and sheer force of scale. It feels like it could ultimately engulf us. However, our contemporary awareness of the climate and biodiversity crisis is forcing us to embrace a radical change in how we envisage the future and rethink our technical approach to consider mixed, rather than mono-culture. We need to capitalise on emergent properties of algal ecosystems as the basis for our building-integrated biotopes – highly dynamic, diverse and vigorous. Our urban environments will depend and thrive on the augmentation of such polycultures and resulting ineffable aesthetics of biofouling so that we can ultimately realise the vision of a future photosynthetic city.

References

Amin, S.A., Green, D.H., Hart, M.C., Küpper, F.C., Sunda, W.G., Carrano, C.J., 2009. Photolysis of iron-siderophore chelates promotes bacterial-algal mutualism. Proc. Natl. Acad. Sci. U. S. A. 106, 17071–17076.

Bhayana, S., Hagopian, J., Mohite, S., Lintong, C., Stoffels, L., Giannakopoulos, S., Beckett, R., Leung, C., Ruiz, J., Cruz, M., Parker, B., 2020. Robotic Extrusion of Algae-Laden Hydrogels for Large-Scale Applications. Glob. Challenges 4, 1900064.

Bixler, G.D., Bhushan, B., 2012. Biofouling: lessons from nature. Philos. Trans. R. Soc. A Math. Phys. Eng. Sci. 370, 2381–2417.

Bloomfield, S.F., Rook, G.A., Scott, E.A., Shanahan, F., Stanwell-Smith, R., Turner, P., 2016. Time to abandon the hygiene hypothesis: new perspectives on allergic disease, the human microbiome, infectious disease prevention and the role of targeted hygiene. http://dx.doi.org/10.1177/1757913916650225 136, 213–224.

Bolker, J.A., 2019. Selection of Models: Evolution and the Choice of Species for Translational Research. Brain. Behav. Evol. 93, 82–91.

Borchert, E., Hammerschmidt, K., Hentschel, U., Deines, P., 2021. Enhancing Microbial Pollutant Degradation by Integrating Eco-Evolutionary Principles with Environmental Biotechnology. Trends Microbiol.

Buhmann, M.T., Schulze, B., Förderer, A., Schleheck, D., Kroth, P.G., 2016. Bacteria may induce the secretion of mucin-like proteins by the diatom Phaeodactylum tricornutum. J. Phycol. 52, 463–474.

Burke, E., 1757. A Philosophical Enquiry into the Origin of Our Ideas of the Sublime and Beautiful. Printed for R. and J. Dodsley, London.

Burlew, J.S., 1953. Algal culture from laboratory to pilot plant. Algal Cult. from Lab. to Pilot plant.

Carney, L.T., Lane, T.W., 2014. Parasites in algae mass culture. Front. Microbiol. 0, 278.

Cavaliere, M., Feng, S., Soyer, O.S., Jiménez, J.I., 2017. Cooperation in microbial communities and their biotechnological applications. Environ. Microbiol. 19, 2949–2963.

Cavicchioli, R., Ripple, W.J., Timmis, K.N., Azam, F., Bakken, L.R., Baylis, M., Behrenfeld, M.J., Boetius, A., Boyd, P.W., Classen, A.T., Crowther, T.W., Danovaro, R., Foreman, C.M., Huisman, J., Hutchins, D.A., Jansson, J.K., Karl, D.M., Koskella, B., Mark Welch, D.B., Martiny, J.B.H., Moran, M.A., Orphan, V.J., Reay, D.S., Remais, J. V., Rich, V.I., Singh, B.K., Stein, L.Y., Stewart, F.J., Sullivan, M.B., van Oppen, M.J.H., Weaver, S.C., Webb, E.A., Webster, N.S., 2019. Scientists' warning to humanity: microorganisms and climate change. Nat. Rev. Microbiol.

Craggs, R., Park, J., Heubeck, S., Sutherland, D., 2014. High rate algal pond systems for low-energy wastewater treatment, nutrient recovery and energy production. https://doi.org/10.1080/0028825X.2013.861855 52, 60–73.

Croft, M.T., Lawrence, A.D., Raux-Deery, E., Warren, M.J., Smith, A.G., 2005. Algae acquire vitamin B_{12} through a symbiotic relationship with bacteria. Nature 438, 90–93.

Cruz, M., 2021. Design for Ageing Buildings: An Applied Research of Poikilohydric Living Walls. In: Duanfang Lu (Ed.), The Routledge Companion to Contemporary Architectural History. Routledge.

Cruz, M., Coletti, M., 2016. Alga(e)zebo: Bartlett Research Folio.

Cruz, M., Parker, B., 2021. From the Anthropocene to the Biocene – novel bio-integrated design as a means to respond to the current biodiversity and climate crisis. In: Harriss, H., House, N. (Eds.), Intersectional Space-Architecture of the Post Anthropocene. RIBA Publications, London.

Cueva, M.E., Horsfall, L.E., 2017. The contribution of microbially produced nanoparticles to sustainable development goals. Microb. Biotechnol. 10, 1212–1215.

Davey, G.C.L., 2011. Disgust: the disease-avoidance emotion and its dysfunctions. Philos. Trans. R. Soc. B Biol. Sci. 366, 3453.

Day, J.G., Gong, Y., Hu, Q., 2017. Microzooplanktonic grazers – A potentially devastating threat to the commercial success of microalgal mass culture. Algal Res. 27, 356–365.

Douglas, M., 1966. Purity and danger: an analysis of concepts of pollution and taboo/Mary Douglas. Routledge and Kegan Paul, London.

Eco, U., 2011. On Ugliness (originally published as Storia della Bruttezza). Rizzoli International Publications, New York, NY.

Flandroy, L., Poutahidis, T., Berg, G., Clarke, G., Dao, M.-C., Decaestecker, E., Furman, E., Haahtela, T., Massart, S., Plovier, H., Sanz, Y., Rook, G., 2018. The impact of human activities and lifestyles on the interlinked microbiota and health of humans and of ecosystems. Sci. Total Environ. 627, 1018–1038.

Flemming, H.C., Wingender, J., Szewzyk, U., Steinberg, P., Rice, S.A., Kjelleberg, S., 2016. Biofilms: An emergent form of bacterial life. Nat. Rev. Microbiol. 14, 563–575.

Forty, A., 1986. Objects of Desire: Design and Society Since 1750. Thames and Hudson, London.

Fraunhofer, 2016. Monitoring Results from the BIQ House.

Gaylarde, C.C., Gaylarde, P.M., 2005. A comparative study of the major microbial biomass of biofilms on exteriors of buildings in Europe and Latin America. Int. Biodeterior. Biodegradation 55, 131–139.

Goulden, C.E., 1969. Developmental Phases of the Biocoenosis. Proc. Natl. Acad. Sci. 62, 1066–1073.

Guillitte, O., 1995. Bioreceptivity: a new concept for building ecology studies. Sci. Total Environ. 167, 215–220.

Hughes, W., Punter, D., Smith, A., 2015. Encyclopedia of the Gothic. Wiley.

Jeffrey, S.W., Wright, S.W., Zapata, M., 2011. Microalgal classes and their signature pigments. Phytoplankt. Pigment. 3–77.

Jones, C.G., Lawton, J.H., Shachak, M., 1994. Organisms as Ecosystem Engineers. Source: Oikos 69, 373–386.

Jüttner, F., Watson, S.B., 2007. Biochemical and ecological control of geosmin and 2-methylisoborneol in source waters. Appl. Environ. Microbiol. 73, 4395–4406.

Kearney, T., 1979. The Poetry of the North: A Post-Modernist Perspective. Crane Bag 3, 45–53.

Kristeva, J., 1982. Powers of horror: an essay on abjection / Julia Kristeva; translated by Leion S. Roudiez, European perspectives. Columbia University Press, New York ; Chichester.

Kuczynska, P., Jemiola-Rzeminska, M., Strzalka, K., 2015. Photosynthetic Pigments in Diatoms. Mar. Drugs 13, 5847.

Landrigan, P.J., Fuller, R., Acosta, N.J.R., Adeyi, O., Arnold, R., Basu, N. (Nil), Baldé, A.B., Bertollini, R., Bose-O'Reilly, S., Boufford, J.I., Breysse, P.N., Chiles, T., Mahidol, C., Coll-Seck, A.M., Cropper, M.L., Fobil, J., Fuster, V., Greenstone, M., Haines, A., Hanrahan, D., Hunter, D., Khare, M., Krupnick, A., Lanphear, B., Lohani, B., Martin, K., Mathiasen, K.V., McTeer, M.A., Murray, C.J.L., Ndahimananjara, J.D., Perera, F., Potočnik, J., Preker, A.S., Ramesh, J., Rockström, J., Salinas, C., Samson, L.D., Sandilya, K., Sly, P.D., Smith, K.R., Steiner, A., Stewart, R.B., Suk, W.A., van Schayck, O.C.P., Yadama, G.N., Yumkella, K., Zhong, M., 2018. The Lancet Commission on pollution and health. Lancet.

Lee, S.M., Ryu, C.M., 2021. Algae as New Kids in the Beneficial Plant Microbiome. Front. Plant Sci. 12.

Louca, S., Jacques, S.M.S., Pires, A.P.F., Leal, J.S., Srivastava, D.S., Parfrey, L.W., Farjalla, V.F., Doebeli, M., 2016. High taxonomic variability despite stable functional structure across microbial communities. Nat. Ecol. Evol. 1, 15.

Marris, E., 2011. Rambunctious Garden: Saving Nature in a Post-Wild World. Bloomsbury Publishing, New York, NY.

McFall-Ngai, M.J., 2015. Giving microbes their due--animal life in a microbially dominant world. J. Exp. Biol. 218, 1968–73.

Miller, W.I., 1997. The anatomy of disgust / William Ian Miller. Harvard University Press, Cambridge, Mass ; London.

Monroe, D., 2007. Looking for Chinks in the Armor of Bacterial Biofilms. PLOS Biol. 5, e307.

Nowak, M.A., 2006. Five rules for the evolution of cooperation. Science (80-.).

Parker, B., Rawat, D., Malik, S., Vilatte, A., Cruz, M., 2022. A Design Manifesto for Bioremediation: 10 Principles for the Creation of Sustainable Systems for Environmental Benefit. In: AR#0 Bioremediation. Atelier Luma, Arles.

Pereira, H., Páramo, J., Silva, J., Marques, A., Barros, A., Maurício, D., Santos, T., Schulze, P., Barros, R., Gouveia, L., Barreira, L., Varela, J., 2018. Scale-up and large-scale production of Tetraselmis sp. CTP4 (Chlorophyta) for CO2 mitigation: from an agar plate to 100–m3 industrial photobioreactors. Sci. Reports 2018 81 8, 1–11.

Pruvost, J., Le Gouic, B., Lepine, O., Legrand, J., Le Borgne, F., 2016. Microalgae culture in building-integrated photobioreactors: Biomass production modelling and energetic analysis. Chem. Eng. J. 284, 850–861.

Pulz, O., Scheibenbogen, K., 2007. Photobioreactors: Design and performance with respect to light energy input. In: Bioprocess and Algae Reactor Technology, Apoptosis. Springer Berlin Heidelberg, pp. 123–152.

Quagliarini, E., Gregorini, B., D'Orazio, M., 2021. An empirical failure model to predict biofouling growth on fired bricks due to microalgae. J. Build. Eng. 44, 102965.

Queller, D.C., Strassmann, J.E., 2009. Beyond society: The evolution of organismality. Philos. Trans. R. Soc. B Biol. Sci. 364, 3143–3155.

Ratti, C., Claudel, M., 2015. Open Source Architecture. Thames & Hudson Ltd.

Rawat, D., Sharma, U., Poria, P., Finlan, A., Parker, B., Sharma, R.S., Mishra, V., 2021. Iron-dependent mutualism between Chlorella sorokiniana and Ralstonia pickettii forms the basis for a sustainable bioremediation system. bioRxiv 2021.06.15.446916.

Reford Gardens, 2011. ALGAEGARDEN | Heather Ring, Brenda Parker, Synnøve Fredericks [WWW Document]. URL https://www.internationalgardenfestival.com/ALGAEGARDEN/ (accessed 10.17.21).

Robinson, J.M., Jorgensen, A., 2020. Rekindling old friendships in new landscapes: The environment – microbiome – health axis in the realms of landscape research. People Nat. 2, 339–349.

Rook, G.A.W., Martinelli, R., Brunet, L.R., 2003. Innate immune responses to mycobacteria and the downregulation of atopic responses. Curr. Opin. Allergy Clin. Immunol.

Sanmartín, P., Miller, A.Z., Prieto, B., Viles, H.A., 2021. Revisiting and reanalysing the concept of bioreceptivity 25 years on. Sci. Total Environ.

Scarano, G., Morelli, E., 2003. Properties of phytochelatin-coated CdS nanocrystallites formed in a marine phytoplanktonic alga (Phaeodactylum tricornutum, Bohlin) in response to Cd. Plant Sci. 165, 803–810.

Scholz, M., Lee, B.H., 2007. Constructed wetlands: a review. http://dx.doi.org/10.1080/00207230500119783 62, 421–447.

Schrödinger, E., 1948. What is life? University Press, Cambridge.

Seymour, J.R., Amin, S.A., Raina, J.-B., Stocker, R., 2017. Zooming in on the phycosphere: the ecological interface for phytoplankton – bacteria relationships. Nat. Microbiol. 2, 17065.

Sheehan, J., Dunahay, T., Benemann, J., Roessler, P., 1998. Look Back at the U.S. Department of Energy's Aquatic Species Program: Biodiesel from Algae; Close-Out Report. Golden, CO.

Simon, H.A., 2019. The Sciences of the Artificial, 3rd ed. ed., The Sciences of the Artificial.

Snoeck, D., Roigé, N., Manso, S., Segura, I., De Belie, N., 2022. The effect of (and the potential of recycled) superabsorbent polymers on the water retention capability and bio-receptivity of cementitious materials. Resour. Conserv. Recycl. 177, 106016.

Tamuli, P., Salmane, A., Jotanovic, N., Cruz, M., Parker, B., 2021. Engineered Living Materials UK Patent Application 2113331.9. UK Patent Application 2113331.9.

Todisco, E., 2019. Microalgae Growth Optimization in Biofaçade Photobloreactors. http://www.theses.fr.

Torres, E., Cid, A., Herrero, C., Abalde, J., 1998. Removal of cadmium ions by the marine diatom Phaeodactylum tricornutum Bohlin accumulation and long-term kinetics of uptake. Bioresour. Technol. 63, 213–220.

Tran, T.H., Govin, A., Guyonnet, R., Grosseau, P., Lors, C., Damidot, D., Devès, O., Ruot, B., 2013. Avrami's law based kinetic modeling of colonization of mortar surface by alga Klebsormidium flaccidum. Int. Biodeterior. Biodegradation 79, 73–80.

Vert, M., Doi, Y., Hellwich, K.-H., Hess, M., Hodge, P., Kubisa, P., Rinaudo, M., Schué, F., 2012. Terminology for biorelated polymers and applications (IUPAC Recommendations 2012)*. Pure Appl. Chem 84, 377–410.

Vigarello, G., 2008. Concepts of Cleanliness: Changing Attitudes in France since the Middle Ages. Cambridge University Press, Cambridge, UK.

Villa, F., Pitts, B., Lauchnor, E., Cappitelli, F., Stewart, P.S., 2015. Development of a Laboratory Model of a Phototroph-Heterotroph Mixed-Species Biofilm at the Stone/Air Interface. Front. Microbiol. 6, 1251.

Wagemans, J., Feldman, J., Gepshtein, S., Kimchi, R., Pomerantz, J.R., van der Helm, P.A., van Leeuwen, C., 2012. A century of Gestalt psychology in visual perception: II. Conceptual and theoretical foundations. Psychol. Bull. 138, 1218–1252.

Watkins, H., Robinson, J.M., Breed, M.F., Parker, B., Weinstein, P., 2020. Microbiome-Inspired Green Infrastructure: A Toolkit for Multidisciplinary Landscape Design. Trends Biotechnol. 38, 1305–1308.

Wigley, M., 1996. White Walls, Designer Dresses: the Fashioning of Modern Architecture. MIT Press, Massachusetts.

Wilson, E.O., 1984. Biophilia / Edward O. Wilson. Harvard University Press, Cambridge, Mass.

Wimpenny, J., Manz, W., Szewzyk, U., 2000. Heterogeneity in biofilms. FEMS Microbiol. Rev. 24, 661–671.

Wurm, J., Pauli, M., 2016. SolarLeaf: The world's first bioreactive façade. Archit. Res. Q. 20, 73–79.

Zallen, D.T., 1993. The "Light" Organism for the Job: Green Algae and Photosynthesis Research. J. Hist. Biol. 26, 269–279.

CHAPTER 12

Phytofictions and Phytofication

Julia Lohmann

Abstract

In 'Phytofictions and Phytofication', designer, researcher and educator Julia Lohmann introduces her practice-led research into seaweed as a material for making. In her work, macro-algae are material, method and muse in one. Lohmann makes a case for speculative and co-speculative design approaches to biomaterial development with an empathic mindset towards regenerative practices. She advocates a shift in the role of designers from individual authors to enablers of communities of practice that envision less harmful multi-species relations, set against the backdrop of the climate crisis. The 'Department of Seaweed', a community of practice Lohmann founded at the Victoria & Albert Museum London, demonstrates how museums can expand their role as repositories of artefacts into becoming spaces for multisensory material engagements and learning. Lohmann explains how 'phytofication' – actively embracing the material agency of macroalgae and treating it as a co-designer – enabled the development of biomaterials and objects that communicate the potential of seaweed to diverse publics. These in turn sparked 'phytofictions': conversations on how we might use algae and other biomaterials in the future. Julia Lohmann believes that working with algae, through phytofictions and phytofication, can help us shift our mindset from extraction towards regeneration – if we, as a species, learn from algae.

Keywords

design – regenerative practices – biomaterials – macroalgae – co-speculative design – material value systems – design ethics – museums – community of practice – ocean literacy

•••

You are mystical multitudes, supple and strong, thriving whilst generously supporting other beings. You are umami, the fifth taste that deepens the other four and completes the experience. Our silent saviour in times of malnutrition and war, of crisis and collapse:

© JULIA LOHMANN, 2024 | DOI:10.1163/9789004683310_014
This is an open access chapter distributed under the terms of the CC BY-NC 4.0 license.

Kelp, may we humbly ask you to help us again in this moment of collective despair?

∴

1 Introduction

Since my first conscious encounter with Japanese kelp on the fish market in Sapporo, Japan in 2007 I have been exploring kelp as a sustainable material for making. In this text I am introducing seaweedness, the inherent properties of macroalgae and describe how it guides the creative process. I describe how we co-imagined seaweed futures (phytofictions) with diverse publics through hands-on seaweed encounters in the Victoria and Albert Museum. I ponder how relating to the algae's marine origin and its agency in its eco-system helps us reflect and understand our own human agenda and mindset. In the process of working with kelp we aspire to become more like kelp. We are becoming phytofied. Can working with algae transform our actions and reflections towards a less destructive, perhaps even regenerative practice?

In 2007 I was a designer in residence at the Sapporo Artist in Residence Centre (S-AIR) in Hokkaido, Japan. I was investigating our human relationships with the sea. On the local fish market, I came across Japanese kelp and was enthralled by its materiality. I immediately saw its leather-like material qualities and imagined it as a potential leather substitute that did not involve the death of an animal, nor the chemicals, time and labour necessary to transform its hide. In this moment I thought that Japanese culture, deeply connected with kelp and known for its outstanding handicrafts, must bear extensive knowledge on kelp craft that I was simply unaware of. I imagined there might be book covers, wall hangings, scrolls, shoji screens, slippers, lamps, vessels, armchairs and dresses made from kelp, and I asked my Japanese friends what they make from seaweed. 'We eat it!' was the answer. 'And?' I probed. 'And …? that's it. We eat it.' They replied. I realised that the kelp craft I had imagined did not exist yet, it was a fiction, a phytofiction. I decided to try and bring it into existence.

This encounter with kelp revealed to me that cultural framing can define the usage of organisms like algae and their applications as strongly as their inherent material properties. If, for generations, a material is known as food, it is difficult to realise that it might also have other uses and applications. Our cultural framing limits the scope of our imagination. Or was it simply that

FIGURE 12.1 Early kelp experiments during the exhibition *Panta rei*, Galleria Nilufar, Milan, Italy, 2008
PHOTO: GERO GRUNDMANN

people had tried to work with kelp and found that it was not possible to utilize it? I took some dried kelp (*Saccharina japonica*) back to Europe with me and began to explore working with it.

In parallel I set out to research the impact and role of brown kelp species on their eco-systems, historical uses for kelp and harvesting and farming techniques. I found a few examples from native cultures, most remarkably the model of a bull kelp water carrier made by Australian Aboriginals.

In Japan, the only non-food related references to kelp as a material for making I could find was the use of kelp to slow down the drying time in traditional lime plastering techniques. These early crafts had not yet developed into contemporary material applications and my initial tests showed promising results.

I began working based on the assumption that most of the things we will have made of kelp in the future simply do not exist yet. What might we make? Who might hold the knowledge needed to develop materials for making? If we succeeded in finding new applications for kelp, what would this increase in harvest / production mean for our marine ecosystems? I shifted

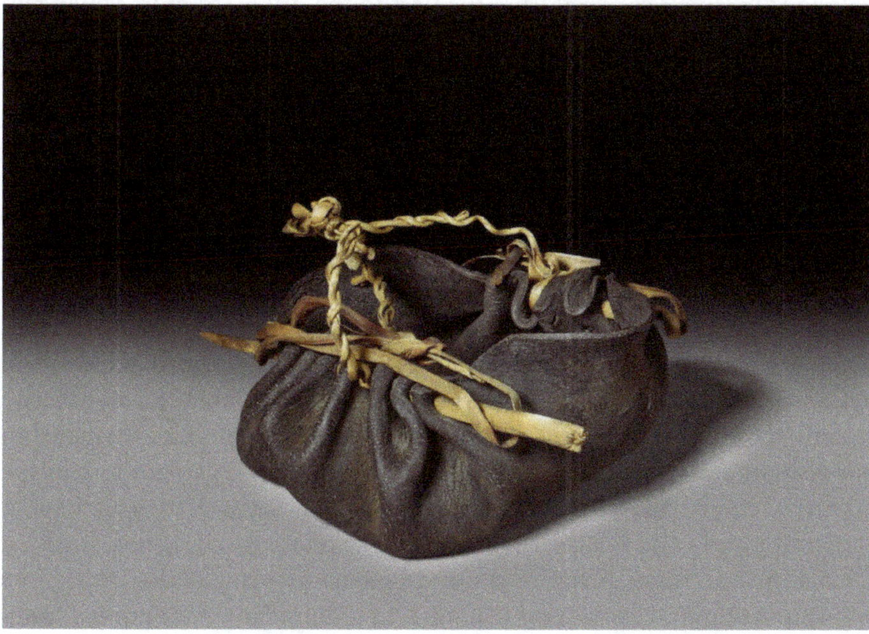

FIGURE 12.2 Model of a water-vessel made of bull kelp, Aboriginal Australian, ca. 1850
PHOTO: © THE TRUSTEES OF THE BRITISH MUSEUM, LONDON, UK

FIGURE 12.3 Seaweed experiments, Stanley Picker Research Fellowship, Kingston-upon-Thames, UK, 2009

my thoughts about seaweed from the past to the future, from imagined and discovered histories to speculative futures: What might a world in which kelp is an established making material look like? How would we need to shift our own mindset, to make this future not just possible, but sustainable and desirable? When MoMA design curator Paola Antonelli asked me about the future of design I said 'algae' but I did not envision that I would still be working with macroalgae 15 years later.

Now, in 2022, the interest and research into the use of algae has grown rapidly and reaches across the whole spectrum of material production, from bioplastics to cosmetics, from food supplements to fuel, from chemical compounds to bathing experiences. The parties engaged in it range from individual entrepreneurs to large international research consortia, from the oil industry to activists with venture capital. Some aspects of the futures we imagined in our encounters with algae are becoming probable as we gather more of the expertise required to realise them. However, at the heart of all these new applications for algae also lies the threat of overexploitation and further stress on the marine ecosystem. If we are to use seaweed, it is essential that we explore any cultivation and harvesting activities in a considered, science-led manner that is location-, case- and species-specific. It depends on what type of seaweed is grown with which technique in which specific location. When I presented the initial idea of developing materials for making from kelp to Prof. Juliet Brodie, the phycologist of the Natural History Museum, she said: "Oh." When I asked her whether or not she thought we could make objects from seaweed she made a statement that has guided my project ever since:

> Brown algae already suffer from warming oceans, habitat-loss, acidification and other anthropogenic changes and impacts on the environment. I am worried that, with all these new human uses emerging for it, we will mess up these algae and the ecosystems they support even more, just as we have impacted on so many of the organisms that we focussed our human interest on.[1]

We urgently need biomaterial alternatives that are healthy for us and the environment; materials with a global impact so that we can turn away from unsustainable fossil-derived materials. Algae can be grown in ways that do not

1 Personal conversation with Prof. Juliet Brodie in 2010, more detailed insights on seaweed harvesting by Prof Brodie can be found in this testimony to Scottish Parliament about the harvesting of wild kelp: https://macroalgalresearchgroup.com/2019/07/05/reflections-on-the-mechanical-harvesting-of-kelp-science-environmental-change-wider-thoughts-and-a-way-forward/.

harm, but benefit the marine ecosystem, capture carbon and reduce eutrophication (nutrient overload) that otherwise leads to systemic imbalances. These benefits can be lost or enhanced through the decisions we make. Our material transformation has to bring forth a shift in our mindset: When we scale algae-production we need to be aware of and scale the environmental benefits without scaling the negative impacts. We need a regenerative mindset that supports the eco-systemic health of the locality we engage in. By employing a bio-inclusive ethical frame[2] that considers the needs of the non-human stakeholders we affect through our actions we can grow algae in a way that supports its agency and benefits ocean health.

2 Seaweedness

In 'Vibrant Matter' Jane Bennett speaks about a 'thing-power' when she refers to objects as actants that are able to influence situations through their materiality.[3] I describe this vibrancy of seaweed, its inherent character and the agency that I aim to support through my creative processes as seaweedness.

The algae I encountered in Japan, *Saccharina japonica*, became the central species for my algae experiments, since I was able to import it as a food ingredient and because there was an established harvesting, drying and classification protocol that guaranteed that my experiments could be repeated and that I was able to reliably source materials. Later, I expanded my practice to European species of brown kelp such as *Saccharina latissima*, *Laminaria digitata* and *Laminaria hyperborea*.

All of these types of kelp continue to possess seaweedness, a dynamic material agency, long after they have been harvested and dried. As a material, kelp by no means feel dead. Taken out of water, it will dry, becoming hard and brittle, at times, depending on the humidity conditions of the room and type of algae it develops a fine white patina of mannitol sugars, fatty acids and salts dispersing from inside the blade. It remains highly sensitive to moisture and humidity and will become soft and supple again if the water content of the air increases. The Victorians knew about this and used fronds of seaweed as a tactile tool to forecast the weather. Touching the seaweed creates a sensory connection between our body and the macroalgae, through our touch we get a deeper understanding of the environmental factors we are surrounded by

2 Veselova, E., & Gaziulusoy, I. (2019). Implications of the Bioinclusive Ethic on Collaborative and Participatory Design. *The Design Journal*, *22*(sup 1), 1571–1586. https://doi.org/10.1080/14606925.2019.1594992.

3 Bennett, Jane Vibrant Matter. Durham: Duke University Press, 2010.

FIGURE 12.4 Humorous Victorian postcard referencing the algae's weather-forecasting qualities. Late 19th century

and even an instinctive glimpse of conditions to come. We cannot feel such connectedness when we touch the glass tube of a thermometer or other, more predictable materials derived from land.

When we are working with seaweed, we engage with it in a kind of dialogue, or correspondence as Tim Ingold calls it.[4] "Making", writes Ingold (2013, p. 31),

4 Ingold, T. (2013): *Making: Anthropology, Archaeology, Art and Architecture*. Routledge, London.

> is a process of correspondence: not the imposition of preconceived form on raw material substance, but the drawing out or bringing forth of potentials immanent in a world of becoming. In the phenomenal world, every material is such a becoming, one path or trajectory through a maze of trajectories.

The kelp I work with has been removed from this natural, eco-systemic context into the human-made world. However, it has not yet settled into its final framing, form or function. It is on the brink of the change that occurs in the shift from one system to another, a transformative process that changes its value, agency and meaning. The multi-sensorial presence of this biological 'actor' pulls the discourse back to its marine origin, which is in turn embedded in people's lived experiences with nature.

As part of my PhD at the Royal College of Art I had the chance to run a six-month residency at the Victoria and Albert Museum in London (V&A). Together with a group of other practitioners I founded the 'Department of Seaweed', an open collective of people wanting to transform algae into a sustainable material for making. We worked together in the V&A and would welcome museum visitors into our workshop, sharing with them the processes, tools, samples and prototypes we were working with. The smell of algae was wafting through the building and the residency space became a multi-sensorial experience both for the museum visitors, for staff and for our collective. These public encounters with seaweed became the context in which we co-imagined phytofictions – scenarios for future uses of algae and their potential benefits and pitfalls.

Most of the visitors to the Department of Seaweed at the Victoria and Albert Museum during our residency in 2013 had prior algae-encounters at the seashore and were reminded of these experiences by its scent. Our conversations often started with these very personal anecdotes and the physical presence of algae enabled a holistic engagement.

The strength of the seaweed lies in its ever-changing and ephemeral nature. To work with it and realise its potential as a material, we must let go of ideas of permanence and uniformity. We need to question prevalent industrial concepts of repeatability and perfection. No two things made of seaweed will be the same. Each frond of kelp has its own characteristics and qualities, its own reactivity to the environment – just as we do – and these qualities change over time. This can be illustrated by the colours of kelp. Its lush greens fade into a translucent parchment-like colour as the chlorophyll is gradually degraded by light, while the fucoxanthin that gives the brown kelps its distinctive colour, remains. This process can take weeks or a few months, depending on species. The artefacts made show the seaweed's characteristics so that encounters with them equally become encounters with seaweedness.

FIGURE 12.5 Julia Lohmann working in the *Department of Seaweed* studio in the Victoria and Albert Museum, 2013
PHOTO: PETR KREJCI

Translucency, shrinkage in drying, natural changes of colour, organic forms, the width and length of the seaweed's blade and the awe for life we feel in experiences of nature were the guiding factors in the creative processes at the Department of Seaweed. Vague in function and legible or rather experienceable in its materiality and form, we created the sculpture Oki Naganode, made of kelp blades stretched onto rattan frames.

Oki is the Japanese word for 'big' and can also be used as a first name, Naga refers Naga-kombu, the type of Japanese kelp the sculpture is made of and node is both synonym of knot and a Japanese syllable ending a family name. Consequently, the Oki Naganode is both the 'big Naga kelp knot' and could also be read as a Japanese name, much like that of a co-designer – which reflects the way I view kelp.

The Oki Naganode was followed in 2020 by two other large structures. The first was Hidaka Ohmu (2020), a seaweed pavilion that I named after the Hidaka kombu used, and the similarly shaped 'Ohmu' creatures from the post-apocalyptic, environmentally themed Japanese anime film 'Nausicaä of the Valley of the Wind'. This was then joined by the Kombu Ahtola (2020), a hanging sculpture I made in Helsinki, Finland. It takes its name from the

FIGURE 12.6 The *Oki Naganode* seaweed and rattan sculpture, built in the Victoria and Albert Museum during the *Department of Seaweed* residency in 2013
PHOTO: PETR KREJCI

FIGURE 12.7 The sunlight-bleached *Oki Naganode* seaweed and rattan sculpture, shown in the exhibition *Earth Matters* at Artipelag, Stockholm, Sweden, in 2015
PHOTO: JEAN-BAPTISTE BÉRANGER

FIGURE 12.8 *Hidaka Ohmu*, seaweed and rattan pavilion, *Partnering With Nature* exhibition, 50th World Economic Forum Davos-Klosters, 2020
PHOTO: WORLD ECONOMIC FORUM

'Ahtola', the underwater residence of the sea gods and marine beings of Finnish mythology.

To some extent, these sculptures design themselves. Considering the attributes of seaweedness, I am proposing their form and they complete it. The kelp I am gluing onto the three-dimensional rattan frames of the sculptures is treated to remain supple, and I attach it while it is wet. It twists and manipulates the rattan rods as it dries, deforming straight lines into concave curves. The seaweed skin stretches taut, into the smallest surface area possible within the framework. This form-finding and shaping correspondence between the two materials makes some of the material's physical attributes legible in the final sculptures. I am embracing and am accommodating this shape-shifting process with rattan frames that are rigid enough to hold the seaweed and weak enough to be shaped by the kelp.

The outcome is an organic-looking, tensile structure that is responsive to light and humidity. In this way, I have shifted the kelp's weather forecasting

FIGURE 12.9 *Kombu Ahtola*, seaweed and rattan sculpture, Annantalo, Helsinki, Finland, 2020

qualities to the inside of buildings, giving an indication of how healthy and habitable interiors are for humans. If the humidity in a space is within normal parameters, the seaweed is supple and strong. If it is too low, often caused by ventilation and air conditioning systems seeking to prevent the spread of mold and other microbes, the seaweed will tense up until it eventually rips. If we realise that the kelp mirrors our own physical responses to such a space, it becomes clear that we too will dehydrate and the ecosystems that are our bodies will degrade if we remain in such an unhealthy manmade environment. Being with kelp raises the question why we should value architectural permanence over human wellbeing.

FIGURE 12.10 Interior view of the *Kombu Ahtola* seaweed and rattan sculpture and its skin on frame structure. The hanging sculpture opens up at the bottom, so that viewers can experience it from the inside and outside.

3 Phytofictions

In the Department of Seaweed, we invited the public of the museum into the process of working with algae. The visitors came into the studio and encountered the seaweed's evocative, multi-sensorial presence within the context of an ongoing design process, not just an exposition of finished artifacts. Neither still in its ecosystem, nor yet fixed into a defined human sphere of reference, sharing process allowed our thoughts to travel along its trajectory of transition both into its multiple potential futures and into its ecological past. The agency and appearance of aliveness in algae also reminds us of their vastly greater agency in the ocean, an ecosystem that we as humans cannot inhabit but depend on for our survival, as do countless other species. Safeguarding the ability of algae to fulfil their habitat-forming, oxygen-producing, nutrient, pollutant and carbon-capturing role – and if possible, expanding it – should be at the core of any algae-related work.

From a maker's perspective, the properties of macroalgae suggest cultural references and material analogies that reach from wood to leather, plastic, lacquer, glass, paper, textile, to the more exotic snake- or dragon-skin, rubber, latex, resin. None of these analogies completely match the material properties, agency and applications of kelp. Still, comparing and contrasting kelp with established materials opens up opportunities to experiment with crafts and processes linked to matter that is alike. We can build on often extensive bodies of knowledge and a wealth of cultural points of reference, past and present. For instance, if we wanted to develop the leather-like qualities of kelp, we could look into tannery, the furrier's trade, leather crafts, surgery, skin-on-frame kayak building, as well as museum conservation techniques, to name but a few fields. On the other hand, studying the differences between kelp and other materials hints at what may be qualities unique to algae and potentially rich in experiences, knowledge and applications.

Unlike for example leather or wood, algae are not yet established as materials for making in a design context. There is no formal algae aesthetic and clearly defined contexts of use for algae beyond the food industry and less visible uses of algal chemical compounds. Even though research into kelp has increased in the past decade, kelp's trajectories into future human contexts are still open to speculation, experimentation and rich in possibilities. These conditions enable us to imagine freely and with few constraints of established applications we are already familiar with. Many of the uses of algae that already exist, apart from foodstuffs, focus on compounds derived from the organisms and do not aim to retain the physical integrity of the algae. They are largely based on the same mindset that humans have used to extract other types of matter from

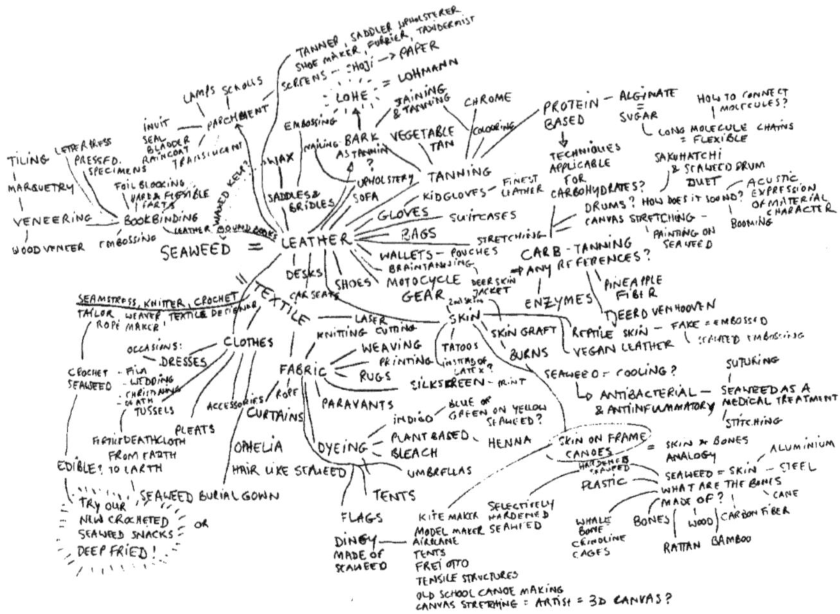

FIGURE 12.11 Mindmap of material analogies of seaweed and leather, craft, cultural heritage and applications that may aid the development of algae as a leather-like material. Algae do not equal leather, but material analogies inspire processes and applications to test with algae, and material differences suggest unique algal qualities and agencies.

the Earth, classifying them as 'natural resources' rather than 'co-inhabitants'. In the Department of Seaweed, we engage with algae by establishing a physical relationship with the organisms, body to body, by letting seaweedness guide our way of working. This type of engagement with algae is sensually rich, evocative through touch and smell and taste as well as visually. The matter itself is versatile and changeable, hyper-reactive to the conditions it is placed in, changing dramatically in its interactions with light, moisture and the treatments applied.

Which phytofictions algae stimulate in people depends equally on how they encounter the material and on their own viewpoint, interest and cultural frame. In the context of food, for instance Japanese dashi soup stock, this ability of algae to enhance the flavour of other ingredients is called *umami*. Within the assemblage of the Department of Seaweed I observed similar *umami*-like behaviours of kelp: It enabled us to imagine through hands-on exploration. Kelp in a material context is highly evocative and relational, meaning that it stimulates the imagination of most people who encounter it and it inspires them to relate it to their own frames of reference. Its vibrancy helps us to see

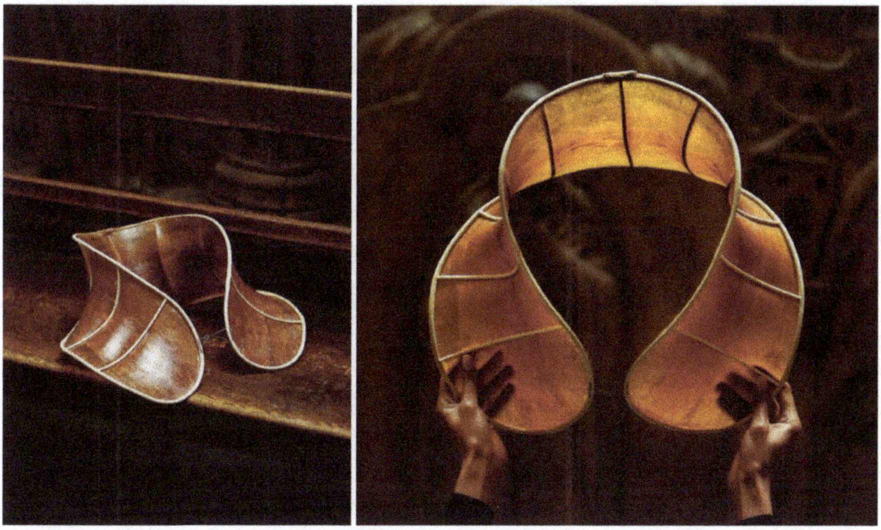

FIGURES 12.12 (LEFT) AND 12.13 (RIGHT) The appearance of a collar made from brown kelp in changing light conditions
PHOTOS: PETR KREJCI, 2013

our own lives in relation to it and to imagine futures that relate to our own lived experiences. With our visitors we discussed a broad spectrum of potential future algae applications and the dialogues proved to be multi-faceted, tying together personal, poetic and scientific questions. In discussing them critically together, we also explored which pathways were merely probable and which might actually lead to meaningful, preferable futures. These collective encounters with the material generated many more ideas, pathways and aspects of possible futures than I could ever have imagined myself. They helped us in articulating our own values, understanding our own biases and helped us to seek and establish a community of practice to achieve our aims. The algae taught us an important lesson: alone we achieve very little, in multitudes we thrive. If we all start at square one with inventing a new material and do not share the knowledge we gain in an open-source way, we will simply repeat these first steps time and again.

I became conscious of the fact that what I had actually been doing at the V&A was to craft multisensorial encounters that enable collective dialogues and expressions of imagination. On this basis, I developed methods for co-speculative design in my PhD.[5]

5 **Lohmann, Julia**, 2018, Thesis, *The Department of Seaweed: co-speculative design in a museum residency* PhD thesis, Royal College of Art. https://researchonline.rca.ac.uk/3704/.

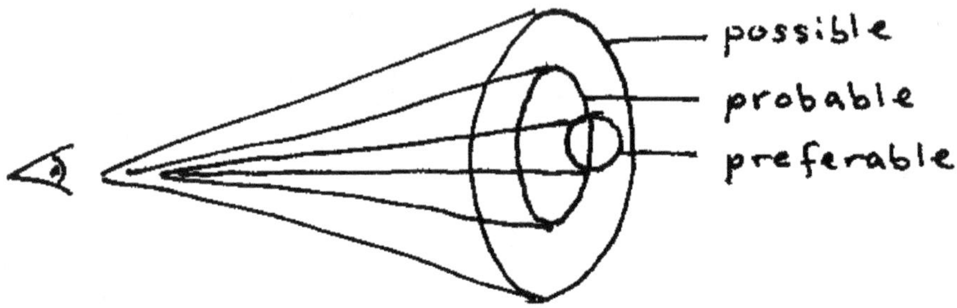

FIGURE 12.14 A *future cone* diagram, used to explain probing possible, probable and preferable future applications for kelp

Co-speculation proposes a shift in roles and mindset from individually authored practices and scenarios to the facilitation of communities of practice that engage in collective speculation and practice-led research centered around a shared concern – in my case seaweed. A 'future cone' diagram consisting of nested or intersecting cones emanating from a point of origin is often used to represent speculative visions. From the centre towards its edges, it may feature cones representing possible, plausible, probable and preferable futures.

However, a single cone only stands for a single perspective, or author. Co-speculative practices generate a multitude of future cones, a field of visions, resulting in a much wider, more diverse scope of speculation around a given concern. This shift from 'Me' to 'We' is also an opportunity to examine which values the community of practice shares and use these to focus on meaningful, preferable futures.

Gradually in the process of working with kelp, our Department of Seaweed community of practice has come to understand the importance of 'listening' to our material. We yield to the macroalgae we work with and relegate the authorship and control over some elements of form-giving to the seaweedness. We learn from kelp, come to anticipate the kelp's reactions and attune ourselves to its material properties. Our maker-and-material relationship becomes reciprocal, the kelp serves us, and we serve it in return; it expands its role from being our material to gradually also becoming our method and our muse. In *Conversations on Plant Sensing*, Natasha Myers discusses the 'affective and kinesthetic entanglements' of plants and the plant scientists who grow and observe them in their research. She asks: "Is it possible that practitioners' sensoria get 'vegetalized' over the long duration of their experimental inquiry?"[6] Even though the algae we work with is not a living organism any more, we feel connected to what it once was and what it can do in its ecosystem and – through

6 N. Myers, Conversations on Plant Sensing.

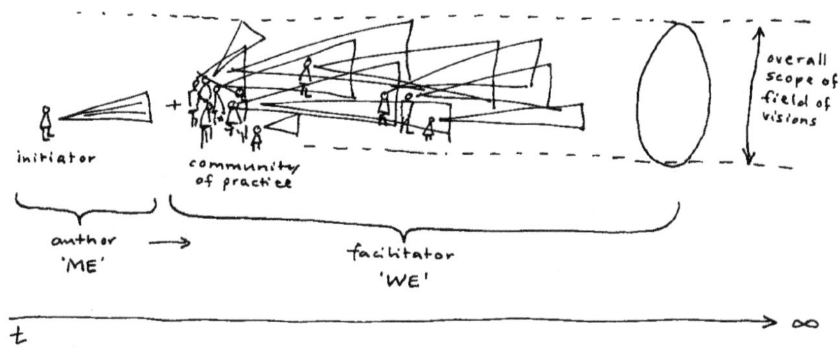

FIGURE 12.15 The open-ended process of *Co-speculative Design*, with its shift from *me* to *we*, from an individually authored vision to the co-speculative field of visions generated by a community of practice

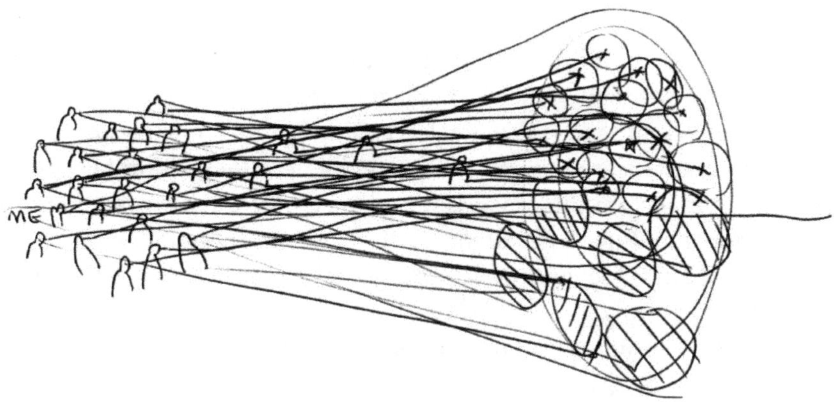

FIGURE 12.16 A *Co-speculative future cone*: the diverse positions from which we imagine our phytofictions enriches the scope of ideas articulated. Shared values enable us to discuss not just possible and probable futures, but also to probe which ones are desirable and undesirable

its evocative material properties – also within the framing of material culture. As practitioners we are humbled by what we learn about, from and with algae. We are starting to aspire to be like seaweed, we are being phytofied.

4 Phytofication

In Phytofiction I focused on the material's trajectory into kelp as a part of human material cultures. In this section I will describe how we can and why we should take a biomaterial's organismic past into account when designing and thinking with it. Both directions of thought on the trajectory from the organism kelp to the materials and artefacts we make from it invite a deep engagement, fuel curiosity and entangle us through haptic, conceptual, olfactory and intellectual engagement.

Algae are marine organisms that thrive in an eco-system that we humans cannot inhabit. The living past of the algae is mystical, alien, enthralling and full of unknowns, even more so when marine biologists, who have studied the ocean for years, contribute to our dialogues. Our understanding becomes less sure-footed, our knowledge patchy; after all, so much of the oceans is still unexplored, 95% of it is still unknown.[7] The bodily encounters we have with algae in its own eco-system when swimming, diving, rock-pooling can be unsettling, often experienced as a loss of control, unwanted touches, an immersion into an unknown world, putting into perspective our perceived superiority as a species, the frames of reference we often unthinkingly apply when we are in our own eco-system.

When I research the marine past of my biomaterials, I encounter our human impact as a major force destabilizing the marine eco-system. Even the natural kelp beds around Hokkaido, Japan, that have been harvested sustainably for generations are now declining rapidly because of warming oceans.[8] Our human impact on the ocean is immense, whilst the ocean's 'eco-system services' buffer and mask the dramatic shifts we are setting in motion. If we disrupt the ocean's ameliorating action any further, the Earth's ecosystems will suffer catastrophically.

Kelp is ideal to illustrate how we need to reframe our thinking from an extractive towards a regenerative mindset because growing it can improve the health of the local marine ecosystem.[9] From an extractive perspective, we might see kelp as another natural resource to be consumed, a foodstuff, a material. We know that it grows afresh in two-year cycles and can be harvested time and again. We may even communicate that it is 'sustainable' or 'regenerative' and tout it as the 'green gold' of the 'blue economy' of the ocean. However, the danger of this lens lies in humans thinking and asking – to paraphrase the US president John F. Kennedy – what kelp can do for them, rather than

7 https://www.marine-ed.org/ocean-literacy/principle-7, accessed on October 10th 2021.

8 https://english.kyodonews.net/news/2020/03/fba75aa19a61-feature-global-warming-wreaks-havoc-on-japanese-edible-kelp.html.

9 https://www.sciencedirect.com/topics/agricultural-and-biological-sciences/seaweed-farming.

what they can do for kelp. An extractive mindset that reduces the immense complexity and agency of the ocean down to human economic principles is an eco-systemic insult. At the level of kelp, taking it out of its natural context and perceiving it in isolation means intentionally or unintentionally disregarding the role it plays in the sea. Within this mindset we would only explore what can be done with kelp, rather than considering what kelp does and is. Whilst extraction is appropriate in some cases, in others we might deplete yet another swathe of species and as a consequence also those that depend on them, their dependants, and so on. Instead of these negative cascades of intervention, our designs must be focused on supporting the agency of algae and ensure their continued existence. We could grow algae in areas that otherwise support very little life, over sandy sea bottoms or hanging above deeper waters to create neopelagic ecosystems that give shelter to fish and offers zones of protection. Of course, for this kind of algae farm to be beneficial to the other organisms we would have to consider their needs and the systemic interactions in our decision-making processes. Their needs would have to be considered when defining where the farm should preferably be located (Marine Spatial Planning), when and how harvesting takes place and which species to grow so that our human impacts on the environment can be alleviated.

Micro- and macroalgae are valuable indicators of marine eco-systemic imbalances. They may be either disappearing from their established habitats or appearing in new ones due to changing climatic conditions or human interference. Algal blooms or beach-cast biomass are other tell-tale signs of off-kilter systems. At the same time kelp forests are some of the most biodiverse places in the ocean and offer shelter and food for a multitude of other species. Kelp is vital to ocean food webs. It shelters fish, becomes a nursery for their eggs and young, as well as habitat and food source for a multitude of other organisms. It grows by removing nutrients from the water, often excess nutrients that impact the ocean because of human activities such as fertilisers and fish farm faeces. Kelp thereby contributes to improving water quality and at the same time it produces oxygen through photosynthesis. At the end of its lifespan, wild kelp sinks to the seabed, taking with it the CO_2 it has captured. If it washes up on land, it can be used as fertilizer and feed for livestock, with recent studies showing it may significantly reduce the methane excretion of cattle. Depending on which type of algae we utilise, the role of the species in its eco-system should direct our methodology.

There are species of algae that are in decline because of human activity and others that thrive and grow excessively, at the expense of the other members of their eco-system. For example, in my local context the Baltic Sea, the brown algae *Fucus vesiculosis* is a keystone species in decline. A keystone species is a species that "relative to its abundance, has a disproportionately large effect on its environment. It plays a critical role in maintaining the organization and diversity of its ecological community, and changes in its abundance and distribution thereby affects many other organisms in the food web."[10]

When working with a keystone species of any eco-system it is obvious that we cannot apply an extractive mindset that might deplete the keystone actor. If we harvest from the wild, the methods we apply need to follow strict sustainability guidelines and cannot exceed the regrowth pattern which needs to be monitored and re-evaluated constantly. In the case of *Fucus vesiculosis* this means that we should not promote any useage of wild stocks at all. At the same time an increase in *Fucus vesiculousis* could benefit the marine ecosystem. Could it be farmed in places where it might not thrive otherwise? Could it be farmed in ways that can nurture its agency as a keystone species as well as provide biomass for human uses? When farming brown kelp, we have to make sure not to negatively affect the wild stock and we should aim to develop practices that support the kelp's potential to benefit its local ecosystem. We need to grow the keystone's agency, not just its biomass.[11] We can achieve this by understanding the keystone species as a key stakeholder or the 'client' of our design process. When we engage with a client, we employ a service mindset that is led by the desire to encounter, come to know and understand their point of view. We begin by aligning ourselves with our client's position as closely as possible, with the aim to understand. Empathy, listening, connecting, research, immersion and supplementary knowledge are the ingredients of this phase of the design process. At the same time, we complement this aligned viewpoint with our own insights as an outsider, our 'fresh eyes'. As such, we engage in stereo with a heightened sensibility for any discrepancies, design insights, in what we encounter. The questions we ask our client are those we ought to pose when we are working with a keystone species: *What is your role here? What do you need to thrive? Who might help me understand you? Who are your potential allies and competitors? How can* I *help you with my skill set, connections and point of view? Are there possibilities for expansion that* I *can help you unlock? How can we both benefit from collaborating? What is a sustainable long term strategy?* Expansion here is not to be understood as a tilting of the balance towards

10 https://stockholmresilience.org/download/18.416c425f13e06f977b14a55/BalticSTERN_State+of+the+Baltic+Sea.pdf.

11 Finnish startup company Origin by Ocean sustainably grows and harvests Baltic algae: https://originbyocean.com/, the Sea Farm in Sweden grows brown kelp *Saccharina latissima* on lines hanging in the ocean to study the environmental benefits of seaweed farming, http://seafarm.se/.

solely benefitting the keystone species, but as a way towards an increased carrying capacity of a biodiverse and thriving ecosystem.

There are also kelp species that only grow in the wild. When we discover properties in these types of algae we need to consider whether promoting them would lead to inevitable exploitation. Perhaps we should then refrain from proposing these types of kelp for use all together.

In the case of algae species that are growing excessively[12] or invasive species that endanger the balance of the local ecosystem,[13] the key aim of the design process could be to utilise the biomass to reduce the environmental impact, ideally in a way that simultaneously addresses its cause and generates biomass that funds the continuous rebalancing efforts. A science-led, extractive mindset within a system-oriented regenerative framing might be an appropriate method for this type of engagement.

Algae-led design processes should be science-led, site specific inquiries. Are we for instance, making algae a viable material to help reduce over-abundant species and extract excess nutrients from the sea? Or are we developing kelp crafts to inspire empathy and engage a community in marine stewardship? Do our proposals have the possibility to benefit the local ecosystem?

The algae's role and agency in its ecosystem defines whether or not we should employ it as a material for making in our human made world, and if so, how. Questions that cannot be answered without a transdisciplinary understanding of the complexity of the matters we design with that includes the material's eco-systemic origins and its future in our material culture. A regenerative mindset involves a systems-level understanding of the situation. It is case- and site-specific and considers the role and impact of the algae in its local marine ecosystem. The designerly approach should be informed by bio-inclusive ethics[14] and shaped by local environmental insights as much as by material affordances, constraints and potentials. It can include methods of empathy and connection as well as extractive practices, always with the aim to regain balance and stability in the socio-ecological system. It recognizes that kelp is more than matter to be shaped by humans – that it can be a marine ecosystem builder, a keystone species or a profiteer of human activities, a destabilizing force or an ecosystem service provider that could reduce our human impact on

12 For instance Sargassum: https://climatecleanup.org/sargassum/.

13 See for instance the case of Caulerpa taxifolia https://plants.ifas.ufl.edu/plant-directory/caulerpa-taxifolia/.

14 Veselova, E., & Gaziulusoy, I. (2019). Implications of the Bioinclusive Ethic on Collaborative and Participatory Design. *The Design Journal, 22* (sup 1), 1571–1586. https://doi.org/10.1080/14606925.2019.1594992.

the sea. It is itself embedded in complex webs of life that we need to consider, to know how to engage in the situation.

Kelp helps us shift our mindset and define a regenerative designerly approach. It inspires us to become more empathic, opens possibilities for community-building dialogues and it raises our awareness of being just one of many species co-inhabiting this world.

CHAPTER 13

Seaweed as the Denizens of the New Commons in the Anthropocene

Soo Jung Ryu and Cintia Organo Quintana

Abstract

The article presents a case for coastal cities to re-envision the current dualistic boundary between city and sea, human and nonhuman, in light of the sea-level rise and storm surges. The aim is to go beyond the traditional defence-driven, mechanical handling of water to engage more meaningfully with the marine world as the water enters our cities. We present seaweed as the representative of the marine world, as a connector between the urban and the marine realm due to its capacity to support life, mitigate climate change, strengthen coastal resilience and impact the local culture and education due to its various ecosystem services. These findings are derived from learnings from three state-of-the-art case studies in Denmark. First, is a design competition incorporating a marine nature-based solution to rethink storm surge protection in Vejle; second is a marine restoration project of a former coastal lagoon in Gyldensteen Strand in Fyn and an artificial underwater reef by the art collective in Copenhagen. The research is based on trans-disciplinary discussions on the value and the role of inviting seaweed as a welcomed key resident of a new coastal commons that fundamentally transform how we live with our watery neighbour, the sea.

Keywords

seaweed – macroalgae – blue urban commons – coastal adaptation – nature-based solutions – blue urbanism – Denmark – marine art – marine restoration – urban seascaping

1 Introduction

This chapter is a call to stakeholders of our urban realm such as planners, developers, designers, ecologists and citizens to grasp the opportunity to re-envision our current business-as-usual waterfront areas in light of a rise in

© SOO JUNG RYU AND CINTIA ORGANO QUINTANA, 2024 | DOI:10.1163/9789004683310_015
This is an open access chapter distributed under the terms of the CC BY-NC 4.0 license.

sea levels and frequent storms. We are researchers from two different fields of urban landscape architecture and marine biology concerned with the worsening biodiversity crisis and the continual degradation of coastal ecosystems in the age of the Anthropocene. This is especially the case in coastal cities where the current business-as-usual waterfront developments fail to acknowledge their responsibility for the ongoing degradation of coastal ecosystems. These are critical habitats for various life forms and are an important meeting place between humans and marine life. Furthermore, with the onset of climate change resulting in a continual rise in sea level and frequent storms, coastal ecosystems play an important role in climate change mitigation and strengthening coastal resilience. We seek to highlight the inadequacy of our current coastal cities to address the challenges of climate change and question the dominant anthropocentric rationale that perpetually excludes the marine world from the urban realm. As the sea seeps further into our coastal cities, we propose that there are transformational opportunities for a new form of commons at our urban shorelines that can reshape the current exploitative relationship with the sea in ways that foster equity for both human and marine life. Therefore, our research seeks to investigate the unexplored potential of a critical coastal ecosystem, seaweed, as a connector between the urban and the marine realm by inviting it to occupy the critical zone of transformation as a new form of commons. This hypothesis will be supported by our learnings from our involvement in a design competition, analysis of state-of-the-art projects and a marine restoration project in Denmark.[1] We hope that the discussions presented in this chapter will stimulate the recognition of the importance of our engagement to the marine world through the lens of the seaweed and inspire the possibility of transformative changes that fundamentally re-envision how coastal cities approach their watery neighbour, the sea.

2 The Plight of Seaweed, the Forgotten Actor

> If we do not change the current course of action to address climate change, an increasingly homogenous biosphere, with empty seas, a world with less diversity of sounds, layers, textures, living colours and perceptible differences, could be the landscape of the future.
>
> KATE ORFF from SCAPE Studio New York (Orff 2016)

1 We will be looking closely at the Danish context, nevertheless, lessons learned could be applied to other similar contemporary coastal cities.

Macroalgae, more commonly known as seaweed, is a forgotten actor in coastal cities despite occupying the crucial land-sea transition space (i.e. intertidal zone) that provides a range of critical ecological and socio-economic services. Predisposing its value as worthless by being named the "weed" of the sea, they do not get the attention of land-based ecosystems and this is despite their important role in tackling negative impacts of anthropogenic climate change. For instance, seaweed filters the water by retaining fine sediment particles and nutrients (i.e. agricultural fertiliser runoffs). They are also one of the most crucial habitats supporting biodiversity and food webs in marine ecosystems worldwide (Krause-Jensen and Duarte 2016). Furthermore, kelp (seaweed) forests, a specific type of brown macroalgae, have the capacity to provide coastal protection from physical disturbances such as waves and storms (Løvås and Tørum 2001; Marine Scotland Directorate 2016; Smale et al. 2013; Zhu et al. 2021). Seaweed may even have an important role in climate mitigation by not only sequestering high amounts of carbon but transferring carbon to the deep sea, where it can be potentially stored for millennia (Boyd, n.d.; Krause-Jensen and Duarte 2016; Krause-Jensen et al. 2018; Nellemann et al. 2009).

Unfortunately, seaweed as a life-giving, life-supporting is under threat of decline (see Figure 13.1). It faces an unprecedented and uncertain future that needs to be protected from the negative impact of human activities while requiring proactive human management to survive (Orff 2016). As such, continual exploitation of the marine world, negative impacts of pollution and accelerating climate change are responsible for the global degradation of 66% of marine environments (IPBES 2019). Furthermore, coastal waters in places like Denmark are in poor ecological conditions due to the effects of eutrophication,[2] which is a barrier towards restoring the ecological health of coastal waters (Miljøstyrelsen n.d.) (see Figure 13.1).

Moreover, the degradation of coastal habitats like seaweed is further exacerbated by human activities stemming from increasing patterns of urbanisation, especially with the rapid increase of harbourfront and waterfront developments from the "land hunger" period of the late 19th century to present times (Firth et al. 2016; Stenak 2005) (see Figure 13.2). Coined "ocean sprawl" (Duarte et al. 2013), the process of reclaiming land from the sea in the name of urban

2 Eutrophication is a process of pollution that occurs when a water body has excess nutrients in the water from runoffs from fertilisers, pesticides and wastes from animals and humans. As a consequence, it becomes overgrown in algae and other aquatic plants that decompose and rob the water of oxygen, killing marine animals (European Environmental Agency, n.d.). Furthermore, increased rainfall predicted in the latest IPCC assessment report (2021) will further carry excess nutrients from agricultural runoffs to the coastal waters, contributing to worsening water qualities for marine life and decreasing coastal waters' salinity levels.

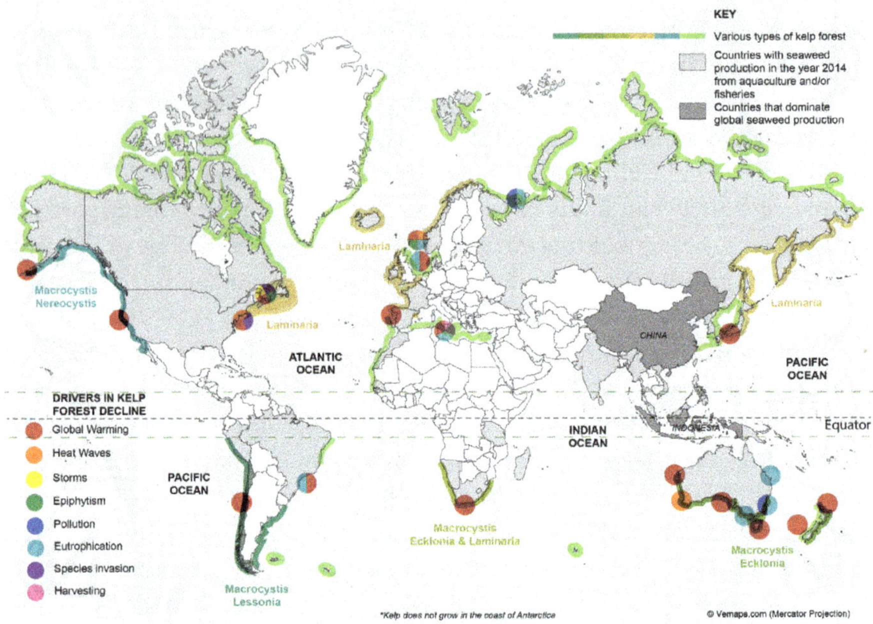

FIGURE 13.1 Various drivers in global kelp forest decline. Note: Unlike kelp, other seaweed types can grow on the equator
MAP ILLUSTRATED BY SOO RYU COMBINING DIFFERENT KELP DECLINE MAPS FROM VARIOUS SOURCES (FILBEE-DEXTER AND WERNBERG 2018; FROEHLICH ET AL. 2019; GUNDERSEN ET AL. 2017; STENECK ET AL. 2002)

development leads to losses of coastal and marine ecosystems by removing structures such as boulders, rocks, stones and shells with crevices and grooves seaweed used to attach their holdfast. Furthermore, intensive fisheries in coastal waters are responsible for further removing habitat-forming reefs and boulders, resulting in homogenised and lifeless sea bottoms that have impacted the marine food chain (Bishop et al. 2017). Reclaimed land is typically filled with concrete to create hard, smooth edge conditions with limited surface area to the water providing poor substrate for marine life (see Figure 13.2). Coastal cities that formed their waterfront areas via land reclamation now face the challenge of having to protect these vulnerable areas due to the onset of more frequent storm surges and rise in sea level throughout this century. And, to make matters worse for the coastal ecosystem like seaweed, the current prevailing coastal protection strategy is a "hard approach" (Faragò et al. 2018), which are engineered infrastructures that work as defence systems to control and contain the water while severing the city's ecological connection[3] to the water (ibid). The current reliance on hard approach protection systems such as dykes, levees and seawalls not only provide a false sense of security[4] at the coast but perpetuates business-as-usual (BAU) waterfront developments that are vulnerable to flood risks and expensive to protect. Furthermore, the developments are not designed to benefit seaweed that lives around it (see Figure 13.2).

3 Anthropocentric Waterfront

To understand some of the key reasons behind the lack of response to the silent loss of the forests of the sea, one must look no further than the current dominant thinking that has influenced our dualistic, dismissive and exploitative relationship with the marine world. Prevailing ideologies such as human exceptionalism, anthropocentrism (Braidotti 2019; Haraway 2016), global capitalism (Campling and Colas 2021; Claudet, Amon, and Blasiak 2021) and terrestrial bias (Dobrin 2021; Jue 2020) are partly responsible for the conceptual barrier that conceives the marine world as something separate from humans and the urban condition. Terrestrial bias is especially evident from our dismissal of the protection of ecologically significant coastal ecosystems at the expense of prioritising developing waterfront areas on reclaimed land (Galland et al. 2012; Filbee-Dexter and Wernberg 2018). Furthermore, our dualistic conception of the relationship between nature (the sea) and culture (the city) does not reflect the reality of the complex entangled network of

3 Ecological connectivity refers to how the movement of living organisms and the flow of nutrients are facilitated or impeded by the presence of man-made or natural landscapes/seascapes (Bishop et al. 2017).

4 This is because these seawalls (levees) are constructed at a certain height level based on a prediction of how high sea level rise and storm surges might be in the future. Past storm surge events around the world (e.g. Hurricane Katrina in New Orleans) have shown that these seawalls and levees can be overtopped due to unpredictability of storm levels thus run the risk of underestimation (UC Berkeley News 2005). Furthermore, a "levee paradox", is a phenomenon to describe the irony of the presence of levees (seawalls, dykes) that leads to less awareness of the flooding risks and, in turn, increased development in the so-called "protected" risk area (Smith, 2002).

FIGURE 13.2 (Top Left) Typical urban coastal edge condition – hard, concrete edges that sever more tactile connection with the water and its life forms such as seaweed. The author took the photo in Lisbon, Portugal, in October 2021. (Top Right) A typical form of urban waterfront commons in Aarhus in Denmark made via land reclamation (i.e. ocean sprawl) is often preoccupied with B-A-U developments. It is a concrete paved area for humans for mainly recreational use and a small space for terrestrial plants in planter boxes for aesthetic purposes. The author took the photo in October 2021. (Bottom) A photo of current waterfront development models in the coastal city of Aarhus in Denmark. The residential and commercial development on Aarhus Docklands started in 2007 on a former container terminal that was reclaimed land. It is 100,000 km^2 with plans to house over 10,000 residents. The Docklands is elevated 2.5m above the previous normal water level based on a future increase in sea level. However, the Docklands is not only considered a storm surge risk area but run the risk of being underestimated to deal with future predictions of sea-level rise and storm surges by the end of the century (Aarhus Kommune, n.d.; Klimatilpasning, 2015)
THE AERIAL PHOTO CREDIT OF AARHUS DOCKLANDS: LASSE LINDKVIST HANSEN (AGERHOLM, 2020)

interdependent connections[5] (Prominski 2014) that make up the urban waterfront. This dualism is exemplified in the physical manifestations of the striking delineation between land and sea in many contemporary coastal cities. As shown in Figure 13.2, the land border's edge on the coast is often constructed from concrete with a 1–1.5m drop to the water, creating a clear separation in between. This particular spatial composition translates into a rationale that designates the watery world as a space that ought to be tamed 'down there' so that the dry urban environment and its inhabitants can reign defiantly from above. As a result, the way coastal cities are designed creates a physical and perceptual barrier for people to engage with the intricacy of the entanglements with the marine world. Moreover, the divided physical conditions at the urban coastal edge highlight the tendency for the cities to only engage with the sea from an instrumental, recreational or idealised way propelled by the waterfront property development boom over the past decades. The contemporary urban waterfront area represents a certain lifestyle, marketed by the promise of constant visual access to mere aesthetic snapshots of the sea (see Figure 13.3), devoid of conveying its complex agency and entanglements. Furthermore, the waterfront has increasingly become an area for a particular group of humans, ones who are privileged enough to afford to live by the water.

4 Learning from the Past

Fortunately, there have been efforts to transcend the current nature-culture divide in the field of landscape architecture and urban planning, reflected in the movements of ecological urbanism (Mostafavi et al. 2016) and blue urbanism (Beatley 2014). These movements refocus on an ecological approach to urbanism that marries ecological health with a mutually beneficial design for both humans and other species (Mostafavi et al. 2016). One of the points of entry into moving past nature-culture dualism is first to recognise that

5 Life on land and sea are closely connected by a network of estuaries and coastal ecosystems which are strongly dependent on each other. Marine life forms move back and forth and exchange energy and materials that make up the ecological connection of this boundary zone between land and sea (Belletti et al. 2020). The ecological connectivity among streams and coastal waters as well as among coastal habitats such as continuous zones of salt marshes, seaweeds, and seagrasses is critical for species to feed, reproduce, distribute over large spatial scales and assure recruitment of next populations (Bishop et al. 2017).

FIGURE 13.3 Marketing narratives for a high-rise apartment complex in the Docklands (reclaimed land) called "The Lighthouse" in the city of Aarhus, Denmark (finished construction in 2022). This development is another form of anthropocentric waterfront area from reclaimed land with hard edges. The landscaped greenery are often add-ons for marketing visualisations and designed for aesthetic and recreational pleasure for people. What is missing is the better integration of the connection to the sea nor the consideration of providing physical conditions conducive to marine life forms such as seaweed. (Translated by the author from Danish)
PHOTO CREDIT: 3XN ARCHITECTS (LIGHTHOUSE N.D.)

baseline ecological conditions of the past (without significant human interference[6]) can never be reached again. Nevertheless, we can still refer back to the past environmental conditions to understand how the coastal landscape/seascape is influenced by the deep geological structures of the land and sea (i.e. via geomorphology, topography and bathymetry). These deep structures tell a story of the movement, distribution and connectivity of materials, species, energy and water that influence the different types of species likely to thrive. Moreover, learning from the past conditions is particularly relevant in coastal cities to restore coastal and marine habitats (destroyed by ocean sprawl) to provide resilience against the negative impacts of climate change. These "former" conditions can give clues and inspiration to understand what the optimal nature-based solution[7] (NbS) could be in a particular context with specific environmental conditions (See Figure 13.4). The aim of coastal cities should therefore be to navigate the precarious scenarios of supporting the activities of human populations[8] while simultaneously regenerating and conserving marine ecosystems under threat from anthropogenic climate change. Thus, the decisions we make now to develop our coastal cities to prepare for the challenges in the future need to address the marine life in crisis and leave room to re-envision new possibilities for the future.

---

6 This is a big discussion among researchers and academics about when the start of major human interference on the so-called "natural world" began. Some researchers claim that it began from the onset of agricultural revolution that made significant changes to the land from the mid-17th century onwards and others argue that significant human interference started after the industrial revolution towards the end of the 19th century (Clarke 2015). For coastal cities (especially in Denmark), it is clear that drastic alteration to the shorelines in the form of land reclamation started from the onset of technological advancement such as pumps in the late 19th century (Miljøstyrelsen, n.d.; Stenak 2005).

7 The official EU definition: Nature-based solutions are "inspired and supported by nature, which are cost-effective, simultaneously provide environmental, social and economic benefits and help build resilience. Such solutions bring more, and more diverse, nature and natural features and processes into cities, landscapes and seascapes, through locally adapted, resource-efficient and systemic interventions. Hence, nature-based solutions must benefit biodiversity and support the delivery of a range of ecosystem services" (EU commission n.d.).

8 Anthropogenic activities are predicted to increase throughout the 21st century from global population growth and increasing urban densification (United Nations, Department of Economic and Social Affairs, and Population Division 2019).

FIGURE 13.4 End of 19th-century drawing and painting of the city of Vejle facing the Vejle Fjord in Denmark. They give clues to a former "soft" edge condition that consisted of beach meadows, salt marshes, seagrass and seaweed that occupied the shorelines that protected the old city from storm surges. These coastal ecosystems worked like a sponge by soaking up the water and mitigating the strength of the waves. These images give clues about what type of species are likely to grow in this area with its local environmental conditions. (Left): A drawing of the city of Vejle in a river valley during the 19th century (Dansk Center for Byhistorie n.d.). (Right): A painting of the Vejle Fjord's soft shoreline conditions by Vilhelm Kyhn (Kyhn 1862).

5 In Search of New Residents of the Urban Commons

Coastal zones are places of connections and solidarity between peoples, places and more-than-humans.[9] It is also a place of opportunity for exposure, awareness and appreciation of the forgotten world below that we are all intertwined with. We are departing from this type of interconnected thinking to explore the scope for new urban shoreline infrastructures that integrate coastal seascapes[10] to enhance human and more-than-human interaction that can develop over time. We propose to revive seaweed from a forgotten actor in our cities to an integral part of place-making and identity for coastal cities in an increasingly

9 "More-than-human" is a term that gained popularity due to the feminist scholar Donna Haraway (2007; 2016) to refer to the entangled multispecies reality that goes beyond the human. The term is used to illuminate new ways of thinking about agency and power of the different living forces. In this chapter, the term will be used to refer to the marine world, in particular the coastal ecosystems such as seaweed as the main representative marine species.

10 Seascape is a relatively new term derived from landscape ecology. Seascapes are defined as spatially heterogeneous and dynamic spaces that can be delineated by a wide range of scales in time and space (Pittman 2018). Seascapes consist of patches or mosaics of interacting ecosystems, such as salt marshes, seagrasses, seaweeds, and soft sediments supporting great biodiversity.

FIGURE 13.5 Aesthetic qualities of macroalgae inspired by the dry pressing of seaweed from the Victorian era by women. Collage of various dry pressed seaweed from the coast of East Jutland, Denmark, in July 2020.

wet world. Seaweed is used as a representative voice of marine life forms that we need to forge better relations in the Anthropocene. They are chosen for its connection and influence on our cultural practices in various societies across the world throughout human history. These traditions and customs range from traditions of making food as a medium for scientific enquiry and practice of art, to name a few[11] (see Figure 13.5). Our proposition is that these preexisting narratives and practices of seaweed in various human cultures make seaweed a good candidate for representing the coastal ecosystems to forge connections

11 Humans have had a long history of forging relations with seaweed, influencing various cultural practices in different societies across history. For instance, many coastal nations in Europe as early as the 10th century ate and traded seaweed also used as fodders for farm animals. Seaweed is known around the world as a healthy and sustainable food source that does not require any land, freshwater, nor pesticide to grow (Krause-Jensen et al., 2018; Krause-Jensen and Duarte, 2016; Mouritsen 2013). Furthermore, seaweed has also had a cultural impact during the natural history boom of the Victorian period for women who were excluded from the practice of scientific fieldwork. It was socially acceptable for British women to venture out to collect seaweed, called "seaweeding" to dry press seaweed showcasing its beauty (Mouritsen 2013; Trethewey 2020).

between humans and more-than-humans. It is an unexplored area that has the potential to influence how we shape our relations with the more-than-human marine world in the age of the Anthropocene.

The following section sets out to explore the proposition to support the role of seaweed as a connector across various disciplines of knowledge, bridging between various human actors to the more-than-humans in the critical meeting space between land and sea. We propose that coastal waterfront strategies should invite the seaweed as the new denizen of the urban commons to tell a new narrative of the transformation that is to occur in our cities as the sea rises.

6 Seaweed as the Denizen of the New Urban Commons and Seaweed as a Connector between Humans and More-than-Humans, Land and Sea

Sea level rise presents a new opportunity in coastal cities as the transitional space between land and sea is increasingly becoming vulnerable to the forces of the sea (see Figure 13.6). Therefore, for coastal cities, a long term strategy of retreat needs to be carefully considered by various stakeholders to plan a secure future at the end of this century. Since guaranteeing the full protection of our coastal cities from the forces of the sea is not possible, especially without resorting to more drastic measures of higher walls and pumps, there will be areas that will increasingly be high-risk zones that need to be relocated to higher ground. The inevitability for retreat in certain places provides a unique opportunity to experiment with the vulnerable low-lying waterfront areas as critical zones of transformation[12] – the new urban commons. However, what exactly is a new urban commons for coastal cities in the age of the Anthropocene? The term urban commons represents shared material and immaterial resources (i.e. land) that belong to or affect the whole community in an urban setting (Hardt & Negri 2009). It is founded on the guiding principle of equity that fundamentally reconceptualises how we view space as something that belongs to and affects all. Urban commoning is thus a collective socio-cultural practice of mutual sharing, collaboration and responsibility to

12 It is a term borrowed from Bruno Latour and Peter Weibel book 'Critical Zones: The Science and Politics of Landing on Earth' (Latour and Weibel 2020). However we are using it in this chapter to refer to the coastal zones (boundary between land and sea) where it is increasingly at risk of inundation from sea level rise and frequent storm surge. These areas present an opportunity for transformation due to these various environmental factors.

FIGURE 13.6 (Left) Seaweed washed on the beach in Lunkebudgten, Denmark. A connecting ecological element between land and sea. The photo was taken in April 2020 by Lars Skaaning. (Right) Various seaweeds (macroalgae) are visible to the human eye from the coast in Elsehoved Beach, in Fyn, Denmark. The photo shows green seaweed called "Sea lettuce" (Søsalat and Vandhår in Danish) and brown seaweed called "Bladderwrack" (blæretang in Danish). The author took the photo in July 2020.

look after urban spaces that have qualities and resources necessary for liveability (Eynaud et al. 2018; Linebaugh 2008; Huron 2017). The concept of an urban commons has gained a growing interest over the past decade, especially in its transformative power to utilise the collective will to implement initiatives that benefit all. However, the traditional conceptions of a commons do not necessarily include more-than-humans such as coastal ecosystems as part of the collective force for change nor as a recipient of the benefits. Therefore, we propose and support the relatively new approach to commons by making room for not only equity among different human actors of various socio-economic classes but for more-than-humans such as seaweed.

The radical act of including the agency of the sea and its life forms into the coastal cities can inspire designs that reflect how the water weaves through the urban landscapes and provide new connections and opportunities to create softer, more dynamic zones. Furthermore, the new urban commons can be a way of returning some of the vulnerable areas back to the oceans, thereby rethinking the traditional ways cities occupy the water's edge. It is a way to recognise that the city and the sea are integral to the urban environment and influence appropriate regulations and policies that address the current accelerating changes in climate. The most appropriate strategies depend on several complex factors but require the careful collaboration of the right actors in the decision-making process to conceive what these new commons could be.

6.1 *State-of-the-Art Projects: Art that Facilitates Interspecies Living by SUPERFLEX*

Envisioning a new form of urban commons as a space for humans and more-than-human requires a paradigm shift that can think from the perspective of marine life. It requires us to ask questions such as, what is an urban commons for seaweed? How can this meeting place be constructed in ways that will allow it to be inhabited by seaweed and humans? Will the seaweed respond to what we created? These are some of the questions that an artist group based in Copenhagen called SUPERFLEX is investigating from the perspective of marine animals. They are paving the way for innovative state-of-the-art projects to think *with* marine life in the hope of fostering multispecies coexistence in light of rising sea levels. For instance, their animated film project 'Vertical Migration' (2021c) tells the parallel story of vertical migration required by humans due to sea-level rise in the coming centuries and the vertical ascent to the surface of the sea shown from the perspective of a siphonophore (a relative of jellyfish). By relating the story of a siphonophore as our own, SUPERFLEX argues that "we can shift our perspective to recognise that we are connected, that our actions affect each other, and that we share a common fate" (SUPERFLEX 2021c).

The concept of 'Vertical Migration' is an appropriate term for the future where a rise in sea level will bring forth marine life forms such as seaweed into the coastal waterfront areas. It is a story of our future where our coastal cities would eventually need to retreat to higher ground (or higher buildings on stilts) to leave the current waterfront spaces to be transformed into a new urban commons where seaweed can find new homes. It addresses the years of colonising the urban shorelines and extracting stones and rocks vital for seaweeds to grow on. By inviting the seaweed to occupy these former waterfront spaces, we give back the surfaces they need to attach themselves to and grow. This idea of re-occupation of artificial structures by marine life due to sea-level rise leads to their next notable project called the 'Interspecies Assembly,' a proposal of "a physical gathering site that aims to foster friendly relations among species and nurture interspecies living" (SUPERFLEX 2021b). It is a designated space where

FIGURE 13.7 (Left): "Pink Elements" (no.6/Zig Zag Column) is part of the research project called "Deep Sea Minding" by SUPERFLEX. The pink sculpture is built with coral friendly bricks for fish. Installed at Galería OMR, Mexico City, 2019 (Photo credit: Enrique Macías Martínez (SUPERFLEX 2019)). (Middle): A diver installing the pink element to test if the fish would inhabit and interact with the sculpture (SUPERFLEX n.d.). (Right): "Interspecies Assembly" – A drawing of what it means to mark the very first gathering of humans and other marine species on earth, to promote interspecies dialogue and cooperation (SUPERFLEX n.d.).

humans need to set aside their terrestrial biases to pay attention to the agency of the marine life below. The strength behind this project is the idea of a space for the meeting of humans and marine life, which can easily translate into the idea of a new urban commons for all (See Figure 13.7) (SUPERFLEX 2021b). For SUPERFLEX, it is important that we design structures near the coast that will eventually be submerged by the sea to be conducive to the needs and desires of marine life. The fish and seaweeds will be the new occupants of the waterfront areas, and we can give them a helping hand by designing sculptures that function as art for humans to create awareness of the marine world while being habitats for fish and seaweed (see Figure 13.7). These porous and modular sculptural blocks designed for fish and other marine species can be installed as part of the current harbourfront buildings that will go under the water in time or as stand-alone sculptures that form a 'Super Reef', an artificial reef to reinstate and revive the lost stone reefs from harbourfront expansion. The porous modules are made of lots of surface area that marine biodiversity thrives on with materials that can withstand marine conditions and are coloured in pink, a colour scientifically known to encourage coral polyp growth in the tropics (See Figure 13.7) (SUPERFLEX 2019; 2021a). The design decisions are made to cater for the occupation of underwater beings in the future but also as a creative mechanism to tell the story of how they will shape the cities as sea level rises. These projects are all part of "SUPERFLEX's research examining the relationship between humans and other species, proposing a new kind of urbanism that reimagines how we live together" (SUPERFLEX 2021a).

6.2 *State-of-the-Art Project: Bringing Back the Seaweed in Gyldensteen Coastal Lagoon, Denmark*

A restoration project that developed an adaptive and resilient nature-based solution to revive coastal ecosystems such as seaweed as a new resident of the commons is situated in Gyldensteen Beach in Fyn, Denmark (see Figures 13.8). It is an example of restoring a coastal marine habitat as a new form of commons through the deliberate act of reversing land reclamation of the sea. Although this initiative is not an urban context, this blue infrastructural strategy led by marine biologists allowed a planned conversion of the hard coastline to an intertidal marine habitat by breaching the dikes to flood the former farmland purposely. The dikes were part of common practice in the 1800s to drain low lying wetlands and construct dykes to gain arable land for agriculture and cattle grazing (Stenak 2005). However, these marginal areas close to the sea often had problems with damages caused by storms that breached the dykes and accumulated rainwater, ruining the crops and pasture fields. Therefore, the re-flooding initiative was possible due to the increasing financial burden of upkeeping the dikes by the farmers and the low profitable yield of the fields. The former site in the late 1700s before it was converted into farmland was a typical coastal marine area (see Figure 13.8 map of the areas in 1780) with transition zones between land and sea that were constantly submerged by the forces of the sea. Understanding the former wet site conditions before the human intervention of dikes and pumps was critical in realising that in the future, as it made little sense to resist the increasing forces of the sea to keep this area dry. Moreover, to address the challenges of climate change, a long-term sustainable solution was needed that considered various factors and values, such as improving biodiversity beyond mere coastal protection.

The Gyldensteen project is forefront in returning the land to sea by inviting the water to create a new form of commons for both human and more-than-human benefit. However, returning the land to the sea was never a viable option until the climate change and the biodiversity crisis brought the imperative of ecological restoration of ecosystems. With the primary aim to restore habitats for birds and increase public awareness and access to marine nature, this concept of bringing the water back in was adopted and funded by private nature foundations in Denmark (Aage V. Jensen Nature Foundation). Thus, after 140 years of agricultural land use of 214 hectares, Gyldensteen's

FIGURE 13.8 (Top row) Before and After photo of Gyldensteen beach (aerial photo credit: Viggo Lind). (Second row) Map of the transformation of Gyldensteen beach over the past 230 years from coastal marine area to farmland to a marine nature reserve for research, recreation and buffer zone to protect the town of Bogense behind. (Third row, left to right) A photo of the education centre for visitors showcasing information on the various land and marine-based species that live in the Gyldensteen nature reserve (called Naturrum)(Aage V. Jensen Naturfond. n.d.); Photo of seaweed in glass tanks for education purposes taken by the author in June 2019; A photo of birds occupying Gyldensteen nature restoration area (VisitNordfyn n.d.). (Fourth row) Showing the growth of brown macroalgae, close up for seaweed growing in Gyldensteen. The author took the photo in September 2021; A photo of people visiting the area (VisitNordfyn n.d.).

dikes were finally breached in 2014. The new lagoon provided a buffer zone in the event of a future storm surge to protect the nearby town of Bogense from flooding (Klimatilpasning n.d.; Aage V. Jensen Naturfond n.d.). Thus, creating a coastal lagoon resulted in providing several benefits for humans as a recreational space to connect with nature with the added benefit of some levels of coastal protection. It was the first time such a large area was restored at once, and the observed results over the course of 6 years showed the gradual colonisation of seaweeds and invertebrates in the lagoon. The monitored results highlighted that if we return the land to the sea, nature will come back (Valdemarsen et al. 2018) (see Figures 13.8 and 13.9). For instance, the biodiversity of seaweeds is twice as high in 2020 compared with 2014, right after the flooding of the coastal lagoon, as shown in Figure 13.9. The number of species of seaweed found in the lagoon is comparable with other natural coastal lagoons and coastal areas in Denmark. Typical species include the brown algae Fucus vesiculosus, the green algae Chaetomorpha linum, and the red algae Agarophyton vermiculophyta.

Furthermore, the increasing number of visitors (approx. 80,000 visitors in summer) and proactive participation of volunteers shows societal engagement and participatory process in the acceptance of this new commons. Recreational facilities contributed to the appreciation of birds and marine nature as well as improved human well-being. A schoolroom and a historical barn were refurbished to a natural "laboratory" to provide tools and opportunities for school classes to experience and observe nature (see Figure 13.8). An outreach exhibition about the project indicated in a survey (n = 433) that Gyldensteen Coastal Lagoon made people aged between 15–30 years more conscious about the impact of climate change in the Anthropocene and understand the benefits of restoration (Frederiksen 2020).

More importantly, Gyldensteen Coastal Lagoon is an example of how sea-level rise and coastal flooding can influence marine restoration projects in vulnerable areas by allowing the water to come back in, giving space for creating a new commons. With time, the new commons transform as seaweeds come back, creating new opportunities for humans to connect better with the marine world forging closer connections across anthropocentric boundaries of walls. The Gyldensteen project highlights the importance of understanding various interconnected networks, such as the dynamic processes of constant synergy between the city and the water and long-time spans required to consider marine nature restoration and sea-level rise (i.e. 5–50 years). Furthermore, the project indicate that the varying levels of restoration of coastal ecosystems could be part of future planning of coastal cities to ensure a higher resilience to climate change and benefit from richer biodiversity.

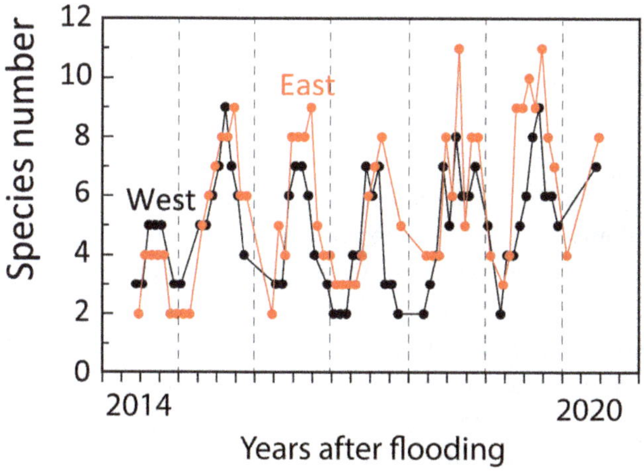

FIGURE 13.9 The return of seaweed biodiversity in the West and East part of Gyldensteen Coastal Lagoon six years after flooding in 2014. Kristensen et al. (unpublished data)

6.3 *State-of-the-Art Projects: City of Vejle's 'Kanten' (The Edge) Design Competition*

The final project that we wish to highlight here involved the integration of seaweed (and other marine life) as a key aspect of a design competition called "Kanten" (called The Edge in English) organised by Vejle Municipality in Denmark from April to October 2020 (Vejle Kommune 2020a). It was a call for architects, landscape architects and artists to collaborate on a holistic proposal that considers nature-based, innovative and recreational solutions that will shape the urban shoreline conditions that protect Vejle from future storm surges and rising seawater (ibid). 'The Edge' here refers to the boundary zone where the city of Vejle meets the sea (Vejle fjord). The competition brief asked the participants to integrate nature not only above the water (i.e. intertidal or coast) but below as well (i.e. subtidal marine life forms). The design brief set a new standard from the usual design competition briefs by asking the participants to find alternative perspectives of the water, emphasising how water and marine nature are integral to the city (Vejle Kommune 2020a).

The winning entries were picked based on their alignment with the brief's core values. One of the few entries that engaged with understanding the intersection between the rules of the land and the rules of the sea was called "The Membrane." It was a proposal that re-envisioned a terrain in the waterfront of Vejle that thrived being above and underwater as sea level rises in this century. Instead of seeing water as something to be kept away from land, they proposed a whole new approach to living close to the water by gradually inviting the water into the city over time. The master plan showed how the waterfront area could be re-developed by designing with the water instead of against water like a cellular membrane that allowed the interchange of land and sea (see Figure 13.10). By allowing the water to enter the waterfront areas and designing to understand the change in tides, the coastal habitats such as seaweeds and saltmarshes could flourish much closer to the city and the residents while providing benefits of associated ecosystem services (see Figure 13.10). The proposed spaces are re-envisioned as natural tidal pools and stones reinstated underwater for seaweeds. It is a strong concept that relates to the entire district of the waterfront to be transformed into a space that allows interconnectivity between humans and more-than-humans.

Other winning entrants, such as "On the edge of Utopia", presented interesting concepts on blending art and marine life as an "underwater botanical garden." Seaweeds hung underneath floating platforms to highlight their important role in filtering the water and showcase their beauty and versatility with various applications (see Figure 13.10). The Kanten/The Edge competition entries highlighting the important role of marine life forms such as seaweed in pushing the creative fields to synthesise the different interdisciplinary knowledge into a holistic outcome.

Our key learnings from being involved as an advisor to the judges were that marine ecosystems such as seaweed created a lot of discussions and scepticism around what the edge conditions (Kanten) should be among different stakeholders with contrasting values (i.e. political, economic, ecological, aesthetic). For instance, judges with political backgrounds had difficulty seeing the benefit of proposals that considered long-term approaches beyond their term of a few years. Moreover, other municipal judges found it difficult to see the importance of giving nature underwater a prominent space on the waterfront as they usually associated these spaces as largely a concrete recreational area for humans. Nevertheless, the multiplicity of values that seaweed represents (i.e. improving biodiversity, water cleaning properties, cultural values as food, beauty etc.) was one of the aspects that convinced the judges to acknowledge them as potential key residents of the new edge condition (Kanten) in Vejle. Moreover, with the expert guidance from Cintia as the marine biologist, the design entries were analysed from the perspective of marine life, to ensure that the physical structures proposed would be conducive for the growth of seaweed. The presence of this alternative marine viewpoint prevented the business-as-usual proposal from being chosen. It is rare to have design competitions involving marine biologists and designers in one room, and it was clear that working with seaweed as a design actor brought together interdisciplinary

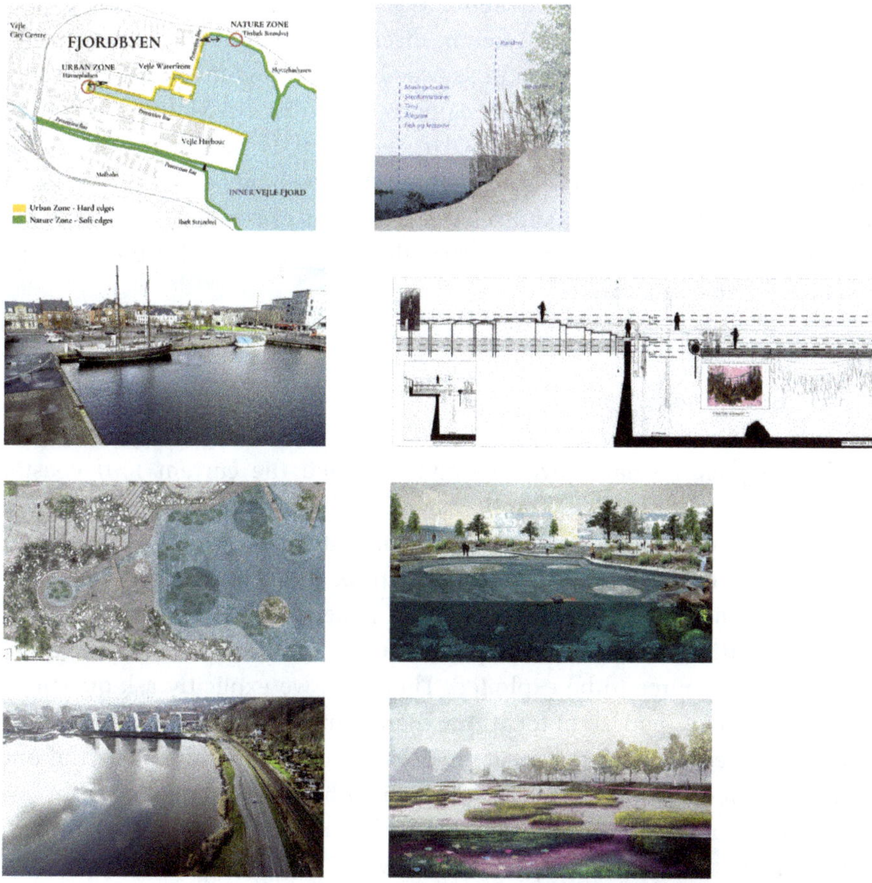

FIGURE 13.10 (Top row) Map of the Kanten (The Edge) design competition by Vejle Municipality. The map of Vejle's waterfront shoreline is mainly made up of hard concrete edges; An illustration of the various species living in the land-water zone (Vejle Fjord's Littoral Zone). (Second row) A photo of the hard edge in the waterfront area called "The Urban Zone" in Vejle and an illustration of a proposal by a Danish architect artist duo, Gamborg/Magnusssen. The proposal is called: 'On the edge of Utopia', where seaweed hangs under a floating system as part of the holistic proposal. (Third row) The winning proposal is called: The Membrane by landscape architects Josephine Philipsen (Sweden), Andrés Hernández Williamson (Colombia) and artist Louisa Brando (Spain) (Vejle Kommune 2020b). The proposal re-envisions natural tidal pools and revives stone reefs on the seabeds for seaweeds. (Fourth row) A photo of the area where the city meets the sandbanks in Vejle called "The Nature Zone" which was re-envisioned by a Danish artist group called SUPERFLEX in collaboration with Baldios. They proposed reviving the habitat such as eelgrass and seaweed for fish by designing elements that provide substrate for seaweed to attach to (Vejle Kommune 2020b).

collaboration among different stakeholders. There was proactive participation and knowledge production of different disciplines in aiming for an outcome that benefited all. The inclusion of seaweed (nature underwater) as a design actor in the design brief allowed design proposals that envisioned alternative scenarios from the status quo that tried to see it from the perspective of marine life forms. The design proposals highlighted the possibility of creating a space for the meeting of humans and seaweeds in light of sea-level rise, spurring potential for new narratives and design realisations of seaweed as the denizens of a new urban commons.

7 Conclusion

In this chapter, we have reviewed and critiqued the current BAU coastal urban development models, emphasising that they are inadequate to handle future water issues due to climate change and its contribution to the degradation of our coastal ecosystems. The health of our coastal ecosystems is of vital importance for the coastal cities that depend on it in numerous ways. It requires various stakeholders to protect it and reverse the trend of treating it as a mere resource to be exploited. Therefore, we explicitly ask questions to understand our persistent terrestrial bias, to move towards thinking from a marine perspective, and to reverse the desire to tame the water as part of our urban coastal development models. This main hypothesis emerged from our involvement and exposure to various blue infrastructural projects in Denmark. For instance, the design competition entries of Kanten/The Edge presented a transformative opportunity at the critical zone between the city and sea. As the increasing impact of water enters the waterfront areas in this century, we suggest that some of these critical infrastructures retreat to higher ground while allowing the waterfront space's transformation to integrate the water into the urban realm better. Winning entries envisioned spaces that allowed citizens to meet and coexist with their new marine neighbours, like seaweed, which is given space to occupy in the new commons. The Kanten/The Edge competition also gave insights to explore new ways of working through the lens of seaweed as a connector between the different disciplines and stakeholders, especially in integrating marine biology with the creative fields in aiding a holistic design outcome.

Furthermore, learnings from the SUPERFLEX projects highlighted the importance of new urban commons for all (humans and nonhumans) and embracing the complex entanglements of humans and the sea in ways that extend beyond our dualistic epistemologies. SUPERFLEX's radical approach resulted in designed structures in the new urban commons conducive to marine life in its choice of material, texture, colour and surface area so that lifeforms like seaweed will come to want to inhabit the spaces as the sea level rises.

Finally, our learnings from the Gyldensteen coastal lagoon restoration project showed that if we invite the agency of the water in, nature will come back. The results of different species of seaweed thriving after six years from its implementation highlighted the resilience of coastal ecosystems if we give them space. These critical zones of transformation are critical for us humans to think with marine ecological awareness and envision fluid futures that transcend the current nature-culture divide.

Seaweed as a denizen of the commons is a potential source of inspiration to birth new narratives required to tackle the challenges of climate change. Therefore, thinking with algae is a call to embrace the agency of the sea and to foster alternative views of the water, influencing different decisions for the future of coastal cities. Seaweed has the potential role to be a connector between various actors (human and nonhuman, urban planners and city developers, marine biologists and politicians) to exercise a collective effort to find creative and alternative ways that can motivate more restorative relations between the city and the sea in the age of the Anthropocene.

We propose that seaweed should no longer be a forgotten actor at our coastal zones but an active denizen of the new urban commons. By turning our gaze below the water, we can start conceiving of alternative ways to live not just by the sea but *with* the sea.

References

Aage V. Jensen Naturfond. n.d. 'Gyldensteen Strand – Aage V. Jensens Fonde'. Accessed 14 October 2021. https://www.avjf.dk/avjnf/naturomraader/gyldensteen-strand/.

Aarhus Kommune. n.d. 'En helt ny bydel skyder op på havnefronten i disse år – Aarhus Ø. En levende og moderne bydel, som allerede i dag er rammen om mange menneskers hverdag og fritid'. Accessed 2 September 2021. https://aarhusoe.dk/om-aarhus-o/historie-og-vision/.

Agerholm D. 2020. 'Bygherrerådgivning på stævnen af Aarhus Ø'. LB Consult A/S (blog). 16 November 2020. https://www.lb-consult.dk/ship/.

Ameel L. 2019. 'Agency at/of the Waterfront in New York City: Vision 2020 and New York 2140'. Textual Practice 0 (0): 1–17.

Beatley T. 2014. Blue Urbanism: Exploring Connections Between Cities and Oceans. 2nd Edition. Washington, DC: Island Press.

Belletti B., C. Garcia de Leaniz, J. Jones, S. Bizzi, L. Börger, G. Segura, A. Castelletti, W. van de Bund, K. Aarestrup, J. Barry, K. Belka, A. Berkhuysen, K. Birnie-Gauvin, M. Bussettini, M. Carolli, S. Consuegra, E. Dopico, T. Feierfeil, S. Fernández, P. Fernandez Garrido, E. Garcia-Vazquez, S. Garrido, G. Giannico, P. Gough, N. Jepsen, P.E. Jones, P. Kemp, J. Kerr, J. King, M. Łapińska, G. Lázaro, M.C. Lucas, L. Marcello, P. Martin, P. McGinnity, J. O'Hanley, R. Olivo del Amo, P. Parasiewicz, M. Pusch, G. Rincon, C. Rodriguez, J. Royte, C.T. Schneider, J.S. Tummers, S. Vallesi, A. Vowles, E. Verspoor, H. Wanningen, K.M. Wantzen, L. Wildman, M. Zalewski. 2020. 'More than one million barriers fragment Europe's rivers'. Nature 588: 436–441.

Bishop M.J., M. Mayer-Pinto, L. Airoldi, L.B. Firth, R.L. Morris, L.H.L. Loke, S.J. Hawkins, L.A. Naylor, R.A. Coleman, S.Y. Chee, K.A. Dafforn. 2017. 'Effects of ocean sprawl on ecological connectivity: impacts and solutions'. Journal of Experimental Marine Biology and Ecology 492: 7–30.

Boyd R. 2021. 'Blue Carbon: An Oceanic Opportunity to Fight Climate Change'. Scientific American. Accessed 5 September 2021. https://www.scientificamerican.com/article/blue-carbon/.

Braidotti R. 2019. Posthuman Knowledge. 1st edition. Medford, MA: Polity.

Campling L., A. Colas. 2021. Capitalism and the Sea: The Maritime Factor in the Making of the Modern World. London; New York: Verso.

Clark T. 2015. Ecocriticism on the Edge: The Anthropocene as a Threshold Concept. Illustrated edition. London; New York: Bloomsbury Academic.

Claudet J., D.J. Amon, R. Blasiak. 2021. Opinion: Transformational Opportunities for an Equitable Ocean Commons. Proceedings of the National Academy of Sciences 118 (42).

Dansk Center for Byhistorie. n.d. Accessed 6 September 2021. Danmarks Købstæder: Vejle – Prospekter http://ddb.byhistorie.dk/koebstaeder/prospekter.aspx?koebstadID=67.

Dobrin, S.I. 2021. Blue Ecocriticism and the Oceanic Imperative. 1st edition. Abingdon, Oxon: New York: Routledge.

Duarte C.M., K.A. Pitt, C.H. Lucas, J.E. Purcell, S. Uye, K. Robinson, L. Brotz, M.B. Decker, K.R. Sutherland, A. Malej, L. Madin, H. Mianzan, J.M. Gili, V. Fuentes, D. Atienza, F. Pages, D. Breitburg, J. Malek, W.M. Graham, R.H. Condon. 2013. Is Global Ocean Sprawl a Cause of Jellyfish Blooms? Frontiers in Ecology and the Environment 11 (2): 91–97.

European Commission. 2021. 'Nature-Based Solutions'. Text. European Commission – European Commission. Accessed 5 September 2021. https://ec.europa.eu/info/research-and-innovation/research-area/environment/nature-based-solutions_en.

European Environmental Agency. 2021. 'Eutrophication – European Environment Agency'. Accessed 5 September 2021. https://www.eea.europa.eu/archived/archived-content-water-topic/wise-help-centre/glossary-definitions/eutrophication.

Eynaud P., J. Maïté, D. Mourey. 2018. 'Participatory Art as a Social Practice of Commoning to Reinvent the Right to the City'. Voluntas: International Journal of Voluntary and Nonprofit Organisations 29 (4): 621–36.

Faragò M., E.S. Rasmussen, O. Fryd, E. Rønde Nielsen, K. Arnbjerg-Nielsen. 2018. 'Coastal Protection Technologies in a Danish Context'. Report. Vand i Byer. https://orbit.dtu.dk/en/publications/coastal-protection-technologies-in-a-danish-context.

Filbee-Dexter K., T. Wernberg. 2018. 'Rise of Turfs: A New Battlefront for Globally Declining Kelp Forests'. BioScience 68 (2): 64–76.

Firth L.B., A.M. Knights, D. Bridger, A. Evans, N. Mieskowska, P.J. Moore, N.E. O'Connor, E.V. Sheehan, R.C. Thompson, S.J. Hawkins. 2016. 'Ocean Sprawl: Challenges and Opportunities for Biodiversity Management in a Changing World'. Aberdeen University Press, Oceanography and Marine Biology: an annual review, 54 (November). https://pearl.plymouth.ac.uk/handle/10026.1/8025.

Frederiksen M.B. 2020. Udvikling og effektanalyse af en nyetableret formidlingsudstilling ved Gyldensteen Strand. Master Thesis, Odense, University of Southern Denmark, 105 pages.

Froehlich H.E., J.C. Afflerbach, M. Frazier, B.S. Halpern. 2019. 'Blue Growth Potential to Mitigate Climate Change through Seaweed Offsetting'. Current Biology 29 (18): 3087–3093.e3.

Galland G., E. Harrould-Kolieb, D. Herr. 2012. The Ocean and Climate Change Policy. Climate Policy 12 (6): 764–71.

Gundersen H., T. Bryan, W. Chen, F.E. Moy, A.N. Sandman, G. Sundblad, S. Schneider, J.H. Andersen, S. Langaas, M.G. Walday. 2017. Ecosystem Services: In the Coastal Zone of the Nordic Countries. Copenhagen: Nordisk Ministerråd. http://urn.kb.se/resolve?urn=urn:nbn:se:norden:org:diva-4719.

Haraway D.J. 2007. When Species Meet. Minneapolis: Univ Of Minnesota Press.

Haraway D.J. 2016. Staying with the Trouble: Making Kin in the Chthulucene. Durham: Duke University Press.

Hardt M., A. Negri. 2009. Commonwealth. Harvard University Press.

Huron A. 2017. 'Theorising the Urban Commons: New Thoughts, Tensions and Paths Forward'. Urban Studies 54 (4): 1062–69.

IPBES. 2019. 'Intergovernmental Science-Policy Platform on Biodiversity and Ecosystem Processes (IPBES) Global Assessment Report'. Media Release: Nature's Dangerous Decline 'Unprecedented'; Species Extinction Rates 'Accelerating'. 5 May 2019. http://ipbes.net/news/Media-Release-Global-Assessment.

IPCC. 2021. Summary for Policymakers. In: Climate Change 2021: The Physical Science Basis. Contribution of Working Group I to the Sixth Assessment Report of the

Intergovernmental Panel on Climate Change [Masson-Delmotte V., P. Zhai, A. Pirani, S.L. Connors, C. Péan, S. Berger, N. Caud, Y. Chen, L. Goldfarb, M.I. Gomis, M. Huang, K. Leitzell, E. Lonnoy, J.B.R. Matthews, T.K. Maycock, T. Waterfield, O. Yelekçi, R. Yu, and B. Zhou (eds.)]. Cambridge University Press.

Jue M. 2020. Wild Blue Media: Thinking through Seawater. Illustrated edition. Durham: Duke University Press Books.

Klimatilpasning. 2015. 'Huse På Aarhus Havn Klarer Høj Vandstand'. 3 December 2015. https://www.klimatilpasning.dk/cases-overview/huse-paa-aarhus-havn-klarer-hoej-vandstand/.

Klimatilpasning. 2021. 'Gyldensteen Strand Gives Tilbage Til Havet'. Accessed 7 September 2021. https://www.klimatilpasning.dk/cases-overview/gyldensteen-strand-gives-tilbage-til-havet/.

Krause-Jensen D., C.M. Duarte. 2016. Substantial role of macroalgae in marine carbon sequestration. Nature Geoscience 9: 737–742.

Kristensen E., C.O. Quintana. (unpublished data) Gyldensteen Strand project, University of Southern Denmark.

Kyhn W. 1862. *Indsejlingen til Vejlefjord.* Oil painting. Randers Kunstmuseum.

Latour B., P. Weibel. 2020. Critical Zones: The Science and Politics of Landing on Earth. Illustrated edition. Cambridge: The MIT Press.

Lighthouse. 2021. 'Lighthouse – Danmarks højeste boligbyggeri'. *Lighthouse.* Accessed 14 October 2021. https://lighthouseaarhus.dk/forside/.

Linebaugh P. "3. The Commodity and the Commons." In The Magna Carta Manifesto: Liberties and Commons for All, 46–68. Berkeley: University of California Press, 2008.

Locher C. 1882. *Der læsses tang ved Hornbæk Strand, 1882, Carl Locher. SMK Open.* Oil painting. https://open.smk.dk/artwork/image/KMS1210.

Løvås S.M., A. Tørum. 2001. 'Effect of the Kelp Laminaria Hyperborea upon Sand Dune Erosion and Water Particle Velocities'. Coastal Engineering 44 (1): 37–63.

Marine Scotland Directorate. 2016. 'Wild Seaweed Harvesting: Strategic Environmental Assessment – Environmental Report'. In. Marine and Fisheries. Scottish Government. http://www.gov.scot/publications/wild-seaweed-harvesting-strategic-environmental-assessment-environmental-report/pages/8/.

Miljøstyrelsen, n.d. MiljøGIS for basisanalyse for vandområdeplaner 2021–2027.

Mostafavi M., G. Doherty, Harvard University Graduate School of Design, eds. 2016. Ecological Urbanism. Revised edition. Zürich: Lars Muller.

Mouritsen O.G. 2013. Seaweeds: Edible, Available, and Sustainable. Translated by Mariela Johansen. Translation edition. Chicago; London: University of Chicago Press.

Nellemann C., E. Corcoran, C. Duarte, L. Valdes, C. Young, L. Fonseca, G. Grimsditch. 2009. Blue Carbon: The role of healthy oceans in binding carbon. A Rapid Response Assessment. United Nations Environment Programme, GRID-Arendal.

Orff K. 2016. Toward an Urban Ecology. Monacelli Press.

Pittman S.L. 2018. Seascape Ecology. Wiley Blackwell.

Prominski M. 2014. 'Andscapes: Concepts of Nature and Culture for Landscape Architecture in the "Anthropocene"'. Journal of Landscape Architecture 9 (1): 6–19.

Smale D.A., M.T. Burrows, P. Moore, N. O'Connor, S.J. Hawkins. 2013. Threats and Knowledge Gaps for Ecosystem Services Provided by Kelp Forests: A Northeast Atlantic Perspective. Ecology and Evolution 3 (11): 4016–38.

Smith D., 2002. 'Understanding and managing large scale flood events'. Geodate 15, 4.

Stenak, M. 2005. De inddæmmede landskaber. Landbohistorisk selskab, Gylling.

Steneck R.S., M.H. Graham, B.J. Bourque, D. Corbett, J.M. Erlandson, J.A. Estes, M.J. Tegner. 2002. 'Kelp Forest Ecosystems: Biodiversity, Stability, Resilience and Future'. Environmental Conservation 29 (4): 436–59.

Superflex. n.d. Instagram. Accessed 6 September 2021. https://www.instagram.com/superflexstudio/.

Superflex. 2019. 'Pink Elements'. 2019. https://superflex.net/works/pink_elements.

Superflex. 2021a. 'As Close As We Get'. 2021. https://superflex.net/works/as_close_as_we_get.

Superflex. 2021b. 'Interspecies Assembly'. 2021. https://superflex.net/works/interspecies_assembly.

Superflex. 2021c. 'Vertical Migration'. 2021. https://superflex.net/works/vertical_migration.

Trethewey L. 2020. 'What Victorian-Era Seaweed Pressings Reveal about Our Changing Seas'. *The Guardian*, 27 October 2020, sec. Environment. https://www.theguardian.com/environment/2020/oct/27/what-victorian-era-seaweed-pressings-reveal-about-our-changing-seas.

UC Berkeley News. 2005. 'Hurricane Katrina Senate Testimony by Prof. Raymond Seed, Ph.D. Professor of Civil and Environmental Engineering, University of California, Berkeley, on Behalf of the NSF-Sponsored Levee Investigation Team, before the Committee on Homeland Security and Government Affairs, U.S. Senate'. 2 November 2005. https://www.berkeley.edu/news/media/releases/2005/11/02_levee_testimony.shtml.

United Nations, Department of Economic and Social Affairs, and Population Division. 2019. *World Urbanization Prospects: The 2018 Revision.* (ST/ESA/SER.A/420). New York: United Nations.

Valdemarsen T., C.O. Quintana, S.W. Thorsen, E. Kristensen. 2018. Benthic macrofauna bioturbation and early colonisation in newly flooded coastal habitats. Plos One 13(4): e0196097.

Vejle Kommune. 2020a. 'Kanten – Åben idekonkurrence program'. *Vejle Kommune.* https://vejle.citizenlab.co/da-DK/projects/idekonkurrencen-kanten/info.

Vejle Kommune. 2020b. 'Vejle Skal Leve Med – Og under – Vandet'. 5 October 2020. https://vejle.presscloud.com/csp/vocast/message.csp?KEY=517943704869993.

VisitNordfyn. n.d. 'Bustur til Gyldensteen Strand'. VisitNordfyn. Accessed 16 October 2021. https://www.visitnordfyn.dk/busture-og-grupper/have-og-natur/besoeg-gyldensteen-strand.

VisitNordfyn. n.d. 'Visit Gyldensteen Strand'. VisitNordfyn. Accessed 16 October 2021b. https://www.visitnordfyn.com/nordfyn/activities/nature/experience-gyldensteen-strand.

Zhu, L., Lei, J., Huguenard, K. and Fredriksson, D.W, 2021. Wave attenuation by suspended canopies with cultivated kelp (Saccharina latissima). Coastal Engineering, 168, pp. 1–20. https://doi.org/10.1016/j.coastaleng.2021.103947.

Being Algae ~ Coda

> It is a curious situation that the sea, from which life first arose should now be threatened by the activities of one form of that life. But the sea, though changed in a sinister way, will continue to exist; the threat is rather to life itself.
>
> RACHEL CARSON, *The Sea Around Us* (1951)

•••

> Today a little more land may belong to the sea, tomorrow a little less. Always the edge of the sea remains an elusive and indefinable boundary.
>
> RACHEL CARSON, *The Sea Around Us* (1951)

∵

We hope this book can be a first stroke at an interchange in contemporary research on algae between botany and anthropology, marine science and art, algae artisans and manufacturers. Transforming the spaces between siloed histories and knowledges allows the liminality of algal being, its ability to transition across borders, to refashion the delineations of disciplines often talking past each other when addressing organisms so radically different from ourselves. By their recalcitrant and excessive tendencies, being with algae requires from us to go beyond the comforts of routinized habits of industrial civilization. Like the aquatic spheres they represent, their foreign medium suggests their transformative potential in engaging with the unexpected philosophical contours their physical being confronts.

Siloed disciplines also speak of the ways modernity has taught us to cultivate both the living and its living culturing techniques. Because 'homogenous', 'industrial', 'standardized' or 'axenic' are not only adjectives we have grown to use when speaking of a mode of relating to the biological realm, these words also describe one way of living the world and telling stories about it. Silos after all collect and store the uniformized harvest within one way of dwelling: that of monocultures. And, as Vandana Shiva would say, monocultures take over the mind beyond land (Monocultures of the Mind, 1993).

© Y. H. HENDLIN, J. WEGGELAAR, N. DEROSSI & S. MUGNAI, 2024 | DOI:10.1163/9789004683310_016
This is an open access chapter distributed under the terms of the CC BY-NC 4.0 license.

As with their agrarian correlates, monocultures of the mind are uniformized fields of reality that in the pursuit of efficiency, specialization and productivity, renounce the possibility of letting diversity emerge and reproduce. Important crafts and knowledges have thus been lost in not complying to those standards. Celebrating interstitial composition and providing sfumatos of legibility into algae alter ways of being, this book recounts Medieval quotidian uses of algae for Arabic medicinal purposes, tales of the revitalization of ancient Danish crafts to make "moss paint," and Japanese ones to fabricate "glue walls," demonstrating algal abilities to engage in unexpected and restorative trajectories. Such stories trouble neat distinctions between the human and the vegetal world and remind us that dwelling beyond anthropo-normativity has been lost but can be refound, reinvented, re-cherished.

These examples show us that algae-human relations can be rich and treasured as conscientious parts of our material and metaphysical world together, rather than instrumentalized for industrialization. More broadly, monocultures as those lived inside biotechnological laboratories, or those written by the tidy lines of high-tech and mute plantations, speak as well to the loss of attention such practices imply. It is precisely such a field of perception that this volume wishes to nurture back, inviting reflections on the 'art of attentiveness' in underwater explorations of algae toxicity, or investigations over eyes' and palates' roles in appreciating algae's beauty and taste. If "ontic monism," which constitutes reducing the essence of life into an increasingly thin unequivocal line, tries to flatten our ability to perceive the world beyond rational schemes, some artists still try to 'phytophy' themselves, tuning their sensing to let kelp's inherent properties guide their algae-based 'making practices.' Learning to live with others means modulating from obsessing over following a painted line of progress and experimenting with turning our gaze to the perfusion of those living around us differently, and to those that have been living here before us, like our ancestors, the algae.

The death of crafts, implicit or tacit praxis and knowledge yoked with sustainably working with one's habitat tolls the death of the range of sensibilities involved. Eventually such amnesia pushes us (un)wittingly to degrade the complexity which allows death and life to exist in the first place: biodiversity at large. Historically, this is far from a purely biological fact. The self-undermining logic of this compulsion to control and obedience according to mechanistic models of life ultimately serves no one: "the destruction of diversity and the creation of uniformity simultaneously implies the destruction of stability and the creation of vulnerability" (Shiva 1993, p. 64). Concretely, biodiversity loss has meant the eradication of millions of vegetal varieties making up the subsistence of farmers all around the world; it has meant, and still means, famines and deaths. In the seas, the loss of relation with algae has led to viewing them as an extractable resource, a CO_2 dump, or a scourge of nature without considering and appreciating their ecological importance at all.

Understanding algal being as the opposite of monoculture, the contributions of this book aim to retrieve polycultures of practice, which in turn means not only providing alternative answers, but most of all, learning to ask oblique questions. For instance, some authors herein task themselves to define moral obligations directed at processual biological entities like algae. They have wondered whether humans could adopt a new ethics of companionship with the living, a sort of natural contract (Serres, 1995), seeing algae not only as another input into our biological machines, but as fluid respondents tracking our reciprocity and capable of supporting us in our mourning practices. Retrieving diversity signifies experimenting with uncomfortable things too, like opening up closed and axenic algae photobioreactors growth systems to new organisms, as well as dealing with the practical consequences of doing it beyond intellectual speculations, creatively finding ecological stratagems to deal with biofouling and decay. In valuing such waste generative processes as part of ecological networks, the Sisyphusian effort of separating value from waste can be reexamined, in hopes of some learning about the ontology of metabolizing.

All this attention continues, without trying to make 'without the trouble,' conscious of the importance of avoiding the trap of getting fooled by an ever-positive savior image of algae, mindful of the negative histories in which they got involved, like the role of Marimo in the construction of a false indigeneity within the Lake Akan communities in Japan, or the Agar soldier's responsibilities during World War II. The efforts made towards acknowledging the historicity of algae perceptions – both negative and positive – comes to our aid here, reminding us of the role that knowledge production and its tools exert on our grasp and instrumentalization of natural beings.

Algae taxonomies contribute to positioning plant, fungi, and bacteria differences as 'difference all the way down', challenging even the uniform animal models we have become accustomed to and have generalized with much violence. Lazy taxonomies such as this, applying a single mode of agency, intelligence, and physical correlates of being everywhere and anywhere, once exposed as dialectical cul-de-sacs, suggest alternative routes of understanding and categorizing. A pluralization of taxonomies, fitting for the capriciously varying domains composing the algal ecosystems, allows entry into apophatic taxonomies, freeing us from recursively cataphatic ones. Rather than affirming algae identity conveniently as 'other' and *ipso facto* 'lesser,' as Linnean or cladistic cataphatic categories have been reinforced, apophatic taxonomies rearrange themselves as understanding is updated, not only according to pre-given categories, but through metamorphosing the categories themselves. Divorce from unitary models of being, with associated hypothesized neural and structural correlates that confer justification for instrumentalization and untoward advantage, discloses an unmistakable being-with fitting for the age of the holobiont. Extending the Evolutionary Synthesis to keep up with the queer processural ontologies of ecological evolutionary developmental biology promises an algalization of human beings, fractaling also into the algalization of knowledge, refiguring our disciplines according to processes and consortia of inquiry.

Approaching algae as engendering new categories and rearranging our understanding of prior ones, this volume aims to contribute to critical plant studies from the science side as much as the arts and humanities. Algae clamor for such non-siloed treatments, and this first attempt at offering a kaleidoscopic approach is no mere apologetics for eclecticism. Rather, it attends to the interbeing of multispecies being, multinatural worlds. The internaturalization of life, through detecting assemblages, clumps, gradations, and textures of affinity, opens scholarship to overcome simulated objectivity via separation. Instead, the auras of community which sustain us, the salinity of the water, the composition of our surrounding media, the macro and micro actors of which we are aware and not aware, get to the paradoxical heart of causality. Can we harness algae without taming them? Can we be with them in cooperation and mutual benefit? We hope this treatment helps readers find motility from industrial culture's mono-cultivated mindset towards the encompassed interbeing of algae.

References

Carson, Rachel (2021), *The Sea Trilogy: Under the Sea-Wind / The Sea Around Us / The Edge of the Sea*, (S. Steingraber, ed.), New York, NY: Library of America.

Serres, Michel (1995), *The natural contract, Studies in literature and science*, Ann Arbor: University of Michigan Press.

Shiva, Vandana (1993), *Monocultures of the Mind: Perspectives on Biodiversity and Biotechnology*, Penang, Malaysia: Bloomsbury.

Index

www.ingramcontent.com/pod-product-compliance
Lightning Source LLC
Chambersburg PA
CBHW071326230926
51849CB00002B/324
9789004683303